AIoT-Driven Sustainable Manufacturing and Operational Initiatives

In an era where industries are tasked with addressing pressing global challenges, ranging from climate change and resource depletion to societal well-being, the fusion of AI and IoT emerges as a powerful catalyst for positive change. The synergistic collaboration between intelligent algorithms and interconnected devices not only enhances operational efficiency but also empowers industries to make informed decisions that align with ecological and social considerations.

AIoT-Driven Sustainable Manufacturing and Operational Initiatives: A Smart Eco-Friendly Manufacturing System offers a comprehensive exploration of the integration of AI and IoT in the context of sustainable industry, covering foundational principles, practical applications, and real-world case studies. It provides practical insights into how AIoT solutions are currently being implemented, offering tangible examples and success stories that readers can relate to and draw inspiration from. Emphasis is placed on the significance of data-driven decision-making, showcasing how AIoT enables industries to harness the power of data for optimizing processes, reducing waste, and achieving sustainability goals. The book includes real-world case studies and examples illustrating the practical impact of AIoT, making complex concepts accessible and demonstrating the transformative potential of these technologies in actual industrial settings.

Tailored to various industries, the book ensures relevance for a diverse audience, including manufacturing, energy, and environmental sectors, making it a versatile resource for professionals across different fields.

Dr. Kali Charan Rath is an associate professor in Mechanical Engineering at GIET University, Gunupur, holds a Ph.D. from NIT Jamshedpur. He is a dedicated academician and researcher with notable contributions to mechanical engineering and sustainable technologies. His research is published in reputed international journals and edited volumes, and he has also served as an editor for several books, demonstrating his commitment to advancing knowledge and promoting academic collaboration.

Dr. Kamalakanta Muduli is an associate professor of Mechanical Engineering at the Papua New Guinea University of Technology, Lae. He holds a Ph.D. from IIT Bhubaneswar and is a fellow of the Institution of Engineers (India), a senior member of the Indian Institution of Industrial Engineering, and an active member of ASME. With strong academic and research credentials, Dr. Muduli contributes significantly to the global advancement of mechanical and industrial engineering.

Advances in Intelligent Decision-Making, Systems Engineering, and Project Management

This new book series will report the latest research and developments in the field of information technology, engineering and manufacturing, construction, consulting, healthcare, military applications, production, networks, traffic management, crisis response, human interfaces, and other related and applied fields. It will cover all project types, such as organizational development, strategy, product development, engineer-to-order manufacturing, infrastructure and systems delivery, and industries and industry-sectors where projects take place, such as professional services, and the public sector including international development and cooperation etc. This new series will publish research on all fields of information technology, engineering, and manufacturing including the growth and testing of new computational methods, the management and analysis of different types of data, and the implementation of novel engineering applications in all areas of information technology and engineering. It will also publish on inventive treatment methodologies, diagnosis tools and techniques, and the best practices for managers, practitioners, and consultants in a wide range of organizations and fields including police, defense, procurement, communications, transport, management, electrical, electronic, aerospace, requirements.

Smart Technologies for Improved Performance of Manufacturing Systems and Services
Edited by Bikash Chandra Behera, Bikash Ranjan Moharana, Kamalakanta Muduli, and Sardar M. N. Islam

Wireless Communication Technologies: Roles, Responsibilities and Impact of IoT, 6G, and Blockchain Practices
Edited by Vandana Sharma, Balamurugan Balusamy, Gianluigi Ferrari, and Prerna Ajmani

Digital Technology Enabled Circular Economy
Models for Environmental and Resource Sustainability
Edited by Bikash Moharana, Bikash Behera, and Kamalakanta Muduli

Industry 4.0, Smart Manufacturing, and Industrial Engineering
Challenges and Opportunities
Edited by Amit Kumar Tyagi, Shrikant Tiwari, and Sayed Sayeed Ahmad

Data Analytics and Artificial Intelligence for Predictive Maintenance in Smart Manufacturing
Edited by Amit Kumar Tyagi, Shrikant Tiwari, and Gulshan Soni

AIoT-Driven Sustainable Manufacturing and Operational Initiatives
A Smart Eco-Friendly Manufacturing System
Edited by Kali Charan Rath and Kamalakanta Muduli

For more information about this series, please visit: https://www.routledge.com/Advances-in-Intelligent-Decision-Making-Systems-Engineering-and-Project-Management/book-series/CRCAIDMSEPM

AIoT-Driven Sustainable Manufacturing and Operational Initiatives

A Smart Eco-Friendly Manufacturing System

Edited by

Kali Charan Rath and Kamalakanta Muduli

CRC Press
Taylor & Francis Group
Boca Raton London New York

CRC Press is an imprint of the
Taylor & Francis Group, an informa business

First edition published 2026
by CRC Press
2385 NW Executive Center Drive, Suite 320, Boca Raton FL 33431

and by CRC Press
4 Park Square, Milton Park, Abingdon, Oxon, OX14 4RN

CRC Press is an imprint of Taylor & Francis Group, LLC

ISBN: 978-1-032-82155-9 (hbk)
ISBN: 978-1-032-82177-1 (pbk)
ISBN: 978-1-003-50332-3 (ebk)

DOI: 10.1201/9781003503323

Typeset in Times
by KnowledgeWorks Global Ltd.

Contents

Preface

AIoT-Driven Sustainable Manufacturing and Operational Initiatives: A Smart Eco-Friendly Manufacturing System explores the integration of Artificial Intelligence (AI) and the Internet of Things (IoT) in advancing sustainable manufacturing practices. The book is structured into 13 chapters, each focusing on a different aspect of Artificial Intelligence of things (AIoT) implementation, including smart systems, energy efficiency, real-time monitoring, and waste reduction. The book aims to provide a comprehensive guide for integrating smart and sustainable initiatives into the manufacturing industry, ensuring both operational efficiency and environmental responsibility.

Chapter 1 introduces the transformative role of the Industrial Internet of Things (IIoT) and the Internet of Things (IoT) in reshaping sustainable manufacturing. It delves into the core technologies and real-world applications of IoT and IIoT, highlighting their impact on enhancing efficiency, productivity, and innovation in modern manufacturing processes.

Chapter 2 delves into the integration of AI with IoT, highlighting how AI technologies like machine learning and natural language processing enhance IoT devices and networks. This chapter explores the transformative capabilities of AIoT in diverse fields, including manufacturing, smart cities, and healthcare, with a focus on sustainable manufacturing practices.

Chapter 3 explores the growing field of smart home IoT deployments, emphasizing the critical factors that influence their success, such as user requirements, device compatibility, and privacy concerns. Through case studies and real-life success stories, this chapter provides valuable insights into the challenges and solutions in achieving effective and secure smart home implementations.

Chapter 4 focuses on the application of computational fluid dynamics (CFD) and the B-spline collocation method to enhance the accuracy and stability of numerical solutions for heat transport problems. It presents a detailed theoretical framework for B-splines, explores various heat transfer applications, and compares the method with traditional numerical approaches, highlighting its potential in optimizing heat transfer system design.

Chapter 5 provides the application of statistical methods to optimize processes in the biotechnology and pharmaceutical industries, focusing on improving efficiency, reducing variability, and ensuring high-quality outcomes. It integrates foundational statistical techniques with modern advancements, offering a comprehensive guide to achieving optimal performance in highly regulated sectors while emphasizing real-world applications and regulatory compliance.

Chapter 6 explains the transformative impact of AIoT-based sustainable engineering on manufacturing, focusing on enhancing efficiency, reducing waste, and promoting sustainability. It explores the integration of advanced technologies such as automation, data analytics, and supply chain management systems to optimize industrial operations, improve decision-making, and achieve significant gains in productivity and sustainability.

Chapter 7 explains the impact of Industry 4.0 technologies, particularly IoT and AI, on revolutionizing Computer Numerical Control (CNC) systems in manufacturing. The chapter presents a unique AI-powered optimization system that leverages real-time data from CNC machines to enhance operational efficiency, reduce waste, and support sustainable manufacturing practices.

Chapter 8 explores the integration of AIoT in predicting the mechanical fracture and thermal properties of Glass Fiber Reinforced Polymer (GFRP) composites, highlighting recent advancements that enhance forecasting accuracy and efficiency for engineering applications.

Chapter 9 explores the impact of self-service business intelligence on customer satisfaction within sustainable manufacturing sectors in Lagos, Nigeria. It highlights how system quality, data accessibility, and user training—moderated by non-technical user independence—shape customer experiences in the era of AI, IoT, and Big Data.

Chapter 10 tells the transformative integration of AIoT in robotics systems, redefining smart manufacturing through autonomy, precision, and intelligent decision-making. It highlights how AIoT enables predictive maintenance, real-time optimization, and adaptive industrial workflows—paving the path toward Industry 5.0.

Chapter 11 provides the convergence of AI and Intelligent Process Automation (IPA) in manufacturing, aiming to streamline complex operations and enhance real-time decision-making. It introduces a hybrid framework that combines AI models with industrial expertise to foster autonomous, adaptive, and sustainable production systems.

Chapter 12 explains the application of lean methodologies in modern manufacturing, highlighting how principles like Value Stream Mapping, Flow, Pull Systems, and Continuous Improvement enhance resource efficiency, reduce waste, and drive competitiveness through sustainable and streamlined processes.

Chapter 13 investigates the impact of solution treatment and aging on the hardness and wear resistance of Ti-15-3 alloy, highlighting how optimized heat treatment enhances its mechanical properties, extends lifespan, and contributes to sustainability by reducing material consumption and environmental impact.

This book explores the integration of engineering applications with advanced technologies to foster a sustainable production environment. By addressing key challenges and innovations across various domains, it provides valuable insights into optimizing efficiency, reducing environmental impact, and enhancing resource utilization. Covering topics such as AI, CFD, lean manufacturing, and material enhancements, the book offers a comprehensive resource for researchers, academicians, and industry professionals. With a strong emphasis on practical applications and emerging trends, it serves as a guide for developing sustainable, high-performance solutions in modern engineering and manufacturing practices.

Contributors

Muhammad Ilman Hakimi Chua Abdullah
Fakulti Teknologi dan Kejuruteraan Mekanikal
Durian Tunggal, Malaysia

Abhinav Aditya
School of Computer Engineering
KIIT (Deemed to Be) University
Bhubaneswar, India

Ch. Mohan Babu
School of Management
Centurion University
Vizianagaram, India

Hadaq Maulana Muji Burrochman
Department of Mechanical and Industrial Engineering
Universitas Negeri Malang
Malang, Indonesia

Adeyemi Oluwatoyin Damilola
Department of Economics and Business Administration
Alethia University
Ago-Iwoye, Nigeria

Dina Darwish
Faculty of Computer Science and IT
Ahram Canadian University
Giza, Egypt

Dipan Kumar Das
Department of Physics
Centurion University of Technology and Management
Odisha, India

Rishav Das
School of Computer Engineering
KIIT (Deemed to Be) University
Bhubaneswar, India

Shashikant Das
Department of Mechanical Engineering
Centurion University of Technology and Management
Odisha, India

Saumya Ranjan Jena
Department of Mathematics,
School of Applied Sciences,
Kalinga Institute of Industrial Technology (KIIT) Deemed to be University,
Bhubaneswar, India

Islauddin Khan
University of Illinois Chicago
Chicago, Illinois, United States

Mallikarjun Koripadu
Labcorp
Karnataka, India

Abdul Munir Hidayat Syah Lubis
Fakulti Teknologi dan Kejuruteraan Mekanikal
Durian Tunggal, Malaysia

Debasis Mishra
Department of Mechanical Engineering
GIET University
Odisha, India

Bhabendu Kumar Mohanta
College of Information Technology
United Arab Emirates University
Al Ain, United Arab Emirates

Gopal Krushna Mohanta
Department of Mechanical Engineering
Gandhi Institute of Engineering and Technology University
Gunupur, India

Debabrata Mohanty
Department of Physics
Chang Gung University
Taoyuan, Taiwan

Hitesh Mohapatra
School of Computer Engineering
KIIT (Deemed to Be) University
Bhubaneswar, India

Kamalakanta Muduli
Mechanical Engineering Department
Papua New Guinea University of Technology
Lae, Papua New Guinea

Siddhartha Mukherjee
School of Computer Engineering
KIIT (Deemed to Be) University
Bhubaneswar, India

OKETOLA Kehinde O.
Department of Business Administration
Osun State University
Osun State, Nigeria

Olubiyi Timilehin Olasoji
Department of Business Administration
Faculty of Management and Social Sciences
West Midlands Open University
Lagos State, Nigeria

Pradeep Padhamnath
Department of Plastic Deformation of Metals and Extractive Metallurgy
Faculty of Engineering Materials and Industrial Information Technology
AGH University of Krakow
Krakow, Poland

Balaji Padhy
Department of Mathematics
Centurion University of Technology and Management
Odisha, India

Lakshmi Prasad Panda
Department of Humanities
Government College of Engineering
Kalahandi, Bhawanipatna, India

Manas Ranjan Panda
Department of Mechanical Engineering
Gandhi Institute of Engineering and Technology University
Gunupur, India

Sasank Sekhar Panda
Department of Mechanical Engineering
Gandhi Institute of Engineering and Technology University
Gunupur, India

Siddharth Parthasarathy
Department of Physics
Centurion University of Technology and Management
Odisha, India

Padmaja Patnaik
Department of Physics
Centurion University of Technology and Management
Odisha, India

Alak Kumar Patra
School of Civil Engineering
PNG University of Technology
Lae, Papua New Guinea

Nirakara Pradhan
Department of Mathematics,
School of Computer Sciences and Artificial Intelligence,
SR University,
Warangal, India

Diki Dwi Pramono
Department of Mechanical and Industrial Engineering
Universitas Negeri Malang
Malang, Indonesia

B. Rabi Prasad
Department of Biotechnology
School of Engineering and Technology
GIET University
Gunupur, India

Ghanishtha Prusty
Life Science, School of Sciences
GIET University
Gunupur, India

Teresa Ptaszynska
Lean Six Sigma Freelancer
Buckinghamshire, United Kingdom

Poppy Puspitasari
Department of Mechanical and Industrial Engineering
Universitas Negeri Malang
Malang, Indonesia

Kali Charan Rath
Department of Mechanical Engineering
Gandhi Institute of Engineering and Technology University
Gunupur, India

Somya Samantaray
Department of Physics
Centurion University of Technology and Management
Odisha, India

Jai Krishna Sanku
Department of Mechanical Engineering
Gandhi Institute of Engineering and Technology University
Gunupur, India

Archana Senapati
Department of Mathematics,
School of Applied Sciences,
Kalinga Institute of Industrial Technology (KIIT) Deemed to be University,
Bhubaneswar, India

Arun Kumar Singh
School of MCS
PNG University of Technology
Lae, Papua New Guinea

Ankur Srivastava
Center of Industrial Electronics
University of Southern Denmark
Alsion, Denmark

Polaki Suman
Department of Biotechnology
School of Agriculture
Centurion University of Technology and Management
Parlakhemundi, India

Adeyemi Omolade Sunday
Department of Business Administration
Christopher University
Ogun State, Nigeria

K. Venkatasubbaiah
Department of Mechanical Engineering
AU College of Engineering
Andhra University
Visakhapatnam, India

1 Sustainability in Manufacturing

The Role of IoT and IIoT Solutions

Siddharth Parthasarathy, Padmaja Patnaik, Dipan Kumar Das, Somya Samantaray, and Debabrata Mohanty

1.1 INTRODUCTION TO IoT AND IIoT

Internet of Things (IoT) and Industrial Internet of Things (IIoT) are two great technical revolutions that redefine the shapes of industries, especially manufacturing. Industrial applications are primarily categorized under IIoT, which is an extension of IoT. With increased connectivity and data analytics, it aims at improving the production process as well as productivity. As part of the much larger IoT scenario, IIoT is important as it allows intelligent industrial operations based on data analytics [Serpanos et al. (2018)]. This would then be considered one of the key shifts for this industrial field, marrying up Operational Technology (OT) with Information and Communication Technology (ICT) to yield better production outcome [Willner (2018)]. IoT, on the other hand, is a number of connected devices that share and exchange data, may conduct direct communication and controlled even remote monitoring and managing [Mathur et al. (2020)]. Both IoT and IIoT have a wide range of applications, from healthcare to smart industries, and are fundamental to the development of Industry 4.0 [Deshpande et al. (2020)].

The traces of historical evolution of IoT and IIoT lie in the early concepts of networked devices and machine-to-machine (M2M) communication. In time, advancements in sensor technology, wireless communication, and data analytics have pushed these concepts into practical applications across various industries. Conceptual framework of IoT and IIoT is the integration of the virtual world with the real one, where collected data from devices is analyzed to make decisions, optimize operations, and create value. In case of IIoT, compared to traditional IACS (Industrial Automation and Control Systems), there has been more connectivity as well as changes in architecture. Drivers of this change are better efficiency of operations, smart identification of objects, and minimization of human intervention in risky environments. The change from the IoT to IIoT was seen with the development of Industry 4.0 and the subsequent acceptance of IIoT across different sectors. On the other hand, the adoption of IIoT poses challenges, including security threats and

DOI: 10.1201/9781003503323-1

the requirement for scalable, interoperable, secure, and reliable architectures. In the case of modern manufacturing, it is impossible to overemphasize the importance and significance of IoT and IIoT. Manufacturers may benefit from data analytics in supply chain efficiency, quality assurance, and predictive maintenance. The sustainable manufacturing industry is becoming increasingly responsive to the rapidly evolving market, mainly because it can collect and analyze huge volumes of data in real time.

The IoT and IIoT are part of the digital revolution, especially in Industry 4.0. Jeschke et al. continue with the challenges and requirements of IIoT, where it is more likely to be applied in Industry 4.0 (2017). Sailaja discussed how IIoT can affect product quality and efficiency and its application in remote monitoring, predictive analysis, and digital twin technology [Axon et al. (2022)]. Willner, specifically in the domain of the IIoT, refers to the necessity to incorporate OT and ICT to reach the goal of Industry 4.0 (2018). Axon et al. identify systemic risk in cybersecurity capability across connected sectors that should be addressed in IIoT [Varga et al. (2020)]. Varga et al. identified challenges in 5G-enabled IIoT such as edge computing, network virtualization, and security [Del Gaudio et al. (2021)].

1.2 TECHNOLOGIES ENABLING IoT AND IIoT

The effective execution of IoT and IIoT in sustainable manufacturing relies on a robust technological foundation. Several key technologies underpin these advanced systems, enabling seamless data collection, communication, and analysis.

1.2.1 Overview of Key Technologies Supporting IoT and IIoT

IoT and IIoT is based on the confluence of different advanced technologies of sensor technologies, communications protocol, cloud computing, and big data analytics. Combining all these components offers networked devices to collect information that is sent and also processes to enable automatic and smart choices in sustainable production.

A variety of enabling technologies underpin IoT and IIoT, from communication standards and messaging protocols to IoT cloud integration platforms. Key enablers in the 5G network and the Semantic Web play a very important role as well. Smart sensors, RFID (Radio Frequency Identification), and M2M are the basic technologies associated with the IoT (Figure 1.1).

1.2.2 Sensor Technologies and Their Role in Data Collection

Sensors are the primary sources of data collection within IoT and IIoT systems. They detect and measure such physical characteristics as motion, chemical composition, temperature, pressure, and humidity, among others. Sensors play a crucial role in these systems because raw data is required for analysis and decision-making.

Cano et al. (2008) discuss safety and efficiency of data acquisition in sensor networks by outlining a method of clustering of sensors in aggregation zones and then a security mechanism based on hash chains. These studies together indicate the multiple applications that have to be considered in sensor technologies employed for data collection. Nowadays, sensor technologies are highly precise, reliable, and also energy efficient (Table 1.1).

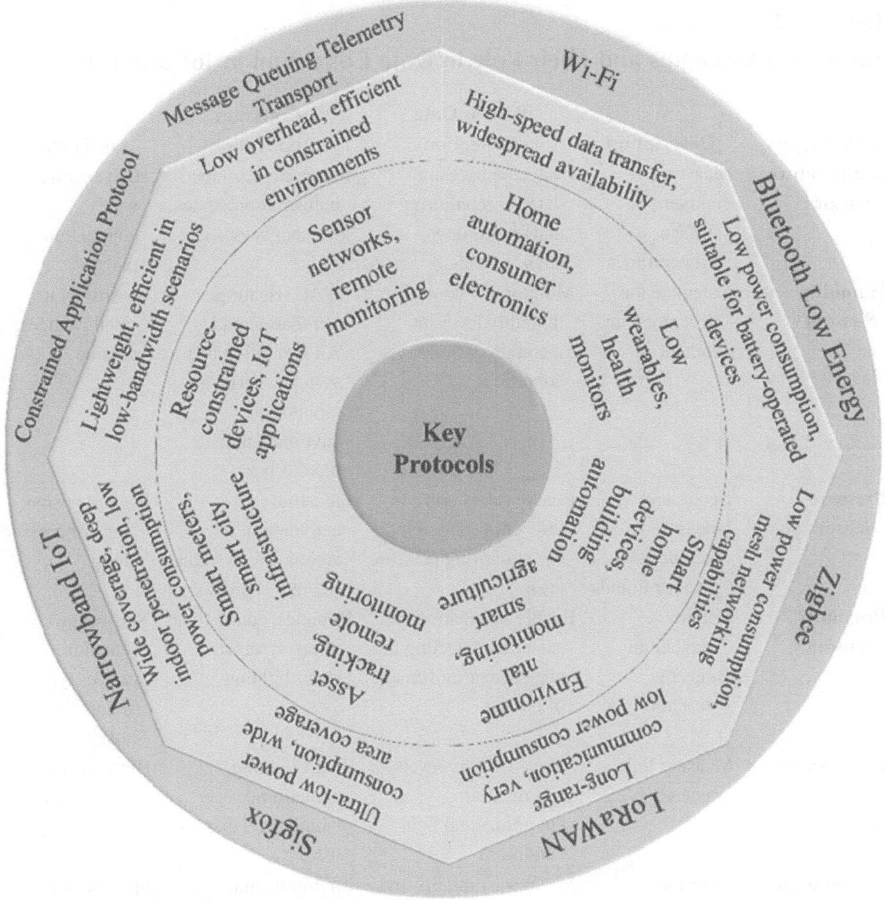

FIGURE 1.1 Types of protocols: its advantages and uses.

1.2.3 Communication Protocols for IoT and IIoT Devices

Effective communication protocols can be assured for reliable data to be exchanged between IoT and IIoT devices. Different types of protocols abound, but each has its weakness and strengths. Choosing an appropriate protocol for the kind of applications is essential, which Jadhav highlighted (2024). Ilchev et al. present a novel protocol targeted for low-cost IoT devices in heterogeneous wireless networks that focus on data encryption, integrity checking, and reliability (2017). Sharma et al. state the necessity of light-weight secure and reliable protocols for the constrained IoT systems by mentioning Constrained Application Protocol (CoAP), Message Queuing Telemetry Transport (MQTT), Extensible Messaging and Presence Protocol (XMPP), Representational State Transfer (RESTFUL) Services, IPv6 over Low-Power and Lossy Networks (6LoWPAN), and Routing Protocol for Low-Power and Lossy Networks (RPL)(2018).

The key protocols in IoT and IIoT are MQTT, a low-bandwidth protocol suitable for environments that have high latency; HTTP/HTTPS, being standard web

TABLE 1.1
Sensor Technologies and Their Role in Data Collection in IoT and IIoT

Sensor Type	Description	Role in Data Collection	Common Applications	Key Features
Temperature Sensors	Measure ambient or surface temperatures	Collect temperature data for monitoring and controlling environments	Smart thermostats, industrial processes, weather stations	High accuracy, wide temperature range
Humidity Sensors	Calculate the air's moisture content	Monitor and control humidity levels to maintain optimal conditions	HVAC (Heating, Ventilation and Air Conditioning) systems, agricultural monitoring, environmental monitoring	Sensitivity to small changes, durable
Pressure Sensors	Detect and measure pressure of gases or liquids	Ensure safety and efficiency in systems by monitoring pressure levels	Industrial automation, automotive systems, weather stations	High precision, range of pressure types
Proximity Sensors	Determine whether an object is present or not	Enable automation and safety by detecting the presence or movement of objects	Robotics, automotive safety systems, smart lighting	Non-contact detection, various types (inductive, capacitive)
Light Sensors	Measure light intensity	Adjust lighting levels and monitor environmental light conditions	Smart lighting, agricultural monitoring, display systems	High sensitivity, broad spectrum response
Gas Sensors	Detect and measure concentrations of gases	Monitor air quality and detect hazardous gases for safety and compliance	Environmental monitoring, industrial safety, HVAC systems	Specificity to gases, real-time monitoring
Motion Sensors	Detect movement within an area	Trigger actions based on movement detection for security and automation	Security systems, smart home devices, fitness trackers	Wide detection range, sensitivity to various movements
Flow Sensors	Measure the flow rate of liquids or gases	Monitor and control fluid dynamics in various applications	Water management, industrial processes, medical devices	High accuracy, various types (mass, volumetric)
Accelerometers	Measure acceleration forces	Track orientation, vibration, and movement for monitoring and control	Wearables, industrial monitoring, automotive systems	High dynamic range, multi-axis detection
Gyroscopes	Measure angular velocity	Track rotational movement for navigation and control	Drones, robotics, mobile devices	High precision, stability over time

(Continued)

TABLE 1.1 *(Continued)*
Sensor Technologies and Their Role in Data Collection in IoT and IIoT

Sensor Type	Description	Role in Data Collection	Common Applications	Key Features
Magnetometers	Measure magnetic fields	Provide directional data for navigation and detection	Compass systems, metal detection, geological surveys	High sensitivity, three-axis measurement
pH Sensors	Measure acidity or alkalinity levels	Monitor chemical compositions in various environments	Water quality monitoring, agricultural soil testing, industrial processes	High accuracy, durable materials
Image Sensors	Capture visual information	Enable visual monitoring and data collection through image and video analysis	Surveillance, medical imaging, industrial inspection	High resolution, wide dynamic range

protocols for internet data transfer; and CoAP which is used to enable devices to interact over the internet and make simple constrained devices operate. Bluetooth and Bluetooth Low Energy (BLE) are meant for short-range device connections and LoRaWAN-low power, long-range Wide Area Network. It is ideal for remote battery-operated devices in most applications (Figure 1.2).

Several works have explored different protocols for IoT applications. One example presented the potential use of Internet Protocol Version 6 (IPv6) over BLE systems through Message Queuing Telemetry Transport for Sensor Networks (MQTT-SN) and CoAP in an IP-based application. Another one compared the two with MQTT being a protocol with lesser delay with a low packet loss, but it produced more extra traffic as opposed to CoAP. CoAP was also suited for media propagation within low-power lossy networks. Additionally, it was tested in Wireless Sensor Networks with Quality of Service Level-2 (QoS 2) identified as the most reliable but slowest publish type and to ensure efficient and safe data transmission for real-time monitoring and control in production processes (Figure 1.3).

1.2.4 Cloud Computing and Data Analytics for IoT and IIoT Applications

Cloud computing plays a critical role in the IoT and IIoT ecosystems in terms of scalable storage and powerful computational resources. In IoT and IIoT applications, it is significant to integrate cloud computing and data analytics, especially to ensure data security and privacy [Kumarage et al. (2016)]. In the context of IIoT, using machine learning (ML) algorithms and predictive analysis is very important for effective data analytics. The proposed application of the concentric computing model, namely sensing systems, gateway processors, and central processors, is proposed for solving

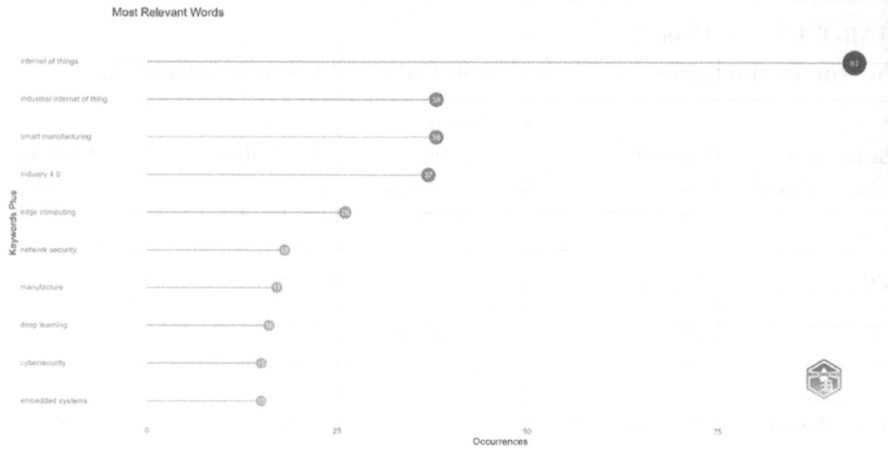

FIGURE 1.2 Most relevant words in IoT and IIoT application for sustainable manufacturing.

challenges such as real-time data analytics and security in big data analytics for IIoT [ur Rehman et al. (2018)].

Key aspects include the following:

- **Data Storage**: Cloud platforms offer vast storage capabilities, allowing manufacturers to store enormous amounts of data produced by IoT and IIoT devices.
- **Data Processing**: Cloud computing enables the processing of this data using sophisticated ML models and algorithms to extract insightful data.
- **Scalability**: Cloud infrastructure can be easily scaled up or down based on the data processing needs, providing flexibility and cost-efficiency.
- **Accessibility**: Access to cloud-stored data is possible from any location., facilitating remote monitoring and management of manufacturing processes (Table 1.2).

FIGURE 1.3 Word cloud in IoT and IIoT application for sustainable manufacturing.

TABLE 1.2
Cloud Computing and Data Analytics for IoT and IIoT Applications

Component	Description	Role in IoT/IIoT	Benefits	Examples
Cloud Storage	Cloud storage services for massive data volumes	Stores IoT/IIoT data securely and scalably	Scalability, data durability, accessibility	Amazon S3, Google Cloud Storage, Azure Blob Storage
Cloud Computing	Delivery of computing power and resources over the internet	Processes and analyzes IoT/IIoT data	On-demand resources, cost-efficiency, flexibility	Amazon Elastic Computer cloud (EC2), Google Compute Engine, Azure Virtual Machines
Data Analytics Platforms	Tools for analyzing large datasets to extract insights	Provides actionable insights from IoT/IIoT data	Real-time analysis, predictive analytics, visualization	Google BigQuery, Amazon Redshift, Azure Synapse Analytics
Edge Computing	Processing information close to its source as opposed to at a centralized data center	Reduces bandwidth consumption and delay through local data processing	Low latency, real-time processing, reduced bandwidth	Amazon Web services (AWS) IoT Greengrass, Azure IoT Edge, Google Cloud IoT Edge
Data Integration Services	Tools for integrating data from various sources	Aggregates and normalizes data from different IoT/IIoT devices for comprehensive analysis	Seamless data flow, unified data view, improved data quality	AWS Glue, Google Cloud Dataflow, Azure Data Factory
Machine Learning (ML) and Artificial Intelligence (AI) Services	Services for building and deploying ML models	Analyzes IoT/IIoT data to predict trends and anomalies	Advanced analytics, predictive maintenance, automation	Amazon SageMaker, Google AI Platform, Azure Machine Learning
Streaming Data Services	Tools for processing real-time data streams	Allows for real-time analytics and data processing for IoT and IIoT applications	Real-time insights, event-driven processing, scalability	Amazon Kinesis, Google Cloud Dataflow, Azure Stream Analytics
IoT Management Platforms	IoT device management and monitoring platforms	Facilitates device management, data collection, and remote control of IoT devices	Centralized management, device security, scalability	AWS IoT Core, Google Cloud IoT, Azure IoT Hub
Security Services	Infrastructure and data security for IoT and IIoT using tools and services	Ensures data integrity, privacy, and protection against cyber threats	Data encryption, access control, threat detection	AWS IoT Device Defender, Azure Security Center, Google Cloud IoT Security

Several research works have targeted using cloud computing to exploit storage, processing, scalability, and accessibility for IoT and IIoT applications. In this, Shukla et al. presented an IoT data storage framework based on the integration of both structured and unstructured data, leveraging Hadoop for multi-types of data (2021). Primananda et al. designed an IoT data center on a model basis that focused on access management and scalability as well (2019). Cloud resources have been considered indispensable for alleviating the strain caused by scarce resources in the IoT smart device, particularly in the areas of low latency, high data rate, quick data access, and real-time analytics of the data. Such a study has the potential for cloud computing along with the others that address the unique challenges of IoT and IIoT data.

Data analytics is critical to data extraction and practical information from the collected data by IoT and IIoT systems. It involves the following:

- **Descriptive Analytics**: Analyzing past data to understand past performance.
- **Predictive Analytics**: Utilizing ML and statistical models to forecast possible problems and future trends.
- **Prescriptive Analytics**: Recommending courses of action to improve manufacturing processes in light of predictive findings.

1.3 IIoT APPLICATIONS IN SUSTAINABLE MANUFACTURING

The integration of IIoT within the production processes in this industry has changed matters and enabled improved efficiency, real-time monitoring, optimal supply chains, and quality assurance. The following shows a number of applications of IIoT in manufacturing, thus showing how this technology transforms industrial processes.

1.3.1 Integration of IIoT in Sustainable Manufacturing Processes

It is a complex field dealing with interoperability, data aggregation, and automated reactions very crucial for the management of information systems and usage of Open Platform Communications Unified Architecture OPC UA devices in the manufacturing sectors as reported by Lee et al. (2019).

The integration of IIoT into manufacturing includes integrating sensors, devices, and intelligent systems into machines and lines of production. The devices collect information and send it over to central systems for processing in a way that can optimize the processes involved. The core aspects of the integration of IIoT include the following:

- **Interconnected Devices**: Machines, tools, and equipment equipped with sensors and communication modules to facilitate data transmission in real time.
- **Data Collection and Analysis**: Constant monitoring of operational parameters and performance metrics, with data sent to cloud or edge computing platforms for analysis.
- **Automation and Control**: Automated systems that adjust operations based on data insights, enhancing efficiency and reducing human intervention.

1.3.2 Real-Time Monitoring and Control Systems

Among the most significant applications of IIoT in manufacturing is real-time monitoring and control, which offer continuous oversight with instant responses to inefficiencies. According to researchers, IoT-based systems are versatile and efficient in monitoring physical parameters and industrial processes. Low-latency communication is the importance of IIoT for power system monitoring, and the development of high-speed, IIoT-based monitoring systems for power substations has been proposed. Moreover, the suitability of IoT-based online monitoring systems for electrical equipment management is discussed. In a nutshell, these studies are some of the essential roles that IIoT plays in enhancing monitoring and control across different industrial applications. Benefits are as follows:

- **Operational Visibility**: Real-time data provides a comprehensive view of the entire manufacturing process, from raw material intake to finished product output.
- **Predictive Maintenance**: By continuously monitoring equipment health, IIoT systems have the ability to anticipate possible problems before they arise, enabling prompt maintenance and minimizing downtime.
- **Process Optimization**: Real-time adjustments to production parameters based on data analytics ensure optimal performance, reducing waste and improving yield.

1.3.3 Supply Chain Optimization through IIoT

Much potential has been explored regarding IoT technology and supply chain optimization, indicating its capability for performance enhancement and revenue improvement. Researchers report that IoT can help increase efficiency within supply chains, with some focusing specifically on e-commerce, as IoT assists with monitoring and managing product flow. Examples of real manufacturing applications have shown how IoT capabilities can be used to save costs, improve quality of goods, and optimize the utilization of resources. These findings, taken collectively, underscore IoT's transformative impact on supply chain management in driving efficiency and effectiveness across various sectors.

By offering visibility and control across the movement of resources and goods, IIoT improves supply chain management. Application includes the following:

- **Inventory Management**: Automated tracking of inventory levels using sensors and RFID tags ensures optimal stock levels and reduces the risk of shortages or overstocking.
- **Logistics and Transportation**: Real-time tracking of shipments and logistics operations improves delivery accuracy and efficiency, reducing lead times and transportation costs.
- **Demand Forecasting**: Data collected from various points in the supply chain enables accurate demand forecasting, allowing manufacturers to adjust production schedules and inventory levels accordingly.

1.3.4 Quality Assurance and Product Lifecycle Management with IIoT

IIoT applications have transformed the sustainable manufacturing of quality assurance and product lifecycle management. Even though the considerations of both software and hardware form the focus, these research studies present a means for managing the quality of products throughout its life cycle. Ondemir et al. (2014) develop an optimization approach to the quality management of product recovery through IIoT, while Tuncer et al. raise the necessity of quality assurance applications to be automated within the framework of IIoT. This shows that there is a capability of IIoT in quality assurance and product lifecycle management.

Key contributions include the following:

- **Enhanced Quality Control**: IIoT-enabled machine vision and sensors detect defects and deviations in real time, ensuring high-quality standards are maintained throughout the production process.
- **Traceability**: Comprehensive tracking of materials and components from suppliers to the end product provides full traceability, facilitating recalls and ensuring compliance with regulatory standards.
- **Lifecycle Management**: Data collected throughout the product lifecycle, from manufacturing to usage in the field, informs design improvements, maintenance schedules, and end-of-life disposal strategies.

Examples of IIoT Applications:

- **Smart Factories**: Smart factories integrate interconnected machines and systems, enabling synchronized operations and automated decision-making. Automotive manufacturers use IIoT to optimize assembly line performance and detect defects early.
- **Food and Beverage Industry**: IIoT systems monitor critical factors such as temperature, humidity, and hygiene in real time, ensuring compliance with food safety standards and reducing the risk of contamination.
- **Pharmaceutical Manufacturing**: Here, precision and compliance have the most significant importance. IIoT allows for strict conditions to be monitored in manufacturing processes and automatically reports back to regulatory bodies, resulting in higher quality and adherence.

1.4 MACHINE VISION AND PREDICTIVE MAINTENANCE USING IIoT

The integration of machine vision and predictive maintenance within IIoT frameworks represents a significant advancement in manufacturing, enhancing both quality control and operational efficiency. What these technologies entail, the strategies used in their integration, and whether they are experienced practically can be seen in the following case studies from real-world applications.

1.4.1 Utilizing Machine Vision for Quality Control in Manufacturing

This permits the real-time inspection of the product and the manufacturing process and has prospects in controlling the quality of manufacture. Bekbayeva et al. (2023)

and Chiou et al. (2009) show that convolutional neural networks are applied for such. Gallagher discusses the barriers to achieving 100% product compliance, including the cost and speed of machine vision systems (1995).

In manufacturing, machine vision systems can:

- **Find Any Flaws**: Identify imperfections like fractures, dents, or surface anomalies on products with high precision.
- **Ensure Compliance**: Verify that products meet specific dimensional, structural, and aesthetic standards.
- **Monitor Production**: Continuously observe the production process to ensure that all steps are performed correctly and consistently.

Machine vision systems automate quality control, reduce manual inspections, and improve accuracy. Integrating into IIoT frameworks reduces waste, lowers rejection rates, and enhances product quality.

1.4.2 Predictive Maintenance Strategies Empowered by IIoT

Predictive maintenance leverages IIoT technologies to anticipate equipment failures before they occur. This proactive approach depends on constant monitoring and data analysis to predict when maintenance should be conducted. It was point out the need for these technologies to enhance fault detection and maintenance prediction, reduce downtime, and optimize maintenance schedules [63–64]. Kliestik et al. also emphasize the change brought by artificial intelligence (AI) in order to further improve maintenance practice and industrial performance (2023).

Key strategies are as follows:

- **Condition Monitoring**: Using sensors to track the health of equipment by measuring parameters such as vibration, temperature, and pressure.
- **Data Analytics**: Examining current and past data to find trends and abnormalities that point to possible malfunctions.
- **Machine Learning Models**: Employing advanced algorithms to predict the remaining useful life of equipment and optimize maintenance schedules.

By implementing predictive maintenance, manufacturers can avoid unexpected breakdowns, extend the lifespan of machinery, and reduce maintenance costs.

1.4.3 Case Studies Illustrating the Effectiveness of Predictive Maintenance in Manufacturing

Case Study 1: Automotive Industry: In an automobile manufacturing plant, IIoT-enabled predictive maintenance for conveyor belts and robotic arms was implemented. Sensors collected parameters regarding the operation of said equipment, and these parameters were analyzed to predict failures in such equipment. This thus meant that the plant could service such equipment on actual conditions prevailing and not based on set schedules. As a result of this strategy, unscheduled downtime of the plant decreased by 20%, while maintenance costs decreased by 15%.

Case Study 2: Food and Beverage Industry: A food processing company integrated IIoT and predictive maintenance in order to monitor its refrigeration units and production lines. Continuous tracking of temperature, humidity, and vibrations in the equipment ensured it would predict when a unit was likely to fail, preventing spoilage and halt in production. Thus, this helped improve overall equipment effectiveness (OEE) by 25% and resulted in great cost savings.

Case Study 3: Chemical Manufacturing: A chemical company was using IIoT sensors to monitor the pumps, compressors, and reactors. Data analytics and ML algorithms predicted when maintenance would be needed based on conditions in real time. It is here that predictive maintenance strategy reduced the reactive maintenance activity by 30%, while production uptime increased by 10%.

1.5 NETWORKING AND EDGE COMPUTING FOR IIoT

The integration of networking infrastructure and edge computing is very important for the effective deployment of IIoT systems. This chapter covers the networking requirements of IIoT devices, discusses the role of edge computing in distributed IIoT systems, and outlines the pros and cons of edge computing in the context of IIoT applications. Foukalas et al. stress the necessity of dependable wireless communications in the IIoT networks and insist on the timeliness and dependability constraints (2019).

1.5.1 Networking Infrastructure for IIoT Devices

A solid networking infrastructure provides the backbone of IIoT systems as a means for seamless connectivity among devices, sensors, and central systems. Essential elements of such an infrastructure are listed below:

- **Wireless Networks**: Technologies such as Wi-Fi, Bluetooth, and LoRaWAN enable wireless communication between IIoT devices. LoRaWAN, in particular, is ideal for long-range, low-power applications.
- **Wired Networks**: Ethernet and other wired connections provide reliable, high-speed communication necessary for critical applications.
- **Communication Protocols**: Protocols like MQTT, CoAP, and OPC UA ensure efficient and secure data exchange between IIoT devices and platforms.
- **Gateways**: These act as intermediaries between IIoT devices and the cloud or edge computing systems, aggregating data and ensuring secure transmission.

1.5.2 Edge Computing and Its Role in Distributed IIoT Systems

Edge computing is computation that occurs closer to the generation of data sources rather than in a central cloud. For IIoT systems, the significance of edge computing is critical in:

- **Reducing Latency**: Edge computing reduces latency in data transmission to and from the cloud through local data processing, allows for real-time control and decision-making.

- **Enhancing Reliability**: The local processing ensures that vital functions can be continued even in case of connectivity loss to the cloud, which enhances the reliability of the system.
- **Optimizing Bandwidth**: This means that raw data can be filtered right at the edge just to prevent so much data flowing into the cloud so as to maximize the utilization of bandwidth and thus cut expenses.

1.5.3 Advantages and Challenges of Edge Computing in IIoT Applications

Advantages:

i. **Real-Time Processing**: Edge computing provides instant analysis and response to data, which is important in applications such as predictive maintenance and real-time quality control.
ii. **Reduced Latency**: Local data processing removed the latency of cloud-based systems, and it enhances system performance along with increased response times.
iii. **Enhanced Security**: Edge computing assures the satisfaction of data privacy legislations and lowers the data breaches risk because of storing the data locally.

Challenges:

i. **Complexity**: Implementing and managing edge computing infrastructure adds complexity to IIoT systems, requiring specialized skills and resources.
ii. **Integration**: Ensuring seamless integration and interoperability between edge devices, cloud platforms, and existing IT infrastructure can be challenging.
iii. **Data Management**: Managing and synchronizing data between edge and cloud environments requires robust data governance and synchronization strategies.

1.6 SECURITY IN IIoT

The IIoT is changing manufacturing and industrial processes through connecting devices, sensors, and systems, all being connected to enable unprecedented levels of automation as well as data analytics. In the same token, connectivity also brings with it various security threats and vulnerabilities that must be addressed in order to protect IIoT ecosystems.

1.6.1 Threats and Vulnerabilities in IIoT Ecosystems

IIoT ecosystems remain vulnerable to several threats and vulnerabilities as a result of highly interconnected and integrated systems. Threats include Distributed Denial of services (DDoS) attacks, malware, ransomware, as well as unauthorized access to industrial processes that result in operation disruptions, sensitive data theft, and manipulation through inherent vulnerabilities [Kebande et al. (2024)].

1.6.2 Security Measures and Protocols for Safeguarding IIoT Devices and Data

Some hazards need to be placed with sophisticated security protocols and safeguards. For this, authentication and authorization of the devices are critical in order to ensure only authorized devices and users could access the IIoT network. Digital certificates and multi-factor authentication (MFA) can be used for this purpose [Mishra et al. (2023)]. Encryption is necessary when the data is in motion and when it is at rest to prevent private data from being intercepted and accessed in unauthorized ways. Patch management and updates of software are essential in fixing known vulnerabilities and stopping the bad actors from exploiting them. Real-time alerts and network traffic monitoring, including activity from intrusion detection and prevention systems, can detect potential attacks quickly. Network segmentation keeps an attack away by segregating vital devices and systems.

1.6.3 Best Practices for Ensuring the Security of IIoT Deployments

There is also a huge potential in securing IIoT deployments by having best practices throughout the lifecycle of a system. It can encompass thorough risk assessments and vulnerability analyses that will discover areas of security gaps; these should rank high on the priority list in terms of mitigative measures. In addition, there should be an extensive security policy uniquely suited to the IIoT environment that needs to be implemented and enforced. The policy should define procedures for secure device provisioning, access controls, and incident response procedures. Awareness and training programs among employees are also critical as human error leaves the security defenses weak [Kumari (2024)]. Coordination with industry standards bodies and compliance with relevant laws add further strength to the security posture of IIoT deployments [Laszka et al. (2020)]. Finally, a culture of continuous learning and keeping up with emerging threats and new technologies ensures that security measures change with the challenges.

1.7 CASE STUDIES OF IIoT IMPLEMENTATION

Significant contributions and concrete developments across the board in various industries are properly implemented in real life through real-time IIoT in manufacturing. Under this section, successful implementation cases, lessons learned, and the positive impacts on productivity, efficiency, and cost savings are included.

1.7.1 Real-World Examples of Successful IIoT Deployments in Manufacturing

Various case studies provide evidence of successful IIoT deployment in manufacturing. For instance, the use of IIoT-related practices practiced by Zenitaka Corporation, Tsuchiya-Gousei, Toyota, and Hitachi has boosted operational efficiency with less intervention from the workforce. Therefore, the role of the production monitoring system using IIoT has been identified as the best way forward to ensure the proper maintenance of manufacturing assets. Finally, various case studies have looked at the role of IIoT in smart factories, for example, designing the next-generation Manufacturing Execution System (MES)/Manufacturing Operations Management (MOM) to improve the responsiveness of the system [Mantravadi et al. (2022)].

Case Study 1: General Electric (GE) – Predix Platform: General Electric developed its Predix Platform as an IIoT-based software that helped to optimize manufacturing processes with higher performance and efficiency. Linking machines, sensors, data analysis tools, and, more importantly, even doing predictive equipment failure in place of maintenance on a better schedule to provide the best possible operational efficiency will ensure data collection. In real-time analysis, less downtime and more production were realized.

Case Study 2: Siemens – MindSphere: Siemens had installed its IIoT ecosystem known as MindSphere in all its plants. This has enabled data-driven decisions for all the plants of Siemens. It had integrated various sensors and machines through this platform to enable Siemens to understand equipment health, monitor production metrics, and optimize resource utilization. Siemens used MindSphere to get the benefits of predictive maintenance to reduce breakdowns and related maintenance costs and increase OEE.

Case 3: Bosch – Connected Industry: Bosch brought its Connected Industry solution to its plant that manufactured automobile parts. This system of the IIoT used sensors, machinery, and analytics that gave real-time visibility into its manufacturing process. Those interconnections provided Bosch with the potential to have better quality control, optimize its supply chain's logistics, and eliminate waste. These Connected Industry solutions further made energy management empowered, and lots of energy savings, then led to lesser operational costs.

1.7.2 Lessons Learned and Insights Gained from IIoT Case Studies

Some of the lessons and insights drawn from these successful implementations of IIoT include the following:

i. **Data Integration and Interoperability**: Data interoperability between different systems and devices that are part of the IIoT implementation is expected to be in place so that a diverse set of sources is integrated. Standardized communication protocols as well as data formats are prerequisites for such integration.

ii. **Scalability and Flexibility**: An IIoT system needs to scale in line with growth in manufacturing operations. It addresses varied business requirements with a flexible architecture that provides for significant increases in sensors, devices, and analytics capability.

iii. **Predictive Maintenance**: In one of the most vital benefits, it predicts the equipment failure even before it happens. Predictive maintenance can significantly reduce the downtime and cost of maintenance. Thus, the productivity will increase.

iv. **Data Security and Privacy**: In IIoT implementations, data security and privacy must be guaranteed. Restrict access, authenticate, encrypt, etc. are some of the significant security protocols that should be followed in the protection of critical manufacturing data against online attacks.

1.7.3 Impact of IIoT on Productivity, Efficiency, and Cost Savings in Various Industries

The impact of IIoT on manufacturing has been profound, leading to measurable improvements in productivity, efficiency, and cost savings across various industries:

- **Increased Productivity**: IIoT gives real-time monitoring and automation of manufacturing operations with effective results in higher production rates and reduced cycle times. There is a higher throughput that continues to keep up with low interaction from people.
- **Enhanced Efficiency**: IIoT helps manufacturers optimize the usage of resources and minimize waste by giving insights into equipment performance and production metrics. It ensures rapid process adjustments through real-time data analysis and avoids downtime, thus maximizing efficiency.
- **Cost Savings**: Predictive maintenance driven by IIoT reduces the frequency of repairs and their corresponding costs, which are actually before issues get major. The right kind of supply chain management and energy efficient measures are also a good contributor to this.

1.8 CHALLENGES AND FUTURE DIRECTIONS

1.8.1 Current Challenges and Obstacles Facing Widespread Adoption of IIoT in Manufacturing

The challenges that the general adoption of IIoT in manufacturing faces include the complexity of systems and interoperability between different kinds of devices and systems. Issues include workforce skills, system inefficiencies, and data management in the application of IIoT in manufacturing [Madugula (2021)]. In terms of the electronics manufacturing industry in Bangladesh, the challenge of Return on Investment (ROI) estimation is particularly significant.

1.8.2 Emerging Trends and Technologies Influencing the Future of IIoT

It's a rapidly evolving sector for improving operational efficiency and productivity within the industrial systems. A smooth connection of people with each other and objects among each other is expected for making more informed decisions for management of assets in future, and thus, an innovation of IIoT becomes a major driver for adding value economically while creating momentum toward the transformation of such conventional industries.

1.8.3 Strategies for Addressing Challenges and Maximizing the Potential of IIoT in Sustainable Manufacturing

To fully exploit IIoT in manufacturing, manufacturers must have a well-thought-out strategy, invest in cybersecurity, apply AI and ML for predictive maintenance, adapt 5G and edge computing for real-time processing of data, and create an environment of continuous learning and innovation [Das et al. (2024)]. Collaboration with technology providers, industry consortia, and standards bodies are important for interoperability and standardization. Security audits and vulnerability assessments need to be done periodically.

1.9 IMPLICATION OF THE STUDY

These studies presented in this chapter depict how IoT and IIoT can change sustainable manufacturing through increased productivity, quality, and creativity. Through a combination of modern sensor technology, real-time monitoring, and data analytics, manufacturers can enhance the quality of production, minimize downtime, and boost operations. Proactive maintenance and quality control also provide significant efficiency improvements and cost savings in operations through IIoT-based machine vision and predictive maintenance.

Critical roles for the edge computing and robust networking infrastructure in the IIoT system are enabling real-time data processing, minimizing latency, and providing reliability of the system. However, the studies show some of the major challenges associated with these, such as legacy integration, cybersecurity threats, and a need for standardization. Challenges like these would need proper security measures, periodic updates, and following best practices.

Emerging trends such as AI, 5G, and edge computing are expected to further revolutionize the applications of IIoT. These technologies will make the sustainable manufacturing processes more agile, responsive, and environmentally friendly, thus significantly improving the productivity and impacts on the environment.

The implications are deep in terms of manufacturing in the future. This will involve all stakeholders in manufacturers, providers of technologies, policymakers, and investors having to work in tandem to ensure adoption and investment in IIoT technologies for sustainable growth, innovation, and competitiveness in the new industrial landscape that is emerging so fast. The studies called for prompt action for maximizing the full potential of IIoT to ensure the sustainable manufacturing industry can survive the digital era.

1.10 CONCLUSION

This chapter unfolds the deep impact of IoT and IIoT on sustainable manufacturing, focusing on key technologies, applications, and strategies required for their effective implementation. This book starts with basics about IoT and IIoT, underlining their importance in modern sustainable manufacturing. The discussion then took an off-on tangent about the enabling technologies: sensor technologies, communication protocols, cloud computing, and data analytics. Both support the functionality and scalability of IoT and IIoT systems.

It further investigates IIoT applications in sustainable manufacturing: real-time monitoring, supply chain optimization, quality assurance, and product lifecycle management. Integration of machine vision and predictive maintenance improved the quality control and operational efficiency in case studies showing significant reductions in downtime and cost savings. Networking infrastructure and edge computing were also covered, where the role of such was discussed in providing real-time processing and less latency in order to allow for improved system reliability. Security was considered the most significant concern, with special regards to threats and vulnerabilities in the IIoT ecosystem. To protect data, therefore, a strong security measure was required in terms of protocol and best practices. It identified current and future challenges within this chapter: some of the mentioned existing challenges are lack of integration, absence of standards, and cyberattacks. There exist current trends which include emerging futures related to AI, 5G, and edge computing for defining future IIoT trends.

The implications of IoT and IIoT for the future of manufacturing are deep. These technologies promise to revolutionize industrial operations, enabling more efficient, flexible, and responsive manufacturing processes. Manufacturers can reach unprecedented levels of productivity, quality, and innovation. The adoption of IoT and IIoT will not only enhance operational efficiency but also drive sustainable growth by reducing waste, optimizing resource utilization, and improving overall environmental impact. Therefore, it is indispensable for the stakeholders – be they manufacturers, technology providers, policymakers, or investors – to embrace and invest in IIoT. Manufacturers must formulate complete IIoT strategies, emphasize cybersecurity, and integrate emerging technologies to survive. Technology providers and industry consortia should look forward to standardization and interoperability in a way that allows effortless integration and scalability. Policymakers need to establish supporting regulatory frameworks, in this case, that facilitate and promote innovation but are vigilant about data privacy protection. Investors need to realize the impact of IIoT and sustain fundings that promote technology and industrial modernization.

REFERENCES

Axon, Louise, Katherine Fletcher, Arianna Schuler Scott, Marcel Stolz, Robert Hannigan, Ali El Kaafarani, Michael Goldsmith, and Sadie Creese. "Emerging Cybersecurity Capability Gaps in the Industrial Internet of Things: Overview and Research Agenda." Digital Threats: Research and Practice 3, no. 4 (2022): 1–27.

Bekbayeva, R., Ye. Ospanov, K. Bekbayev, D. Kozhakhmetova, B. Zhapar, Ye. Zhanuzakov, and B. Mussayev. "Automation of quality control in manufacturing using machine learning and computer vision: System development and performance analysis." Bulletin D. Serikbayev of EKTU, December 2023, 77–91. https://doi.org/10.51885/1561-4212_2023_4_77.

Cano, Cristina, Manel Guerrero, and Boris Bellalta. "Secure and efficient data collection in sensor networks." In International workshop of the EuroNGI network of excellence, pp. 37–48. Springer Berlin Heidelberg, 2008.

Chiou, Richard, Prathaban Mookiah, and Yongjin Kwon. "Manufacturing e-Quality through Integrated Web-Enabled Computer Vision and Robotics." The International Journal of Advanced Manufacturing Technology 43 (2009): 720–730.

Das, Dipan Kumar, Priyambada Mallick, Siddharth Parthasarathy, Santosh Ku Satpathy, and Padmaja Patnaik. "The transformation of manufacturing by artificial intelligence." In Industry applications of thrust manufacturing: Convergence with real-time data and AI, pp. 42–73. IGI Global, 2024.

Del Gaudio, Daniel, and Pascal Hirmer. "The IoT Vision: Challenges and Research Gaps." International Journal of Organizational and Collective Intelligence (IJOCI) 11, no. 4 (2021): 1–12.

Deshpande, Sujit N., and Rashmi M. Jogdand. "A Survey on Internet of Things (IoT), Industrial IoT (IIoT) and Industry 4.0." International Journal of Computer Applications 175, no. 27 (2020): 20–27.

Foukalas, Fotis, Paul Pop, Fabrice Theoleyre, Carlo Alberto Boano, and Chiara Buratti. "Dependable wireless industrial IoT networks: Recent advances and open challenges." In 2019 IEEE European Test Symposium (ETS), pp. 1–10. IEEE, 2019.

Gallagher, Paul K. "Vision systems for quality control." In IEEE Technical Applications Conference and Workshops. Northcon/95. Conference Record, pp. 381–388. IEEE, 1995.

Ilchev, Svetozar, and Zlatoliliya Ilcheva. "Internet-of-Things communication protocol for low-cost devices in heterogeneous wireless networks." In Proceedings of the 18th International Conference on Computer Systems and Technologies, pp. 272–279. 2017.

Jadhav, Ashish Madhukar. "Communication Protocols for IoT Applications." International Journal for Research in Applied Science and Engineering Technology 12, no. 4 (April 30, 2024): 5875–5877. https://doi.org/10.22214/ijraset.2024.61338.

Jeschke, Sabina, Christian Brecher, Tobias Meisen, Denis Özdemir, and Tim Eschert. Industrial internet of things and cyber manufacturing systems. Springer International Publishing, 2017.

Kebande, Victor R., and Ali Ismail Awad. "Industrial Internet of Things Ecosystems Security and Digital Forensics: Achievements, Open Challenges, and Future Directions." ACM Computing Surveys 56, no. 5 (2024): 1–37.

Kliestik, Tomas, Elvira Nica, Pavol Durana, and Gheorghe H. Popescu. "Artificial Intelligence-Based Predictive Maintenance, Time-Sensitive Networking, and Big Data-Driven Algorithmic Decision-Making in the Economics of Industrial Internet of Things." Oeconomia Copernicana 14, no. 4 (2023): 1097–1138.

Kumarage, Heshan, Ibrahim Khalil, Abdulatif Alabdulatif, Zahir Tari, and Xun Yi. "Secure Data Analytics for Cloud-Integrated Internet of Things Applications." IEEE Cloud Computing 3, no. 2 (2016): 46–56.

Kumari, Anusha. "A Comprehensive Survey of Security Challenges and Solutions of Industrial Internet of Things (IIoT) Networks." International Journal of Scientific Research in Engineering and Management 08, no. 04 (April 24, 2024): 1–5. https://doi.org/10.55041/ijsrem31337.

Laszka, Aron, Waseem Abbas, Yevgeniy Vorobeychik, and Xenofon Koutsoukos. "Integrating Redundancy, Diversity, and Hardening to Improve Security of Industrial Internet of Things." Cyber-Physical Systems 6, no. 1 (2020): 1–32.

Lee, Chao-Hsien, Zheng-Lin Wu, Yun-Ting Chiu, and Vi-Shu Chen. "Heterogeneous industrial IoT integration for manufacturing production." In 2019 International Symposium on Intelligent Signal Processing and Communication Systems (ISPACS), pp. 1–2. IEEE, 2019.

Madugula, Lavanya. "Applications of IoT in Manufacturing: Issues and Challenges." Journal of Advanced Research in Embedded System 8, nos. 1 & 2 (2021): 3–7.

Mantravadi, Soujanya, Charles Møller, and Reto Schnyder. "Design Choices for Next-Generation IIoT-Connected MES/MOM: An Empirical Study on Smart Factories." Robotics and Computer-Integrated Manufacturing 73 (2022): 102225.

Mathur, Puneet. "Overview of IoT and IIoT." IoT Machine Learning Applications in Telecom, Energy, and Agriculture: With Raspberry Pi and Arduino Using Python (2020): 19–43.

Mishra, Amit Kumar, and Mohammad Wazid. "Design of a cloud-based security mechanism for Industry 4.0 communication." In 2023 Third International Conference on Secure Cyber Computing and Communication (ICSCCC), pp. 337–343. IEEE, 2023.

Ondemir, Onder, and Surendra M. Gupta. "Quality Management in Product Recovery Using the Internet of Things: An Optimization Approach." Computers in Industry 65, no. 3 (2014): 491–504.

Primananda, Rakhmadhany, Reza Andria Siregar, and Mochamad Atha. "Cloud-Based Data Center Design as a Data Storage Infrastructure on Internet of Things." Journal of Information Technology and Computer Science 4, no. 2 (2019): 185–192.

Serpanos, Dimitrios, Marilyn Wolf, Dimitrios Serpanos, and Marilyn Wolf. "Industrial Internet of Things." In Internet-of-Things (IoT) systems: Architectures, algorithms, methodologies, pp. 37–54. 2018.

Sharma, Cheena, and Naveen Kumar Gondhi. "Communication protocol stack for constrained IoT systems." In 2018 3rd International Conference on Internet of Things: Smart Innovation and Usages (IoT-SIU), pp. 1–6. IEEE, 2018.

Shukla, Ankita, Priyatam Reddy Somagattu, Vishal Krishna Singh, and Mala Kalra. "IoT-based data storage for cloud computing applications." In Advances in artificial intelligence and data engineering: Select proceedings of AIDE 2019, pp. 1455–1464. Springer Singapore, 2021.

ur Rehman, Muhammad Habib, Ejaz Ahmed, Ibrar Yaqoob, Ibrahim Abaker Targio Hashem, Muhammad Imran, and Shafiq Ahmad. "Big Data Analytics in Industrial IoT Using a Concentric Computing Model." IEEE Communications Magazine 56, no. 2 (2018): 37–43.

Varga, Pal, Jozsef Peto, Attila Franko, David Balla, David Haja, Ferenc Janky, Gabor Soos, Daniel Ficzere, Markosz Maliosz, and Laszlo Toka. "5G Support for Industrial IoT Applications—Challenges, Solutions, and Research Gaps." Sensors 20, no. 3 (2020): 828.

Willner, Alexander. "The industrial internet of things." In Internet of things A to Z: Technologies and applications, pp. 293–318. 2018.

2 AIoT Application in Smart Eco-Friendly Manufacturing System

Arun Kumar Singh and Alak Kumar Patra

2.1 INTRODUCTION

Artificial Intelligence of Things (AIoT) has the prospective to significantly improve sustainability, optimize production processes, and minimize environmental impact when applied to intelligent, environmentally friendly manufacturing systems. By incorporating Internet of Things (IoT) sensors to track machine performance in real time, AIoT can facilitate projecting maintenance of manufacturing equipment by analyzing sensor data; machine learning algorithms can identify early indicators of equipment failure or degradation, and this enables proactive maintenance to minimize unplanned downtime and save energy. With the help of energy pattern analysis and detection of opportunities for optimization, energy management with the help of AIoT systems may leverage energy consumption in manufacturing facilities. To predict the energy consumption, how production is to be planned, and proposing ideas and alterations with regard to energy saving like process or equipment changes, a historical record is used by the machine learning algorithms. It means that observing such failures and inefficiencies in production lines, AIoT-based waste reduction systems may prevent wasting excessive amounts of materials in industrial processes that take place in industries. Innovative solutions based on artificial intelligence (AI) analyze data from IoT gadgets to detect improvement and waste saving potentials. These sensors gather details concerning the quality of the products to be manufactured, the material to be used, and other properties of the manufacturing process. In sustainable and advanced industrial processes, the AIoT helps provide supply chain transparency and sustainability. Some of the IoT sensors detect the movement of components, finished products, and raw materials within the supply chain. Then this data is evaluated by AI algorithms to ensure that sustainability needs have been met in terms of inventory and emission transport. In smart manufacturing operations, AIoT facilitates conformity to the concepts of autonomy and flexibility in production, as well as the ability to alter the production process according to the environmental conditions, available resources, or even the customers' requirements. Internet-connected devices gather real-time information on the performance of production line, ambient conditions, and market information, while artificial intelligence (AI) processes and controls the schedules and resources and parameters to yield the highest output and the lowest environmental effect. Potential environmental parameters consist of air quality, water consumptions, and waste generation for manufacturing companies to meet environmental standards while reducing their environmental footprints with the help

DOI: 10.1201/9781003503323-2

of AIoT systems. IoT sensors measure various environmental variables that are then analyzed by AI that predicts environmental problems, patterns, and needed corrective measures to sustain environmentally friendly operations. Through the data collection and analysis of the environmental cost at each phase in the product's life cycle, the AIoT can help with product life-cycle assessment. IoT sensors observe product consumption, maintenance processes, and end-of-life processes and AI algorithms analyze environmental impact of product in terms of carbon foot printing, water usage, energy usage for eco-designing, and sustainability management. AI and IoT equalized with intelligent and environmentally friendly production systems offer a lot of opportunities to increase productivity, minimize losses, and enhance sustainability in value chains of the industrial sector. Using AIoT, manufacturers could possibly maintain market competitiveness in an ever-shifting market, implemented with compliance to the environmental regulations.

AIoT is of great help for sustainable engineering and through the connection of the dots by AIoT, firms can monitor production inputs and outputs in real time. By which cause significant cost reductions and equivalent energy savings as well as efficiency enhancements essentially occur and in addition, firms can enhance their supply chain management using AIoT. AI algorithms are then used to analyze data collected from IoT sensors applied in firms' supply chain in order to determine problem areas. Identification is another step that can be followed by educative action and there are many benefits that a business can get from it such as time delivery, minimal wastage, and transportation cost. AIoT is used in sustainable manufacturing for predictive maintenance whereby manufacturers are able to identify possible breakdowns of their equipment and take appropriate measures in a bid to avoid productivity losses that may be costly by using AI algorithms to analyze data that is collected from IoT sensors. This leads to reduced maintenance costs, higher equipment reliability, and low wastage with the integration of these technologies; it enhances the usage of energy by varying the settings of the equipment through real-time data obtained from the IoT. AIoT's capability to enhance product design is a plus, which is least considered. By extracting customer feedback as well as patterns in product usage, manufacturers might design better products that are environmentally sound to meet client needs at the same time. This is a step toward sustainability but at the same time it retains a company's focus on its customers.

2.2 SMART ECO-FRIENDLY MANUFACTURING SYSTEM

The "smart eco-friendly" approach integrates state-of-the-art technologies, eco-friendly methods, and creative solutions to minimize ecological damage and enhance ecological sustainability in a number of areas: smart eco-friendly approach in flips and integrates smart technologies, data analytics, AI, and IoT. These technologies enable real-time monitoring, analysis, and optimization of resource use, energy consumption, and environmental impact. The intelligent environment-friendly plan has to set in order the energy efficiency by putting into place energy-saving strategies such as smart meters, Light Emitting Diode (LED) lighting, energy-efficient appliances, and automated systems that maximize the energy usage in productions, transportation, and building operations. Renewable sources of energy need to be put into place in order to reduce dependency on fossil fuel and cut down greenhouse gas emissions. This requires integrating renewable energy technology into infrastructure, transportation, and building systems, reduction of waste, and conservation of resources – all under the smart eco-friendly strategy. That means applying the ideas of the circular economy: to design products that will be more durable, repairable, and recyclable and to encourage all industries to reduce waste and develop reusing and recycling programs. Investing in green infrastructure – like green buildings, green spaces, and schemes for sustainable transport – helps improve the quality of life both in urban and rural regions, reduces environmental impact, and improves air quality accordingly. Examples of green infrastructure solutions include, but are not limited to, green tops, permeable pavement, urban trees, and public transportation systems powered by renewable energy. Conscious eco-friendly transportation strategies encourage walking, cycling, public transportation, and electric cars or electric vehicles (EVs). Smart mobility technologies can be integrated to bring down traffic congestion, air pollution, and carbon emissions. These include ride-sharing platforms, intelligent transportation systems (ITS), and EV charging infrastructure. It would be enabled in real time by the environmental monitoring systems: soil health, biodiversity, water quality, and air quality. IoT devices with remote-sensing technologies can collect, analyze, and provide visualizations of environmental data to support proactively managing the environment with well-informed decisions. In sustainable culture development and behavior change, stakeholder and community involvement is imperative. Awareness, outreach, and education provide the public with knowledge of environmental concerns and give communities and individuals the tools they will need to take action toward a sustainable future. Strong policy frameworks and efficient governance processes are key to systemic changes in sustainable development. It is a major responsibility of governments, legislators, and regulatory agencies to provide the necessary legislation, regulations, and incentives for encouraging wise environmental activities that hasten the shift to a sustainable society.

The whole world, nowadays, is toward ecological, social, and corporate governance concerns. Thus, businesses are constantly being asked to be more ecologically and socially responsible. Indeed, today many buyers and consumers presently rate and select suppliers, manufacturers, and service providers on their respective Environmental, Social, and Governance (ESG) scores. It is the social and environmental movements of today affecting every company and organization, yet the compliance weight now seems to fall on manufacturing. Sustainability has then rapidly become one of the major commercial drivers. With the enabling power of modern manufacturing technology, sustainability measures can empower business to create actionable information that can be utilized in supply chain optimization, initiatives related to the environment, and better productivity targets for higher profit margins. Automation, productivity, and operational efficiency are concepts commonly thought of in relation to Smart Manufacturing or Industry 4.0 or the Industrial Internet of Things (IIoT). Although this is, of course, the case, the advantage of smart manufacturing technology involves watching the consumption of both the processes and the resources which are needed throughout the entire productive process. This would also include waste, water, energy, and other materials. Because most metrics of sustainability evaluation are plotted at the plant level, combining the sustainability indicators within a particular plant or plant line is quite useful. Normally, it allows the producers to:

- Lower cost and waste while improving operations
- The ability to turn to or find more clients in order to extend the competitive advantage

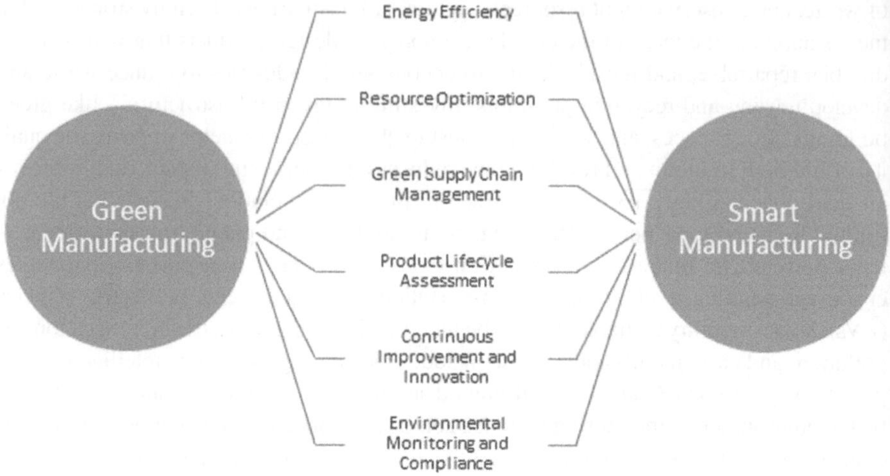

FIGURE 2.1 Smart–green approach.

- Build and increase public trust in order to promote brand image
- Provide long-term performance and survival of the business performance
- Use or address regulatory constraints and use regulatory advantage (Figure 2.1)

Cutting-edge technologies are used by an intelligent, environmentally friendly manufacturing system to streamline operations, lessen their negative effects on the environment, and enhance sustainability. By utilizing technology like LED lighting, energy-efficient machinery, and automated systems that reduce energy usage during production operations, smart manufacturing systems prioritize energy efficiency. To further lessen carbon footprint, renewable energy sources like solar or wind power may also be incorporated. Utilizing sophisticated analytics and optimization techniques, these systems maximize the utilization of resources including energy, water, and raw materials. IoT sensors gather resource usage data, which AI systems then evaluate to find inefficiencies and suggest optimization techniques to reduce waste and optimize resource use. Smart manufacturing solutions aim to decrease waste generation and promote recycling throughout the production process. This may require implementing waste reduction techniques, such as waste segregation, recycling programs, and lean manufacturing principles, in order to lessen landfill garbage and conserve resources. Eco-friendly manufacturing techniques support green supply chain initiatives by extending sustainability efforts outside of the manufacturing floor. This includes finding items from sustainable sources, promoting environmentally friendly packaging and logistical practices, and optimizing transportation routes to reduce emissions. Environmental monitoring and assessment: these systems make use of environmental monitoring technologies to monitor and control environmental performance metrics like waste generation, greenhouse gas emissions, and air and water quality. Ensuring environmental compliance involves both taking corrective action and routinely observing conformity with environmental laws and standards.

Work in partnership with stakeholders and industry partners, as well as involving staff members in sustainability projects to promote change. To establish a production environment that is both socially and environmentally responsible, a smart eco-friendly manufacturing system combines cutting-edge technologies, sustainable practices, and ongoing improvement initiatives. These systems line up sustainability, which not only minimizes its impact on the environment but also increases operational effectiveness, lowers costs, and boosts competitiveness in the global market. A smart eco-friendly manufacturing system that takes part these elements and features produces a more resilient, competitive, and sustainable manufacturing environment that strikes a balance between environmental responsibility and economic growth. Smart, environmentally friendly manufacturing technologies will be crucial in determining the direction of manufacturing in the future as companies place a greater emphasis on sustainability and environmental stewardship [Peter et al. (2022); Ruben et al. (2017); Saxby et al. (2020)] (Table 2.1).

TABLE 2.1
Breakdown of a Smart Eco-Friendly Manufacturing System

Aspect	Description
Energy Efficiency	• Integration of equipment and energy-efficient machinery. • Application of LED lighting and smart sensors to optimize energy usage. • Acceptance of automated systems for energy management and optimization. • Integration of renewable energy sources such as solar or wind power.
Resource Optimization	• Deployment of advanced analytics and AI algorithms to optimize resource usage. • Observing and minimizing waste generation throughout production processes. • Putting into practice of circular economy principles to promote resource efficiency and waste reduction.
Green Supply Chain Management	• Acceptance of green supply chain practices such as sustainable sourcing and ethical procurement. • Optimization of transportation routes and logistics to reduce emissions. • Advancement of eco-friendly packaging and recycling initiatives throughout the supply chain.
Environmental Monitoring and Compliance	• Application of environmental monitoring technologies to track and manage environmental performance metrics. • Agreement with environmental regulations and standards through proactive monitoring and corrective actions. • Incorporation of IoT sensors and AI algorithms for real-time environmental monitoring and analysis.
Product Lifecycle Assessment	• Steering lifecycle assessments of products to evaluate environmental impact from cradle to grave. • Evaluating environmental metrics such as carbon footprint, water usage, and energy consumption throughout the product lifecycle. • Enlightening eco-design decisions and sustainable product development strategies based on lifecycle assessment findings.
Continuous Improvement and Innovation	• Adopting a culture of continuous improvement and innovation to drive sustainability initiatives forward. • Engaging cross-functional teams in identifying opportunities for improvement and implementing sustainable practices. • Capitalizing in research and development of eco-friendly technologies and fostering collaboration with industry partners and stakeholders.

2.3 AIoT APPLICATIONS

In the current arena, further integration of AI with the IoT has led to a surge of interest in AIoT. Artificial intelligence, now commonly referred to as AI, is a field dedicated to intelligent machines that have the capability to replicate human intelligence. In short, AI tries to do just what the term says: make the computer imitate the human senses, thinking, reasoning, and other functions. In this regard, the improvement in most industries regarding the production of better goods and services highly relies on the strongly disruptive capabilities of AI by intelligent systems [Muduli et al. (2018)]. IoT represents the integration of devices or physical objects into one system that may be enabled to collect data through devices such as sensors and pass them in real time through software. The major purpose of adopting AI and IoT is enabling an AI engine to analyze the data created and harvested through IoT devices; hence, the networking and data transport character of the IoT is integrated within the learning models in AI systems. That is to say, the integration of AI into IoT systems enhances not only their capacity for data collection and transmission but also processing the data received in as [Abubakr et al. (2020); Ali et al. (2020); Belhadi et al. (2019); Buer et al. (2021)]. Actually, new industries can take more advantage of AI than what the conventional ways provide by using AIoT. Combined IoT with AI, however, has the following advantages: two main benefits can be noted – increase of productivity and reduction of costs. AIoT provides industries with the highest level of recognition of operational efficiency. AIoT-enabled devices have the capability to collect, process, and analyze data and patterns with the use of machine learning methods. In this way, it can swiftly create insights for operations, pinpoint and rectify issues, as well as enhance the degree of automation of tasks that take too much time. As a result, through AI handling routine activities, the business can operate on reduced workforces but still deliver efficient services [Inman et al. (2018); Leong et al. (2020); Morales Méndez & Rodriguez (2017)] (Figure 2.2).

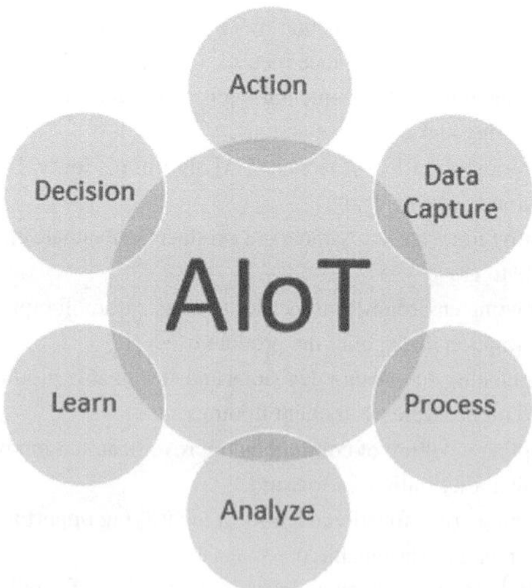

FIGURE 2.2 Elements of AIoT.

Among the IIoT that will prevail, the usage of AIoT systems will also prevail heavily. These smart systems are for monitoring machines in real time and for detecting defected parts among many others in manufacturing. These modern deep learning applications are one by one replacing the generalized machine vision systems of the bygone era used in SO skirmishes named as quality control. Our inspector works around the clock by connecting live to cheap cameras examining Vernier parts in real time, analyzed using fast AI with AI models [Caiado et al. (2020); Cherrafi et al. (2019); Chhotaray et al. (2024); Choudhary et al. (2019)] (Table 2.2).

TABLE 2.2
AIoT (Artificial Intelligence of Things) Applications across Various Domains

Domain	Description
Smart Manufacturing	• Predictive maintenance: AIoT applications forecast equipment breakdown and schedule maintenance timings by processing the data obtained from sensors across industries in real time. • Feature control: The AIoT driving improved efficiency in the inspection and quality control through the utilization of computer vision and deep learning.
Smart Healthcare	• Distant patient monitoring: AIoT devices keep track of the patient's vital signs and overall health condition with an aim of diagnosing any health complications at an early stage. • Adapted treatment: Modern AIoT systems assess patient data for individualized treatments and enhance the quality of people's health.
Smart Agriculture	• Precision farming: The AGRIoT solutions enhance the agricultural practices by keeping track of the environment conditions, the moisture content of the soil, and the health of crops; hence, higher crop yields and better utilization of resources. • Livestock monitoring: Through the AIoT devices streamline animal behavior, health condition, and production, thereby helping farmer in decision-making process for enriching their animals.
Smart Cities	• Traffic management: AIoT systems analyze the flow of traffic and align the signals and routes to ensure that the traffic density is low. • Environmental monitoring: Such AIoT devices as those for environmental control by indicating air and water quality, noise levels, and waste management systems for better environmental and public health.
Smart Energy Management	• Energy optimization: Clarification of AIoT patterns ensures that the usage of energy and consumption and distribution patterns are checked, interpreted, and controlled to avoid wastage and increase efficiency. • Request response: AIoT devices are flexibly tailored so that they may adapt the energy use in accordance with the grid conditions and signals to avoid blackouts.
Smart Retail	• Consumer analytics: AIoT solutions understand and learn the customer behavior, their needs, and their purchasing patterns so as to create effective marketing strategies that offer the right customer experiences. • Record management: Few of the AIoT applications include tracking of inventory levels, identification of stock out, and several supply chain management issues such as pricing and material acquisition costs, among others.

(Continued)

TABLE 2.2 *(Continued)*
AIoT (Artificial Intelligence of Things) Applications across Various Domains

Domain	Description
Smart Homes and Buildings	• Home automation: AIoT devices change home appliances, lighting fixtures, and Heating, Ventilation, and Air Conditioning systems (HVACs) based on preferred choices of users and weather conditions to improve comfort and efficiency. • Building management: Thus, AIoT systems enhance the performance efficiency of buildings by keeping track of energy consumption and occupancy rates as well as environmental factors, signaling energy saving and operating expenses.
Industrial IoT (IIoT)	• Predictive maintenance: AIoT solutions manage the industrial equipment and its expected breakdown requirements from the data collected by the sensors so that the downtime and maintenance expenses are minimized. • Process optimization: AIoT devices enhance the production processes through the analysis of real-time production data in order to highlight areas of improvement.
Autonomous Vehicles	• Self-driving cars: This means that AIoT systems make self-driven vehicle plateau regarding roads, mechanical obstacles, and driving decisions as well as on the basis of sensor data and AI algorithms to enhance road safety and energy transportation. • Connected vehicles: In vehicles, AIoT applications are also connected with each other as well as connected roadsides to provide traffic information and safety measures to avoid collision and efficiently manage the traffic.

2.3.1 AIoT Applications with Manufacturing System

Using AIoT, users can manage to convert massive unorganized IoT data into meaningful information that helps precisely in making a decision inside the system. According to MarketsandMarkets (https://www.marketsandmarkets.com/Market-Reports/ai-in-iot-market-43388726.html), the global AI in IoT market will grow from US$5,1BN in 2019 to US$16.2 billion by 2024. Overall market growth is expected to be primarily driven by the need for efficient handling of huge volumes of data in real time generated from IoT devices so as to reap valuable insights, ensure real-time monitoring, deliver better user experience, and reduce maintenance costs and downtimes (Figures 2.3 and 2.4).

Enterprise companies around the world are increasingly moving from cloud-based to Edge AI in order to reduce latency and costs, as several recent market reports have suggested increasing demand for real-time monitoring. AI boasts investment into many IoT technology businesses aiming to foster new "smart" products, increase corporate productivity, and fully exploit data to determine business-inducing reflections and enrich consumer experience. In addition, the amount of venture capital funding has skyrocketed as well as acquisitions of IoT start-ups with AI at their core solutions. At the same time, implementation of AI functionalities even for these purposes has mostly been limited to well-established general-purpose and industrial IoT platform vendors (including Amazon, IBM, Microsoft-Platform, and Oracle) who strive to supplement their Cloud-Cloud Machine Learning product lines with new offerings including more complex predictive analytics tasks enabled

FIGURE 2.3 AI in IoT market. (Source: https://www.marketsandmarkets.com/Market-Reports/ai-in-iot-market-43388726.html.)

by deep learning methods [Embia et al. (2023); Gaspar et al. (2021); Ghobakhloo and Ching (2019)].

When the groundbreaking integration of AI and IoT is properly implemented, it is changing sectors, improving productivity, and inspiring men and women to be assistance-oriented_advances_people. A clear synergy happens between AI and IoT as the technology etches out, where, therefore, sure to create an emergent market that will upset the global economic picture. Rising trend of AI in IoT Applications This paper explores In IOT industry AI development, In details; challenges for applying AI to IoT and the vision for a kind of Internet referred to as this in the future. As AI emerged to dictate the tech realm particularly by gradual incorporation of IoT devices across various sectors including smart cities, production houses, healthcare facilities, and transport systems, this market was steered toward vast growth, thereby generating higher revenues for those who are developing AI to be integrated to the IoT. As the number of connected devices continued to rise across the globe, there has

Report Metrics	Details
Market size available for years	2017–2024
Base year considered	2018
Forecast period	2019–2024
Forecast units	Billion (USD)
Segments covered	Components, Technologies, Verticals, and Regions
Geographies covered	North America, Europe, APAC, Latin America, and MEA
Companies covered	Google (US), Microsoft (US), IBM (US), AWS (US), Oracle (US), SAP (Germany), PTC (US), GE (US), Salesforce (US), Hitachi (Japan), Uptake (US), SAS (US), Autoplant Systems Pvt Ltd. (India), Kairos (US), Softweb Solutions (US), Arundo (US), C3 IoT (US), Anagog (Israel), Imagimob (Sweden), and Thingstel (India)

FIGURE 2.4 Research report categorizes the AI in IoT market based on component, technologies, verticals, and regions. (Source: https://www.marketsandmarkets.com/Market-Reports/ai-in-iot-market-43388726.html.)

been generation of explosive data which requires to be processed or analyzed in real time using AI. Therefore, any enterprise always seeking to enhance top- and bottom-line efficiencies, trim costs, and harness competitive edge is turning to AI-backed IoT solutions. Specialists from the field predict that the AI in IoT market will keep on developing in the following years. The need for smarter and more self-governing systems is shifting investment in this area, with estimates indicating the market may be valued in tens of billions kind. Some of the factors that have led to the growth of AI in different fields include enhanced AI algorithms, development of cloud computing, and increased utilization of real-time data and information, among others.

2.3.2 Industry Applications

Now AI with IOT is combinedly appearing in many verticals of industries, replacing or developing new feature in the classic process. Some of the smart home products include AI-based IoT devices like voice assistances, smart home security systems, and smart thermostats that have make life relatively easier and reduced energy wastage. Many of these gadgets personalize the interaction that makes day-to-day living simpler because it accommodates the behavior of the user. AI-enabled IoT devices, for example, wearable sensors, smart medical equipment, are changing patient care in the healthcare industry. They are constantly monitoring the pulse, predicting any possible illness, and helping to define perfect treatment. By providing data and information in real time, they are able to assist healthcare givers in making the right decisions needed to improve on the healthcare of patients with less expenses.

Some of the positive impacts seen in the integration of AI technological advancement with IoT have also been experienced in manufacturing industries. AI-based IoT solutions are enhancing industrial processes, estimating when sometime is due for repair and minimizing time off by using analytics techniques. This is the case since better product quality, low operating expenses, and high efficiency is the direction a company wants to be heading in. AI in IoT is also there in agriculture where scarce resources are managed efficiently, irrigation is made automated and crop health is tracked. These are in turn leading to increased yields, minimal wastage, and more environmentally friendly farming practices. Transport is one more sector which is benefiting from the use of AI and IoT to a rather considerable extent. Advanced car safeguarded with AI-based devices and communication technology are making roads safer, traffic smoother, and the transportation system more efficient. Additionally, in the energy domain, AIoT technologies have been developed in controlling energy use in smart grids and in buildings for efficiency, reduction in costs, and for sustainability.

2.4 CHALLENGES AND HURDLES

While the AI in IoT market shows great promise, it encounters several significant challenges. Data security and privacy are primary issues, as IoT devices produce large volumes of sensitive information that need security from unauthorized access and cyber threats. Protection of this data while ensuring user privacy is a complicated endeavor that necessitates strong encryption, authentication, and access control measures. Interoperability grants another challenge, as IoT devices and platforms often created from various manufacturers and utilize different communication protocols. For AI-driven IoT systems to be successful, these devices must be able to communicate with one another and work as a cohesive unit. Moreover, widespread adoption may be hampered by the high costs of integrating AI into IoT systems, especially for small and medium-sized enterprises. Another significant barrier is the complexity of creating and deploying AI-based IoT solutions. These systems require specific knowledge in AI, data analytics, and IoT technologies, which can be expensive and difficult to come by. Reaching the full potential of AI in the IoT market requires overcoming these obstacles.

2.4.1 Future Prospects

Looking onward, the AI in IoT market has a capable outlook. One notable trend is Edge AI, where data processing takes place closer to the source, such as straight on IoT devices, rather than relying on centralized cloud servers. This approach minimizes latency, saves bandwidth, and facilitates real-time decision-making, making it particularly suitable for applications that require speed, like autonomous vehicles and industrial automation. The evolution of IoT will also be greatly impacted by enhanced AI capabilities. IoT devices will becoming more autonomous as AI algorithms develop, able to take in information from their environment, and carry out increasingly complicated tasks without the need for human intervention. Smarter systems that can adapt to shifting circumstances and gradually improve their performance will be the outcome of this progress. Another area where AI in IoT is expected to have a significant impact is sustainability. AI-driven IoT solutions can contribute to the solution of global problems like resource scarcity and climate change by encouraging more effective resource management and reducing waste. Smart grids that employ AI to optimize energy distribution have the potential to reduce carbon emissions and expedite the transition to renewable energy sources.

2.4.2 Advantages of AIoT

- Enriched operational efficiency: AIoT can also analyze operational data in a real-time mode, as well as select specific patterns for evaluation and identification that humans are incapable of perceiving. From this, it can identify operating conditions in real time that yield optimal levels of commerce performance. Therefore, AI has benefits to the workflow and production processes, hence enhancing efficiency and reducing costs of operation.
- Improved risk management: AI may help organizations to determine threats with the help of data as soon as possible and use these findings to optimize business processes in order to reduce risks and increase safety, and make wiser management decisions. It can be applied in such areas as recognizing danger zones in a manufacturing plant to predict mechanical failure on planes to support reducing risk.
- Unique goods and services: The fact that one can get insight and draw sense out of huge volumes of information makes it conceivable to create

heretofore incognizable technologies such as voice, face, and predictive analysis. Such recently acquired skills can be utilized in numerous applications such as voice-activated virtual personal assistants, smart video doorbells, preventive maintenance in automobiles, home automation systems, and use of robots for delivery purposes or for emergency disaster response.
- Diminished unscheduled downtime: For the industrialized industry, this causes breakdown in machines which leads to unplanned downtimes causing a big problem. In relation to the machinery, this kind of maintenance not only enables the scheduler to assess the data concerning possible failure and plan maintenance in advance but also reduces the number and cost of times when equipment fails without any warning.
- AIoT allows personalized recommendation and a better purchasing experience related with the consumer behavior, demographic, and customer analytics in the retail environment.
- Compact product costs: AI reduces the data that is required to be transmitted to the cloud, thus reducing the cloud connectivity and service costs to the edge.

2.4.3 Disadvantages of AIoT

The AIoT associates AI with the IoT to generate intelligent systems that can analyze data, learn from it, and make decisions. Although AIoT has many potential benefits, such as enhanced efficiency, automation, and data-driven insights, it also comes with several disadvantages (Table 2.3).

2.4.4 Applications

What are the types of applications that are benefiting or can benefit from AIoT? Here are some examples.

- Smart Farming with AIoT: AIoT is a game changer in farming, one of the key industries. It uses AI to create a smart system that changes its settings based on temperature, crop and soil conditions, water use, and weather. It assesses sensor data to decide the best crops, fertilizers, watering systems, and pest control methods. AI supports farmers in boosting their harvest, forecasting weather, and seasonal trends for crop planning, and getting the most from their resources. Big farmlands and crops are watched over with computer vision and AI to spot and alert about problem areas when needed.
- Intelligent Helpers – Robots: AI and robots mesh well in making both factory and consumer goods. Robotic vacuums have sensors to collect data about the surroundings and use AI to decide the best way forward. Similarly, robots in industries, food, package delivery, and search and rescues in disaster areas use AI to navigate complex and sometimes hostile spaces, tweaking their behaviors as needed. Robots are even being used in shops because they can recognize faces and human feelings to enhance customer service and manage foot traffic.

TABLE 2.3
Disadvantages of AIoT

Aspect	Description
Complexity and Integration	Technical Complexity: Integrating AI with IoT systems can be technically complex, requiring specialized knowledge and expertise. Integration Challenges: Combining AI algorithms with IoT infrastructure often involves significant integration efforts, which can be time-consuming and costly.
Data Privacy and Security	Security Risks: AIoT systems are vulnerable to cyberattacks, hacking, and data breaches, which can compromise sensitive data and operational integrity. Privacy Concerns: Large amounts of data are being bought and processed by the AIoT devices, and thus privacy concerns arise particularly in relation to personal and sensitive data.
Cost and Resources	High Implementation Costs: The deployment of AIoT systems requires substantial investment in hardware, software, and skilled personnel. Maintenance Costs: Ongoing maintenance, updates, and security management of AIoT systems can be expensive.
Reliability and Dependence	System Reliability: To the user's disservice, AIoT systems can be prone to reliability problem, for instance, device malfunctions, unavailability of connectivity, or software glitches. Over-Reliance on Technology: This implies that the use of AIoT systems can make the overall systems to become oligopolistic by making human intervention to be out of proportion, hence relying so much on technology than human abilities.
Scalability and Flexibility	Scalability Issues: Scaling AIoT systems across large and diverse environments can be challenging due to varying technical requirements and infrastructure limitations. Flexibility Limitations: AIoT solutions may lack the flexibility to adapt to changing business needs or technological advancements without significant modifications.
Ethical and Social Implications	Job Displacement: AIoT brings benefits of automation and increases efficiency at workplaces; this in turn displaces people at their jobs, thus increasing unemployment rates in some industries. Bias and Fairness: When AI is trained on such datasets, it ends up replicating the bias, hence leading to different forms of injustice and discriminations in the decisions arrived at.
Legal and Regulatory Challenges	Regulatory Compliance: Ensuring compliance with data protection laws, industry regulations, and standards can be complex and costly. Legal Liability: Determining liability in cases of AIoT system failures or erroneous decisions can be legally challenging.
Environmental Impact	Energy Consumption: AIoT systems can be energy-intensive, particularly when processing large volumes of data, leading to increased carbon footprints and environmental impact. E-Waste: The proliferation of IoT devices contributes to electronic waste, posing environmental and disposal challenges.

- Smart Factories: Using AI and computer vision upgrades the quality control on assembly lines and helps spot irregularities. AI lends a hand in foreseeing maintenance needs of machines to prevent downtime, prolong machine life, and cut down production costs. Robots can move packages, assist on assembly lines, quality check the products, and handle repetitive but high-precision tasks on the factory floor or in storage.
- Self-Driving Cars: Autonomous vehicles or self-driving cars rely on IoT and AI to ply through traffic, adapt to changing weather, road situations, and traffic flow, and even guess what pedestrians might do next. AI can also predict when maintenance might be needed and assess the car's condition based on gathered use data.
- Building/Home Automation: If consumption and preference data of the building are provided to AIoT, AIoT can help business organizations save energy costs and provide them energy-efficient buildings by changing the lighting and climate control. By using the results of diagnostic data on the condition of the building systems, it allows performing a repair and maintenance when it is necessary rather than spending money on such services in a regular basis and that will save money for businesses. They can also help in varying degree to optimize the various systems and are capable of indicating times of possible system failure ahead of time. By using cameras sensors, AI has the capability of being implemented in the automatic gate control.
- Smart Cities: AIoT can potentially offer fresh ways of building better cities and sustaining municipal property and community public services. To do this, data from a large variety of sensors and IoT devices may be gathered and analyzed, and from this information, you may obtain useful insights that can be applied to real-time adjustments. Useful applications include waste collection, urban parking, and traffic control smart lighting, and many others. For example, drones can be employed in traffic control to track traffic in real time and the data retrieved collected can be used to change traffic signals or the number of lanes to control and reduce traffic in traffic flow – all this done automatically and without the physical assistance of a human being.
- Transportation and Logistics: AI is applied in fleet management in predictive maintenance, which involves monitoring of the fleet in real time with the data discovered from the sensors and the GPS trackers used in the vehicles for proactive maintenance. There are also added advantages to the fleet managers where AI has real-time navigation; it reduces the fuel expenses, tracking of the maintenance, and more crucially detect reckless driving. AI has the potential to benefit retail management in two ways. Retailers can forecast and make precise, data-driven business decisions by using AI and predictive analytics to help collect and analyze vast volumes of data. Through the use of consumer intelligence, demographic information, and behavioral analytics, AIoT may enhance customer service, product placement strategy, retail operations, and user experience overall by offering shoppers individualized recommendations.

- The advancement and application of AIoT particularly in the field of healthcare is diverse. These are inclusive of using imaging data for the identification and characterization of diseases, tele monitoring of patient data using sensors, and subsequently alerting clinicians when deviations from normalcy are observed, identification of drug interactions, and estimation of a patient's disease probability via analysis of the Electronic Health Records (EHRs). In addition, there is the use of robotic surgical systems that allow surgeries such as minimally invasive surgery and may perform or assist highly specialized and accurate procedures.
- AIoT systems use sensors incorporated in manufacturing equipment to detect the health status of the machinery in real time by using the predictive maintenance technique. By analyzing such data, the machine learning algorithms are able to predict faults on the equipment before it occurs. This makes it possible noting that timely maintenance decreases the cost of maintenance and sudden stoppage of the machines.
- Quality Control and Inspection: During manufacturing, the tasks of visual product inspections might be performed through the aid of AIoT-enabled cameras and sensors. Perfect for ensuring that only quality product is produced; machine learning algorithms analyze images and sensor data to identify defects, variations, or any sort of abnormalities.
- Process Optimization: Through the collection of data from the sensors installed along the production line, AIoT systems broaden industrial operations. These information taken through computers and with the help of analytical models or better still AI, it pinpoints areas that need improvement to increase output and optimize resources.
- Supply Chain Management: The AIoT systems contribute to the enhancement of the supply chain velocity and visibility through tracking of the movements of the components, finished goods, and raw materials. Sensors and Radio Frequency Identification (RFID) tags help in real-time data about the inventory of inventories, shipment status, and logistics activity resulting in better decision-making and inventory management.
- Energy Management: While analyzing the patterns, energy consumption, and opportunities to seize, AIoT arrangements ensure rational use of energy in production facilities. Machine learning helps through analyzing data results, which, in its turn, is important for predicting energy consumption and making suggestions concerning production timetable, for example, to save energy, changing processes, or equipment.
- Inventory Management: These AIoT devices allow manufacturers to control inventory conveniently since the devices can track the quantity of inventory and movement of stock in real time. As a result of cutting down on excess stocks, minimizing on stock out situations, and increasing the effectiveness of the supply chain, RFID tags and sensors are placed on inventory goods. These tags and sensors provide transparency of stock condition, its location, and usage pattern.
- AIoT technologies offer additional protection and ergonomics for the working environment as well as monitor areas of possible risks. To spot risks

such as repetitive strain injuries or risky work patterns, wearable devices capture the worker's motion, posture, and physiological data. The information helps to plan healthy preventive measures and safety changes.

- Demand Forecasting: AIoT systems predict the quantity of products and their components that is to be produced and supplied to the clients based on data from sensors integrated with the systems, clients' orders, and market trends. To meet the demand of consumers more accurately, the businesses can integrate supply chain and apply machine learning algorithms to analyze the data of demand to decide the timing and distribution of the product.

AIoT will help IoT understand its full potential by opening up new use cases and applications. Submissions of AIoT can be found in a wide range of industries, including smart homes, medical, industrial automation, smart cities, and agriculture. AI-enabled IoT endpoint applications will only cultivate in number, and a collective number of manufacturers will commit substantial resources to the field of AIoT [Dey et al. (2021); Dhingra et al. (2024); Duarte et al. (2017); Embia et al. (2023)].

2.5 CONCLUSION

According to the AIoT concept, the integration of the IoT technology with AI capabilities enhances the activities associated with such devices as networks and objects. AI is a technology that includes but is not limited to computer vision, machine learning, deep learning, and natural language processing which are said to be the game changers in the development of more intelligent IoT devices that can operate and make decisions independently. Smart thermostats with advanced technology such as AI can, for instance, learn the user's behavior patterns and lower the operational costs by optimizing their energy consumption appropriately. In addition, more insights and prediction of future occurrences can be synthesized from the enormous data continuously churned out by the IoT devices with the help of proficient analytical tools in real time. It allows companies to make decisions based on the facts to enhance the optimization operations, foresee the likelihood of failures of the equipment, and identify any unusual occurrences. The management of IoT devices and their control responsibilities can be enhanced with the help of automated structures powered by AI. As an example, AI-based control systems can provide autonomous reorganizations to a product workflow in response to sensor input, manage traffic in the intelligent city environment, and balance the power usage in intelligent houses. The setting up of the IoTs embodies a built-in structure that helps in the communication and the management of the devices in IoT, AI improves the network, decreasing embedment, and improving interoperability. By incorporating this machine to the IoT, the technology can detect and pinpoint cybersecurity threats quicker than before which in turn improves the security and privacy of IoT. AI capabilities can also conceptualize security systems that respond to threats in the network by identifying abnormal activity such as network traffic and preventing data and device intrusion. Certain AI-integrated, Internet-connected household devices may take into consideration a user's likes, behaviors, and environment which may influence how they interact with the devices. For example, smart home products can get to know the behavioral patterns of individuals and provide support such as useful recommendations on tasks depending on the need of the user. IoT devices are connected with higher levels of intelligence, by AI-powered edge computing that makes it possible to compute and analyze data in real time at the edge of the network. This increases the practicality and response of IoT systems by decreasing cloud dependency and increasing latency and bandwidth usage. The usage of these two technologic gadgets presents benefits toward several platforms such as healthcare activities, manufacturing industry, and others through the development of intelligent autonomous self-connected systems that can create new ideas and enhance the living standards.

REFERENCES

Abubakr, M.; Abbas, A.T.; Tomaz, I.; Soliman, M.S.; Luqman, M.; Hegab, H. Sustainable and smart manufacturing: An integrated approach. *Sustainability* 2020, *12*, 2280.

Ali, A.A.; Mahmood, A.; Ikram, A.; Ahmad, A. Configuring the drivers and carriers of process innovation in manufacturing organizations. *J. Open Innov. Technol. Mark. Complex* 2020, *6*, 154.

Belhadi, A.; Kamble, S.S.; Zkik, K.; Cherrafi, A.; Touriki, F.E. The integrated effect of big data analytics, lean six sigma and green manufacturing on the environmental performance of manufacturing companies: The case of North Africa. *J. Clean. Prod.* 2019, *252*, 119903.

Buer, S.V.; Semini, M.; Strandhagen, J.O.; Sgarbossa, F. The complementary effect of lean manufacturing and digitalisation on operational performance. *Int. J. Prod. Res.* 2021, *59*, 1976–1992.

Caiado, R.G.G.; Scavarda, L.F.; Gavião, L.O.; Ivson, P.; de Nascimento, D.L.M.; Garza-Reyes, J.A. A fuzzy rule-based industry 4.0 maturity model for operations and supply chain management. *Int. J. Prod. Econ.* 2020, *231*, 107883.

Cherrafi, A.; Elfezazi, S.; Hurley, B.; Garza-Reyes, J.A.; Kumar, V.; Anosike, A.A.; Batista, L. Green and lean: A Gemba–Kaizen model for sustainability enhancement. *Prod. Plan. Control* 2019, *30*, 385–399.

Chhotaray, P.; Behera, B.C.; Moharana, B.R.; Muduli, K.; Sephyrin, F.T.R. Enhancement of Manufacturing Sector Performance with the Application of Industrial Internet of Things (IIoT). In Smart Technologies for Improved Performance of Manufacturing Systems and Services (pp. 1–19). CRC Press, 2024.

Choudhary, S.; Nayak, R.; Dora, M.; Mishra, N.; Ghadge, A. An integrated lean and green approach for improving sustainability performance: A case study of a packaging manufacturing SME in the U.K. *Prod. Plan. Control* 2019, *30*, 353–368.

Dey, B.K.; Bhuniya, S.; Sarkar, B. Involvement of controllable lead time and variable demand for a smart manufacturing system under a supply chain management. *Expert Syst. Appl.* 2021, *184*, 115464.

Dhingra, S.; Raut, R.; Naik, K.; Muduli, K. Blockchain technology applications in healthcare supply chains – A review. *IEEE Access* 2024, *12*, 11230–11257.

Duarte, S.; Cruz Machado, V. Green and lean implementation: An assessment in the automotive industry. *Int. J. Lean Six Sigma* 2017, *8*, 65–88.

Embia, G.; Mohamed, A.; Moharana, B.R.; Muduli, K. Edge Computing-Based Conditional Monitoring. In Intelligent Manufacturing Management Systems: Operational Applications of Evolutionary Digital Technologies in Mechanical and Industrial Engineering (pp. 249–270). 2023.

Embia, G.J.; Moharana, B.R.; Behera, B.C.; Mohmaed, N.H.; Biswal, D.K.; Muduli, K. Reliability Prediction Using Machine Learning Approach. In Smart Technologies for Improved Performance of Manufacturing Systems and Services (pp. 21–37). CRC Press, 2023.

Embia, G.; Moharana, B.R.; Mohamed, A.; Muduli, K.; Muhammad, N.B. 3D Printing Pathways for Sustainable Manufacturing. In New Horizons for Industry 4.0 in Modern Business (pp. 253–272). Springer International Publishing, 2023.

Gaspar, P.D.; Fernandez, C.M.; Soares, V.N.G.J.; Caldeira, J.M.L.P.; Silva, H. Development of technological capabilities through the internet of things (IoT): Survey of opportunities and barriers for IoT implementation in Portugal's agro-industry. *Appl. Sci.* 2021, *11*, 3454.

Ghobakhloo, M.; Ching, N.T. Adoption of digital technologies of smart manufacturing in SMEs. *J. Ind. Inf. Integr.* 2019, *16*, 100107.

Inman, R.A.; Green, K.W. Lean and green combine to impact environmental and operational performance. *Int. J. Prod. Res.* 2018, *56*, 4802–4818.

Leong, W.D.; Teng, S.Y.; How, B.S.; Ngan, S.L.; Rahman, A.A.; Tan, C.P.; Ponnambalam, S.G.; Lam, H.L. Enhancing the adaptability: Lean and green strategy towards the Industry Revolution 4.0. *J. Clean. Prod.* 2020, *273*, 122870.

Morales Méndez, J.D.; Rodriguez, R.S. Total productive maintenance (TPM) as a tool for improving productivity: A case study of application in the bottleneck of an auto-parts machining line. *Int. J. Adv. Manuf. Technol.* 2017, *92*, 1013–1026.

Muduli, K.; Raut, R.; Narkhede, B.E.; Shee, H. Blockchain technology for enhancing supply chain performance and reducing the threats arising from the COVID-19 pandemic. *Sustainability* 2022, *14*(6), 3290.

Peter, O.; Swain, S.; Muduli, K.; Ramasamy, A. IoT in Combating COVID-19 Pandemics: Lessons for Developing Countries. In Assessing COVID-19 and Other Pandemics and Epidemics Using Computational Modelling and Data Analysis (pp. 113–131). Springer International Publishing, 2022.

Ruben, R.B.; Vinodh, S.; Asokan, P. Implementation of lean six sigma framework with environmental considerations in an Indian automotive component manufacturing firm: A case study. *Prod. Plan. Control* 2017, *28*, 1193–1211.

Saxby, R.; Cano-Kourouklis, M.; Viza, E. An initial assessment of lean management methods for Industry 4.0. *TQM J.* 2020, *32*, 587–601.

3 AIoT-Adopted Sustainable Smart Eco-Friendly Home Systems

Siddhartha Mukherjee, Rishav Das, Abhinav Aditya, Hitesh Mohapatra, and Bhabendu Kumar Mohanta

3.1 INTRODUCTION

Smart homes have become a reality because of the ever-growing technology and now the use of Internet of Things (IoT) technologies in homes enhances the homeowners convenience, comfort, and efficiency. In the application of smart homes under the IoT network, the arrangements of interconnected devices and sensors to enhance the value addition in the daily domestic life has gained much interest. Overall, the awareness of real-life applications of IoT in the growing ecosystem becomes important in order to evaluate its efficiency, its issues, and its presence in the people's lives. The research topic for this study is the exploration of actual smart home success stories and use them as basis for practical lessons on IoT wise home. Hence, this research aims at identifying success factors, emerging trends, and even the merits and demerits of incorporating IoT in smart homes based on real-life references. Through a comparative analysis of various smart home implementations, including energy control systems and home protection systems, this work intends to advance the knowledge of how IoT is changing the human lifestyle and the way that an individual interacts with the environment [1].

Smart home technologies mean extensions of IoT and home automation, which make homes active structures [2]. In this research, five technological disruptors to the traditional residential model are analysed with respect to energy efficiency, convenience, and security: Nest Thermostat, Philips Hue, Amazon Echo, August Smart Locks, and Samsung SmartThings. For instance, the Nest Thermostat adjusts the usage of heating and cooling based on the user habits and preferences for lower energy consumption but the comfort of the house. Likewise, Philips Hue personalizes the lighting, whereby Via can change the appearance and nature of lighting, put in features such as motion sensors and natural light saving options to decrease wastage of power [3].

Smart home is one of the most known applications in IoT environment that provides additional convenience in our daily lives and brings some security threats starting from the insider ones due to the device sharing and numerous social interactions in this context [4]. Modern IoT platforms counter this with what is known as multifactor authentication. Smart gadgets such as Nest Thermostat learns behaviour to optimize energy use, while Philips Hue makes light an experience [5]. Of all the connected devices,

DOI: 10.1201/9781003503323-3

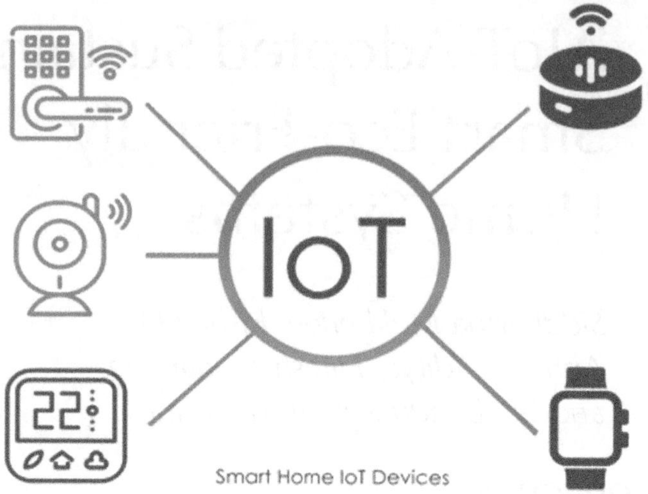

FIGURE 3.1 Smart home IoT devices [7].

the Amazon Echo gives voice-controlled automation a new face, while the August Smart Lock provides a blend of home security and convenience by allowing home access control from a distance. Samsung SmartThings comprises various Appliance and MADE for smart things technologies to create comfort and security experiences.

This research, grounded in case studies and references post-2020, aims to offer a timely examination of these advancements. By analysing how Smart home technologies support energy efficiency, convenience, and security, this research contributes to understanding of how these technologies transform the structure of today's homes. Not only does the research seek to influence policymakers, industry specialists, and scholars but also enable the public to have better control over their lives within the backdrop of an unpredictably advancing Smart Home Landscape [6] (Figure 3.1).

In line with this context, authors identified that the user-relevant customizations of smart home systems considerably improve user satisfaction and usage rates. They pointed out on the suitable interface qualities and provided functionalities that are end-user-centric [8]. Huraj et al. [9] pointed out that integration of different smart devices is significant to the operations of smart homes. They also highlighted that there was no consistency and suggested to implement protocols that would allow products of one firm to correspond with products of another line. Altogether, these studies are aligned to solve fundamental problems of smart home implementations including the aspects of user-centric design, device integration, and data protection. Introducing these solutions enables researchers to improve the performance of smart home systems in real-life application and increase user satisfaction with their use [10].

3.1.1 Research Gap

Sorting through potential IoT applications in smart homes is a treasure hunt of sorts. Every corner is rich with information, and it is somehow concealed; it is a stimulus

TABLE 3.1
Statistical Analysis on Several Ways to Deploy Smart Home

Metric	DIY Installations	Professional Installations	Hybrid Models
User Satisfaction (out of 5)	3.8	4.5	4.2
Initial Setup Cost ($)	500	1500	1000
Long-Term Maintenance Cost (5 years) ($)	200	400	300
Energy Consumption Reduction (%)	10	18	15
Security Incidents (per 100 installations per year)	15	5	8
Interoperability Issues (per 100 installations per year)	20	7	12

to researchers. Current home ownership with IoT devices has helped to change the way people live making life easier and convenient to live in [11]. However, there are still many issues with them, such as a problem of interconnection, compatibility, protection, and convenience of the interface. Although IoT focuses on the creation of more comfort, long-term efficiency and the possible drawbacks, they require further recognition. Some of these issues are emerging best practices and implications of IoT penetration in the environment. It has been noted in Table 3.1 that there are notable voids in smart home research [12].

Of the two short-term impacts of the use of IoT, the main risks include resource requirement and electronic wastage. Development of smart home devices, which include rare materials like the rare earth metals, often means accelerated depletion and thus prompting harmful minification. These devices also have very short product lifecycles, which results in the production of the dangerous e-waste, which negatively impacts ecological systems, and which poses further problems for waste management [13]. He also noted that higher IoT usage at home leads to higher energy consumption and greenhouse emissions. While IoT was developed with an idea in mind of energy efficiency achieved by turning off all unnecessary appliances or consumer devices, millions of these devices connected to the single network could achieve the opposite. Future studies are required to quantify the net energy impact of IoT and seek options to minimize the effects on the ecosystem. Also, little is known about consumers' IoT longitudinal process and what shapes their perception of privacy and security even further. In this case, IoT transforms as well as experiences the risk of adversarial cybersecurity which can negatively affect the security of smart homes [14]. Attacker traffic and listening are one of the threats typically applied by the attackers. They can occur unnoticed and grant unauthorized access to user's information. While approaching these threats, the researchers seek to enhance the security and unyieldingness [15]. The IoT has the benefit of edge computing, which has emerged as a recent trend. Some of the benefits with the integration of Edge Computing (EC) are as follows: real-time decision-making, low latency, and energy saving in a smart home environment. Nevertheless, there is limited research on how these technologies can be employed at their full potential towards extending IoT systems. However, more research still has to be conducted to close these gaps [16].

3.1.1.1 Long-Term Sustainability and Impact

When it comes to IoT deployments in smart homes, the research done has mostly focused on short-term effects with minimal concern for long-term sustainability and wider impact. It would be crucial to analyse aspects such as the lifespan of the device, compatibility with dynamic standards, and resource consumption [17]. Longitudinal surveys following up on how IoT devices perform over time can provide a good basis for drawing conclusions about their potential long-term future use [18].

3.1.2 User Acceptance and Behavioural Aspects

On the other hand, user experience with IoT devices has been studied by some scholars; however, there is a need for further investigation into user acceptance and behavioural patterns over time. In this regard, elements like personal data protection worries, opinions on reliability levels, as well as changeable likes of users should be taken into consideration when determining the design and launch of potential IoT products and services [19].

3.1.3 Security Concerns

Another challenge comes with the growing integration of what is known as the smart home devices which are vulnerable to cybersecurity threats. These include covert monitoring and traffic analysis, among others; this makes smart home systems vulnerable to attacks and hackers without the client knowing. Recommendations based on current literature revealed the need for more empirical studies to discover risks and design firmer securities to protect smart home settings. Integration of edge computing: As edge computing technologies are implemented at an accelerated pace, there is research void regarding their application in smart IoT systems. Real-time decision-making and other benefits of edge computing and edge processing technologies have to be studied in detail on how they apply to IoT devices in a smart home [20]. Scalability and interoperability: When there are more IoT devices in smart home, the problem of scalability and achieving interoperability arise. Further studies must be conducted to address difficulties that are associated with the use of multiple interfaces in a smart home. It is, therefore, important to coordinate these devices. Interoperability between devices of different manufacturers is also another factor that should also be well managed [21].

3.1.4 Human-centric Design

The emphasis of the current literature is set on technological aspects of SHs while there is a research void concerning human-centric approaches to SHs. We have to know all features that their users require, like, and do often. This in return assists in developing enhanced user interfaces of IoT systems. These solutions are, therefore, flexible enough to be implemented depending on the needs of each of the households.

Regulatory and policy frameworks: Since more and more smart homes are adopting IoT technology research on the policies and regulations that need to be implemented to address issues such as data privacy, security, and interoperability. Legal and ethical issues are also important and can be solved by enhancing strong regulatory measures and policy that will encourage innovation in the smart home market [22]. It will be critically important to fill these research gaps to propel further the field of IoT deployments in smart homes, and to enhance the higher condition of innovation and sustainable smart home setting that will benefit the homeowners and strike the goals of smart home technology [23].

3.2 LITERATURE REVIEW

Smart home automation technologies began being developed in the 1980s by the United States, Japan, and Europe. The term "automation" was used by Japanese companies such as Hitachi and Mitsubishi, with the focus being on home systems. The United States initiated the Smart House Project in 1984 by National Research Centre NAHB with the industrial key players [24]. Six important companies supplied the required hardware and software parts to the system. IoT integration has attracted more attention in the recent years with more literature being published on smart home environments [25]. Other papers illustrate how interoperability is achieved between smart devices such as smart thermostats, lighting, and security systems. The examples of real-life cases prove that applications reveal the effectiveness of real-time monitoring in conjunction with machine learning for efficient energy conservation [8].

IoT is established in healthcare by smart homes. Also, there are works explaining how use of sensors and monitoring technologies contribute to health enhancement under Assisted Ambient Living (AAL) [26]. Smart homes using IoT include case studies on energy management smart meters and sensors in homes for monitoring and optimizing the energy used. This has resulted into reduced tariffs and bills, energy consciousness, and enhanced environmental conservation. Another group of works emphasizes the possibility of managing interconnected IoT systems in intelligent homes for health observation. The smart clothing, smart scale, and health platforms monitor the biometrics signs, movement, and sleep. Such systems facilitate long-distance health supervision and m altered individual attention. These salient findings are useful for researchers, practitioners, and policymakers interested in smart home technologies [27].

This is because IoT systems are now embedded in smart homes for sophisticated energy control. Such systems are being receiving much attention from both researchers and users. Essentials of IoT energy includes smart meters, sensors, and others for realizing energy usage. Some of them gather information about electricity, gas, and water usage assists the users in tracking and optimize the consumption [28]. Sensors, for instance, are used in temperature, lighting, and movements, while energy monitors offer apps for real-time data. These technologies interact through the internet making a network that oversees and regulates energy optimally. The system provides information on how and where energy is consumed, notifies about unexpected consumption increases, and controls, for example, lights out in rooms with no occupants.

In fact, some systems offer advice on how energy can be conserved in the building. All in all, IoT energy management creates awareness and offers control to users, and trims cost and emission, thus encouraging environmental sustainability.

Smart meters are one of the mechanisms of technology in today's world to measure the consumption of electricity in houses [29]. REMs offer specific information about energy consumption to the homeowners to analyse patterns and to spend less on energy consumption. IoT systems with smart meters take this several steps further with sensors used to record a number of things including humidity and occupancy. These systems do not only record energy but also assists in maintaining it. What they do in a way is move people from being mere consumers of energy to active and informed savers of energy [30]. Smart home technologies make living environmentally friendly possible and enable residents to achieve a greener living.

A considerable number of works address the topic of controlling the energy consumption in smart home via IoT technology. This includes the use of smart EMS, smart meters, sensors, and energy monitoring devices. It gives the owners of these devices reliable information concerning the energy being used by them and hence can be regulated more effectively. Smart meters are especially useful as they maintain energy usage measurements in households and provide further details to find out wastage. To identify usage pattern and for better improvement in usage, homeowners can use advanced analytics provided to them. Other devices like sensors and monitoring devices also help IoT-based energy management in smart homes.

These sensors installed in all homes are a communication system designed for a radical change in the large-scale management of energy beyond the temperature, humidity, and occupants as mere observers of the domestic environment: they are watchmen. Sub-daily resolution of meteorology: Their abundance offers broad dynamical coverage of energy and captures details of energetic processes that cannot be observed through other ways. Viewing them not as sensors: They are great musicians, searching for ideas and showing how energy dissipates. From simply flickering standby appliances to concealed losses in heating and cooling appliances, these sensors go extra mile. However, their effects are also more widespread featuring the territories of the individual dwellings. They are able to gather data from in numerous households and they show an average picture of energy usage across neighbourhoods. This collective knowledge informs policy adjustment to work towards sustainable and energy efficient developments that are better aligned for planners, urban planners, and alike.

Nevertheless, their goals hit closer to the bone, so to speak. They bring about cultural change through the provision of real-time information and data that enable homeowners make better decisions that are not damaging to the environment. While at the micro level people are charged with the responsibility of managing their energetic choices, this does not turn them into spectators on an energy stage. This new conception gives not only measurable economic benefits but also helps to change perception of responsibility for the environment. In other words, these IoT-based energy management systems are a complete paradigm shift of how we interact with energy. It is not only objects, it is holders of a promise, driving us to a setting where sustainability and effectiveness are not dreams and visions, but the foundational pillars of the environment. Table 3.2 summarizes a tabular form of the findings of all the papers that were reviewed in the course of this study.

TABLE 3.2
Comparison and Analysis of a Few Research Papers

Papers	Objective	Remark
[22]	To provide a comprehensive overview of the vision and challenges in realizing the IoT.	Presented a vision for the IoT and identified key challenges including interoperability, security, privacy, and legal aspects.
[23]	To investigate benefits of IoT in healthcare monitoring and assisted living.	Presented compelling case studies demonstrating the benefits of IoT in enhancing health outcomes and assisted living in smart homes.
[24]	To provide an overview of IoT applications in smart homes and highlight their benefits and challenges.	Reviewed various IoT applications in smart homes and discussed their benefits in enhancing convenience, security, and energy efficiency.
[25]	To explore energy management through real-time monitoring and machine learning.	Showcased substantial energy savings achieved through real-time monitoring and machine learning algorithms.
[26]	To discuss recent advances and challenges in IoT-based smart homes.	Explored recent technological advancements in IoT-based smart homes and discussed challenges related to security, interoperability, and user acceptance.
[27]	To analyse security challenges and propose solutions for IoT-based smart home systems.	Identified security vulnerabilities in IoT-based smart home systems and proposed solutions to address these challenges.
[28]	To demonstrate the scalability and collective benefits of interconnected IoT systems.	Highlighted the potential for interconnected IoT systems to create synergies and enhance efficiency, convenience, and user experiences in smart homes.
[29]	To propose a smart home energy management system leveraging IoT technologies.	Presented a smart home energy management system integrating IoT devices for real-time monitoring and optimization of energy consumption.
[30]	To identify gaps in existing research on IoT deployments in smart homes.	Identified gaps including long-term sustainability, user acceptance, security concerns, integration of edge computing, scalability, human-centric design, and regulatory frameworks.

3.3 PROPOSED STUDY

From the identified gaps of IoT in smart homes, it is crucial that multifaceted approaches will be deemed useful. First and foremost, sustainability means recognizing that it is not business's role to fix the deficit, but rather that the latter must be addressed systemically. To have real-time information concerning lifespan of the devices, flexibility of the applications, and influence on the surroundings, it is mandatory to undertake longitudinal surveys and assessments. Environmental impacts of the IoT devices can be reduced through adopting recycling measures and encouraging green manufacturing functions. Most of the devices being developed should be designed to be durable while the life cycle of these devices should be significantly shorter, environmentally. It is paramount to engage other industry participants and policymakers in the promotion of sustainability, through the IoT, across the entire lifecycle of smart homes. Secondly, one of the priorities to consider for the IoT technologies' broader application is the problem of user acceptance and behavioural factors' influence. More analytical work has to be conducted in respect of user adaptation, privacy issues, and shifting trends. Such studies will help the developers to understand the users' requirements and expectations and thus develop easy to use interfaces. Applying the concept of human-centred design, smart home solutions are not only intelligent solutions but also more suitable to the user's need and preference. This could be achieved through additional and continued contact with the end users or by subsequent cycles of design work, with improved work through the end-user acceptance and work satisfaction among the end users.

Altogether, the solutions that help to eliminate gaps in IoT deployment in smart homes include a complex approach to regulation and policies. Furthermore, there is need for policymakers to engage stakeholders in the industry. In this case, they have to understand that it is high time to define the guidelines and standards. These should include IoT privacy, IoT security, and the issue of IoT compatibility with other devices, systems, and networks. Therefore, by building up a permissive yet constructive legislative framework, legislative authorities are likely to provide for innovation while preventing abusive practices which may be employed by companies targeting smart home users. Also, continuous assessment and review of such regulation measures are crucial since IoT initiatives are dynamic and transformational in the network environment. In addition, there is a need to engage partners at the global level to align the IoT regulations and ease on the implementation of IoT solutions across countries. Through encouraging cooperation between regulatory authorities and promoting the discussion of the prospects for IoT solution implementation for smart homes between policymakers, the best conditions for the worldwide application of IoT solutions in smart homes can be created in order to bring the envisioned positive changes improving people's quality of comfort, convenience, and efficiency in the corresponding living spaces.

Security threats in IoT deployments: From the study, attention is given towards the security risks prevalent in the smart home with the use of IoT systems, much attention being paid to the residents. To address all these risks, the proposed solutions seek to reduce the vulnerabilities in order to highlight how security of IoT deployment in smart home can be improved. The OCTAVE Allegro approach examines the manner in which the user or device in a whole system processes information. It also examines where the information is located or reside and threats that it is likely to encounter. Other important assets are recognized by associating them to the main information asset. Smart home risk assessments are well done and more comprehensive than digital security when using Allegro. The process follows eight steps divided into four main phases such as:

a. Establishing drivers phase
b. Profile assets phase
c. Identify threats phase
d. Risk mitigation phase (Figure 3.2)

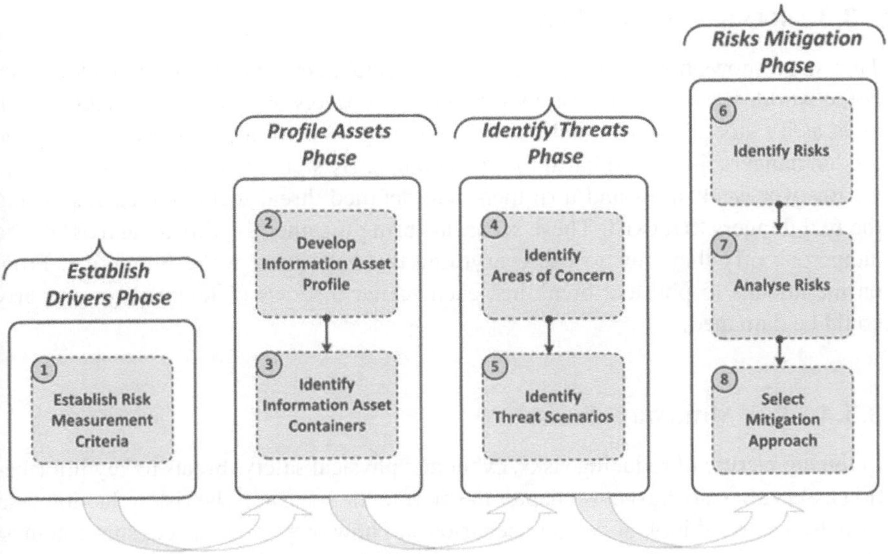

FIGURE 3.2 OCTAVE Allegro methodology flowchart of the eight steps.

3.3.1 Establish Drivers Phase

The purpose of the establish drivers segment goes beyond just starting; it's a key part that helps build how we look at risks in smart homes. This important section is like setting up for a concert where every sound is planned to work together, showing stakeholders how to check data assets and verify their readiness for effective use. At its heart, the establish drivers section aims to make a strong base—a place to grow knowledge—where we can check data assets well. By making a full set of rules for measuring risk, stakeholders can deal with tricky smart home tech more easily and see what might happen next.

3.3.2 Profile Assets Phase

In the big picture of smart home security, the profile segment stands out as a key platform—the base on which the security fortress rests. This layer serves as a melting pot of insights where we examine and dig into vital information assets to spot their weak points and key protections. This part starts by looking at how we find this crucial information—the heart of the smart home system. Through careful study and breakdown, we pick each asset and bring it into focus, weighing how much it matters and how it affects the system's overall safety. With the basic parts of identity in place, we start to profile—making a clear and forward-looking list to set each property's limits and show what's needed for security rules. Here, we spell out each asset's size in its digital, physical, and idea-based forms, and we take a deep look at the tricky ties that connect them to the wider system.

3.3.3 Identify Threats Phase

This stage shows how security risks and key data assets interact in the wider smart home world. It calls for a close look at the many levels of risk, while tracking each asset as it's stored, moved, and used. The main goal is to find security threats—the hidden dangers that could harm the smart home. By studying each property's setup, we uncover weak spots and turn them into detailed threat scenarios, each showing the real dangers that exist. These scenarios bring together the threats and assets—a dangerous mix that tries to use everything to attack what we're protecting. From online attacks to physical break-ins, each scenario shows different ways property could be damaged.

3.3.4 Risk Mitigation Phase

In the big picture of reducing risks, cyber and physical safety threats to key information assets are starting a new chapter in smart home security. The risk reduction field calls for a careful look at danger scenarios and how they might affect smart homes. Here, the push for security is driven by a full review of how each threat scenario can spread through the connected systems in smart homes. As the review goes on, we examine how these threats could impact smart homes—from small problems to big breaches. Through careful study, we give each threat a quality score helping us understand how important it is for smart home security. At its core, risk reduction isn't just about defence, but about taking action against uncertainty and challenges. This shows a strong commitment to protect smart home users' safety and security. It helps them feel confident and at ease as they use online services [13].

One of the problems with IoT setups in smart homes is the absence of standard communication protocols between devices, which causes compatibility issues. To fix this, we could create and use universal communication protocols that ensure smooth connectivity and data sharing among different IoT devices in a smart home setup [14]. The sovereign system gives users a controlled, self-contained smart home answer that runs on current IoT hardware without much extra work. This hints that another problem with IoT setups in smart homes could be the lack of user control and self-containment. To tackle this issue, we could put in place systems like sovereign that put user control first and work well on existing hardware [15]. Another issue with IoT setups in smart homes is that IoT sensors are easy targets for distributed denial-of-service (D-Dos) attacks, which can mess up communication between sensors and IoT devices putting user safety at risk. To solve this problem, we suggest setting up personal assistant systems in smart homes to make user communication with sensors stronger during D-Dos attacks. This way real users can still control and get data from IoT sensors even during attacks. This makes IoT systems in smart homes safer and more reliable [16]. To make IoT systems in smart homes better, we recommend using distributed fog computing (DFC) along with IoT in smart villages. This method tries to provide secure services closer to end devices, which could make smart home IoT systems more efficient and secure [17].

The growing number of IoT devices in smart homes has created big security worries. These gadgets often don't have much storage or processing power, which makes

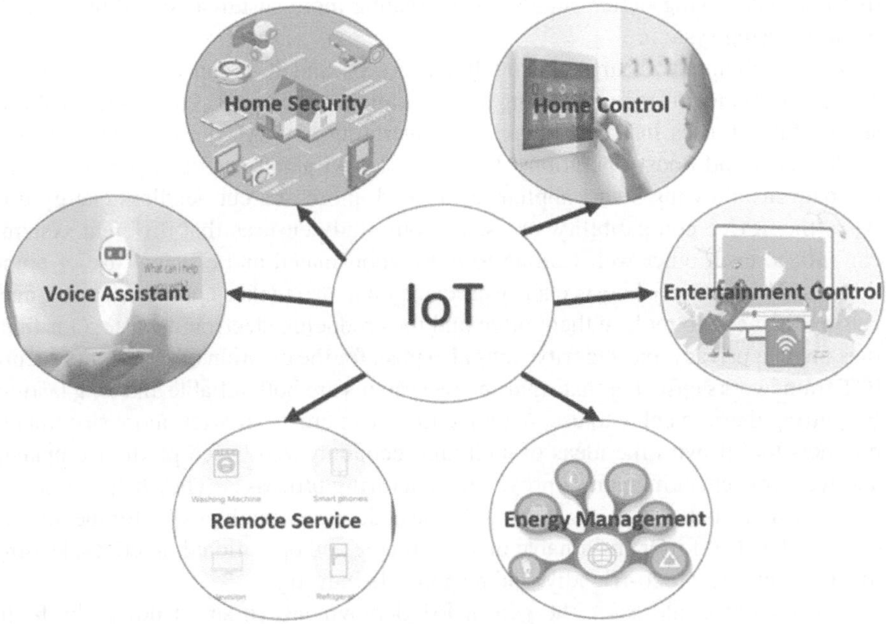

FIGURE 3.3 Application of IoT in smart home [24].

them easy targets for attacks. The article points out that old-school security methods haven't quite cut it when it comes to these special challenges. To tackle these security risks, the writers suggest using blockchain tech in smart home setups. The fix is to use a consortium blockchain where each IoT device keeps its own record. Every smart home would have its own miner to handle blockchain transactions boosting security in smart home settings [18]. Putting IoT devices in smart homes opens up more ways for attackers to get in showing a hole in security measures. To fix this, some ideas include using strong encryption protocols, keeping firmware up to date, and improving network separation to reduce possible threats in smart home IoT setups [19]. Figure 3.3 shows how IoT works in a smart home.

The proposed study has a significant impact on sustainable manufacturing by encouraging production processes that are good for society and the environment. By looking at real-world uses of smart home technologies, the study finds best practices that boost energy efficiency cut down on waste, and make the best use of resources in homes. These findings push manufacturers to create IoT devices and systems that put sustainability first such as appliances that use less energy smart thermostats, and home automation systems that work together to reduce environmental harm. Also, the study stresses how important it is for devices to work together and last a long time, which leads to less electronic waste and more durable upgradable products. By showing the key role of user-friendly design in getting good results, the research also supports production processes that think about the social side of sustainability making sure smart home technologies are easy to use, inclusive, and meet the needs of different users. In the end, this study backs the development of smart home solutions that not make living better but also help create a more sustainable and responsible manufacturing system.

The combination of artificial intelligence (AI) and ICT can boost the sustainable performance of manufacturing and operational processes. Our research shows successful IoT uses in smart homes demonstrating how AI can cut energy waste, slash waste, and boost operations. For example, AI algorithms can predict energy use patterns allowing smart appliances to work more and cut needless energy use. Also, the device compatibility stressed in our study ensures that different systems can talk to each other well leading to more coordinated and efficient use of home devices. This not only boosts energy efficiency but also makes appliances last longer cutting the need to replace them often and thus reducing electronic waste. Our findings on data privacy and security steps also push for the creation of strong and secure ICT frameworks ensuring that smart home systems are both reliable and long lasting. By putting these insights to use, manufacturers can create smarter, more sustainable products that fit with the ideas of a circular economy. AI-driven predictive upkeep and real-time checking can improve manufacturing processes. They help cut downtime and make better use of resources. In the end, our research pushes for the mixing of AI and ICT to foster sustainable manufacturing and operational practices, leading to a more efficient, eco-friendly, and responsible industry.

The reference addresses the gap in IoT deployments in smart homes by highlighting the inherent IT-security risks associated with interconnected IoT devices. The proposed solution, FANE, is a Firewall Appliance for Smart Homes. It reduces risks by automatically learning firewall rules based on network segmentation and IoT device communication profiles. This innovative approach not only enhances security in smart homes but also eliminates the need for network-security expertise from the users, thus providing a practical solution to the identified gap [20]. The reference delves into the time-to-provision evaluation of IoT devices, a crucial aspect in IoT deployments. This study sheds light on the challenges faced in provisioning IoT devices efficiently, which could be a significant gap in smart home deployments. Proposed solutions could include streamlining the provisioning process through automated zero-touch provisioning methods to enhance the deployment speed and reliability in smart home environments [21].

Admittedly, addressing the security issues in the IoT is crucial to building trust and faith among consumers, particularly in the context of smart home deployments. Effective security measures must be employed during every phase, from design to implementation and beyond, in order to effectively avert threats and preserve user data. This needs a tricky approach that brings together cybersecurity experts, partners, and regulators to create and keep up tough security standards and protocols. A key part of making IoT more secure is to set up research and development to handle and predict new threats that might come up. This way, dynamic stakeholders like cyber experts, who are ever evolving in their trade, can at all times detect and eliminate any potential weaknesses before they are taken advantage of. In this proactive approach, it not only boosts the robustness in the IoT environment but also propagates the spirit of security cautiousness among the industry. Moreover, in order to encourage a responsible implementation and adoption of IoT in smart homes, it is important to also consider regulatory frameworks that relate to privacy, security, and interoperability. When collaborating with lawmakers, business people who are giants in the sector of IoT technologies, as well as customer rights groups, transparent and ethical principles can be defined as those which give priority to user privacy and security, while encouraging innovation and market expansion. It is important to address security issues, but on the other hand, integrating edge computing technologies into the IoT systems of smart homes holds the promise of significantly increasing effectiveness and responsiveness.

The fact that data is processed and analysed where it is generated reduces latency, enhances decision-making capabilities in real time, and optimizes resource utilization, leading to overall performance improvement. The furtherance of research, therefore, will necessitate that these studies delve into the unknown promises regarding smart home deployment involving edge computing, such as optimizing energy usage and increasing the satisfaction of individuals. Mutual assistance among educational establishments, enterprises, and technical colleges has become an integral part of advancing towards state-of-the-art approaches and deploying efficient methods to tackle a range of problems that are presented during the residential setting's deployment of this cutting-edge technology. Also, it is important that investment in the infrastructure and technology development to support the diffusion of edge computing within smart homes is implemented.

This means that we need to build edge computing platforms, networking infrastructure, and special devices for homes. So, by investing in the right infrastructure and helping interested parties work together, people at home will get the most out of edge computing and use it to create a new smart living space. To make IoT work well in smart homes, we need to focus on keeping things secure and adding edge computing technology. By putting security first teaming up with industry players and government bodies, and investing in new ideas and growth, we can create a place where IoT tech thrives. This leads to a better life without putting data privacy at risk. Stress how important easy-to-use interfaces and personal experiences are. Suggest doing user surveys and testing how easy things are to use when designing. Push for open standards to make sure different devices and platforms can work together. Think about using protocols like Zigbee, Z-Wave, or Thread to make everything work. Suggest using strong encryption and ways to check who's who to protect smart home systems. Advise regular software updates and security fixes to close any gaps. Look into ways to lower the starting costs of smart home tech, like subsidies, rewards, or bundling services [22]. Point out the money people can save in the long run with energy-saving devices and automation.

The study title plays a key role in improving our grasp and real-world use of IoT tech in homes. As the smart home market grows fast pushed by its ability to boost ease, output, and safety, we need to go beyond theory and laboratory tests to look at actual setups. This study fills a big gap in current research by taking a deep look at smart homes that work well, showing the many things that help them succeed. By zeroing in on what users want how devices work together, and how to keep data safe and secure, this work gives useful insights into the real problems and fixes found in everyday use. These findings are vital for researchers, workers, and rulemakers who want to make smart home tech better so it meets user hopes and fits into daily life. In the end, what we learn from this study will shape future IoT growth, help more people use it, and get the best out of smart home tech for end users in all kinds of settings.

3.4 CONCLUSION

To sum up this journey to the real-world deployments of IoT solutions for smart homes, it can be concluded from proven deployment case studies that such innovative are prevalent and diverse in our growing field. In each of the case studies, we have seen how user needs and sensemaking between devices can either drive positive results with happy users or vice versa; data, privacy issues, and security measures being key to this. A closer look reveals that not only do successful smart home deployments hinge on future technological breakthroughs but also by catering to users' need and facilitating harmonious interaction of increasingly disparate devices. Further, ensuring that data privacy and security are top of mind at all times becomes a key tool in building trust through user buy-in. The complexity and difficulties surrounding the implementation of a real-world smart home provide an insight into why most developers tend to ignore this area. This underscores the importance of stakeholders (researchers, practitioners, and policymakers) to realize that technology is an intimate component in how users meet his/her different needs. This enables the seamless integration of IoT applications and services in residential spaces, therefore enhancing overall end-user satisfaction. It helps to bridge the gap between theory and practice in smart home deployments. Delivering these empirical insights provides useful assistance to stakeholders in the development of IoT deployments. It also underlines the value of iterational review and adaptation to fluctuating user needs and tech developments. In next iterations, more recent trends and technologies in the smart home environment should be studied. Moreover, long-term follow-ups in longitudinal studies to capture real-life behaviour change and quality of life (QoL) data are needed that reflect the dynamics of this field. These results underline the significance of integrated strategies taking into account not only technological but also human factors in enabling smart home technologies. Stakeholders seeking to enhance their decision-making can draw on knowledge from real-world case studies and insights from technology experts, which together guide the progressive adoption of IoT in residential settings through defined milestones.

REFERENCES

1. Catarinucci, L., de Donno, D., Mainetti, L., Palano, L., Patrono, L., & Stefanizzi, M.L. (2015). An IoT-Aware Architecture for Smart Healthcare Systems. IEEE Internet of Things Journal, 2(6), 515–526.
2. Veracini, S., Hasan, M.Z., Capodieci, N., & Georgakopoulos, D. (2021). A Review of Blockchain-Based Architectures for the Internet of Things: A Case Study on Smart Cities. Sensors, 21(7), 2463.
3. Ho, G., Leung, D., Mishra, P., Hosseini, A., Song, D., & Wagner, D. (2016). Smart Locks: Lessons for Securing Commodity Internet of Things Devices. EECS Department, University of California, Berkeley. Technical report no. UCB/EECS-2016-11.
4. Kong, L., Tan, J., Huang, J., Chen, G., Wang, S., Jin, X., Zeng, P., Khan, M., & Das, S.K. (2022). Edge-Computing-Driven Internet of Things: A Survey. ACM Computing Surveys, 55(8), Article 174 (August 2023), 41 pages.

5. Ali, B., & Awad, A.I. (2018). Cyber and Physical Security Vulnerability Assessment for IoT-Based Smart Homes. Sensors, 18(3), 817.
6. Mohapatra, H., & Rath, A.K. (2023). Designing of Fault-Tolerant Models for Wireless Sensor Network-Assisted Smart City Applications. In: Dash, S.R., Das, H., Li, K.C., & Tello, E.V. (eds) Intelligent Technologies: Concepts, Applications, and Future Directions, vol 2. Studies in Computational Intelligence, vol 1098. Springer, Singapore. https://doi.org/10.1007/978-981-99-1482-1_2.
7. Al-Fuqaha, A., Guizani, M., Mohammadi, M., Aledhari, M., & Ayyash, M. (2015). Internet of Things: A Survey on Enabling Technologies, Protocols, and Applications. IEEE Communications Surveys & Tutorials, 17(4), 2347–2376.
8. Zhang, Z., Yu, T., Ma, X., Guan, Y., Moll, P., & Zhang, L. (2022). Sovereign: Self-Contained Smart Home with Data-Centric Network and Security. IEEE Internet of Things Journal, 9(15), 13808–13822.
9. Huraj, L., Šimon, M., & Horák, T. (2020). Resistance of IoT Sensors against DDoS Attack in Smart Home Environment. Sensors, 20(18), 5298.
10. Aljuhani, A., Kumar, P., Kumar, R., Jolfaei, A., & Islam, A.K.M.N. (2023). Fog Intelligence for Secure Smart Villages: Architecture and Future Challenges. IEEE Consumer Electronics Magazine, 12(5), 12–21.
11. Arif, S., Khan, M., Rehman, A., Kabir, S., & Imran, M.A. (2020). Investigating Smart Home Security: Is Blockchain the Answer? IEEE Access, 8, 117802–117816.
12. Kavallieratos, G., Chowdhury, N., Katsikas, S.K., Gkioulos, V., & Wolthusen, S.D. (2019). Threat Analysis for Smart Homes. Future Internet, 11(10), 207.
13. Haar, C., & Buchmann, E. (2019, September). Fane: A firewall appliance for the smart home. Federated Conference on Computer Science and Information Systems (FedCSIS) (pp. 449–458). IEEE.
14. Boškov, I., Yetgin, H., Vučnik, M., Fortuna, C., & Mohorčič, M. (2020). Time-to-provision evaluation of IoT devices using automated zero-touch provisioning. In GLOBECOM 2020-2020 IEEE Global Communications Conference (pp. 1–7). IEEE.
15. Mohapatra, H., Dehury, M.K., Guru, A., & Rath, A.K. (2023). IoT-Enabled Zero Water Wastage Smart Garden. In: Marques, G., Saini, J., & Dutta, M. (eds) IoT-Enabled Computer-Aided Systems for Smart Buildings. EAI/Springer Innovations in Communication and Computing. Springer, Cham. https://doi.org/10.1007/978-3-031-26685-0_4.
16. Ehsanifar, M., Dekamini, F., Spulbar, C., Birau, R., Khazaei, M., & Bărbăcioru, I.C. (2023). A Sustainable Pattern of Waste Management and Energy Efficiency in Smart Homes Using the Internet of Things (IoT). Sustainability, 15(3), 1821.
17. Wei, L. (2023). Research on Smart Home Vision System Based on Computer Artificial Intelligence Technology, 2023 IEEE 3rd International Conference on Electronic Technology, Communication and Information (ICETCI), Changchun, China, pp. 1227–1231. https:doi.org/10.1109/ICETCI57876.2023.10176647.
18. Bhalotia, N., Kumar, M., Alameen, A., Mohapatra, H., & Kolhar, M. (2023). A Helping Hand to the Elderly: Securing Their Freedom through the HAIE Framework. Applied Sciences, 13, 6797. https://doi.org/10.3390/app13116797.
19. Mohapatra, H., & Mishra, S.R. (2024). Unlocking Insights: Exploring Data Analytics and AI Tool Performance Across Industries. In: Singh, P., Mishra, A.R., & Garg, P. (eds) Data Analytics and Machine Learning. Studies in Big Data, vol 145. Springer, Singapore. https://doi.org/10.1007/978-981-97-0448-4_13.
20. Mohapatra, H. (2024). The Future of Smart Agriculture. Nova Science Publishers, Incorporated.
21. Mendes, E.J., Silveira, M.M., Araújo, M.B., Celestino, J., & Gomes, R.L. (2019). Abstraction of Heterogeneous IoT Devices for Management of Smart Homes, 2019 IEEE Latin-American Conference on Communications (LATINCOM), Salvador, Brazil, pp. 1–6. https://doi.org/10.1109/LATINCOM48065.2019.8937985.

22. Sundmaeker, H., Guillemin, P., Friess, P., & Woelfflé, S. (2010). Vision and Challenges for Realising the Internet of Things. Cluster of European Research Projects on the Internet of Things. European Commission, 3(3), 34–36.
23. Catarinucci, L. et al. (Dec. 2015). An IoT-Aware Architecture for Smart Healthcare Systems. IEEE Internet of Things Journal, 2(6), 515–526. https://doi.org/10.1109/JIOT.2015.2417684.
24. Alaa, M., Zaidan, A.A., Zaidan, B.B., Talal, M., & Kiah, M.L.M. (2017). A Review of Smart Home Applications Based on Internet of Things. Journal of Network and Computer Applications, 97, 48–65.
25. Li, M., Li, G.Y., Chen, H.R., & Jiang, C.W. (2018). QoE-Aware Smart Home Energy Management Considering Renewables and Electric Vehicles. Energies, 11(9), 2304.
26. Zielonka, A., Woźniak, M., Garg, S., Kaddoum, G., Piran, M.J., & Muhammad, G. (2021). Smart Homes: How Much Will They Support Us? A Research on Recent Trends and Advances. IEEE Access, 9, 26388–26419.
27. Kumar, P., & Chouhan, L. (2021). A Secure Authentication Scheme for IoT Application in Smart Home. Peer-to-Peer Networking and Applications, 14(1), 420–438.
28. Simeoni, E., Gaeta, E., García-Betances, R.I., Raggett, D., Medrano-Gil, A.M., Carvajal-Flores, D.F., … & Arredondo Waldmeyer, M.T. (2021). A Secure and Scalable Smart Home Gateway to Bridge Technology Fragmentation. Sensors, 21(11), 3587.
29. Al-Ali, A.R., Zualkernan, I.A., Rashid, M., Gupta, R., & AliKarar, M. (2017). A Smart Home Energy Management System Using IoT and Big Data Analytics Approach. IEEE Transactions on Consumer Electronics, 63(4), 426–434.
30. Zaidan, A.A., Zaidan, B.B., Qahtan, M.Y., Albahri, O.S., Albahri, A.S., Alaa, M., … & Lim, C.K. (2018). A Survey on Communication Components for IoT-Based Technologies in Smart Homes. Telecommunication Systems, 69, 1–25.

4 Sustainable Manufacturing with AIoT-Integrated B-Spline CFD Simulations

Archana Senapati, Shashikant Das,
Balaji Padhy, Nirakara Pradhan,
Saumya Ranjan Jena, Ankur Srivastava

4.1 INTRODUCTION

A few examination distributions have shown how computational fluid dynamics (CFD) utilizes the B-spline collocation way to tackle heat transport issues. Partial differential equations (PDEs) have numerous applications in CFD, such as the heat equation, Korteweg-de Vries (KdV) equation, Burgers' equation, advection-diffusion equation (ADE). To stimulate thermal processes in various logical and modern applications, the heat equation should be mathematically settled with accuracy and productivity. B-spline collocation strategies are one of the most utilized mathematical procedures due to their amazing exactness, computational proficiency, and flexibility. Heat conduction equations are approximated specifically by the cubic B-spline approach. This scheme describes the piecewise continuous arrangements across the area by discretizing time derivatives using finite difference schemes and B-splines functions in spatial dimension. The methodology is additionally extended to cover both linear and non-linear fractional initial value problems, showing its flexibility to various problems. Stability and convergence analysis, along with error norm, approve the accuracy and efficiency of the B-spline collocation strategy, demonstrating its better exhibition over existing mathematical methodologies.

Spline collocation techniques have been shown in recent research to be effective in solving PDEs. For example, Kutluay et al. [1] have researched numerical solutions to nature's most significant problem: the heat conduction equation of 1D. They did this with the help of collocation and finite element techniques along with B-spline basis functions. Tassaddiq et al. [2] numerically approximated the heat equation employing the cubic B-spline collocation technique with both conventional and non-classical boundary conditions (BCs). Rabah et al. [3] demonstrated a cubic B-spline interpolation technique for numerically solving linear, non-linear, singular, and non-singular problems. The non-linear heat conduction difficulties were solved by Annasabi et al. [4] using the Kirchhoff transformation and the B-spline method. Trigonometric spline collocation was introduced by Schoenberg [5]. In addition, by creating a quartic order cubic B-spline, Jena and Senapati [6] examine the solution to the problem of heat and advection-diffusion. Shafiq et al. [7] proposed cubic B-spline for the

DOI: 10.1201/9781003503323-4

wave equations with the Mittag-Leffler function. The heat parabolic equation was solved by Hamad and Sabawi [8] using a cubic B-spline collocation. Raslan and Ali [9] proposed the identical method for higher dimensional (3D and 4D) heat equation. Numerous authors have contributed to the conclusion of the heat and ADE, which is the most important issue in nature. Mittal and Rohila [10] developed a fourth-order cubic B-spline colocation approach for ADE; meanwhile, Dwivedi et al. [11] approximated the fractional reaction-diffusion equation by employing the cubic B-spline collocation technique. Umer et al. [12] presented a more extended cubic B-spline for time-fractional ADE. A cubic B-spline with an exponential kernel approach was used by Shafiq et al. [13] for the approximate solution of the wave equation. Jena and Gebremedhin [14] used the octic B-spline collocation method for ADE. Dehghan [15] applied the weighted finite difference scheme for the 1D heat equation. In order to solve the convection-diffusion problem, Mittal and Jain [16] used redefined cubic B-spline collocation, and Nazir et al. [17] proposed cubic trigonometric B-spline, whereas Mohammadi [18] presented an exponential B-spline approach. Gnedin et al. [19] enforced the Courant-Friedrichs-Lewy condition in temporal stepping schemes. Khan et al. [20] employed a cubic B-spline collocation technique combined with a weighted approach to approximate the time fractional ADE, utilizing the Atangana-Baleanu derivative. The cubic trigonometric collocation approach was utilized by Yaseen et al. [21] to solve the temporal fractional diffusion problem numerically. The non-linear time fractional telegraph equation was approximated by Akram et al. [22]. B-spline with a semi-analytical approach was utilized by Reutskiy et al. [23] for the 2D convection diffusion problem. Hidayat et al. [24] proposed local B-spline with finite difference method (FDM) for 2D heat conduction problems, and to find the approximate solution of explicit and implicit heat equations, Jena and Senapati [25] used spline collocation with the Thomas algorithm scheme.

Despite these advancements, there remain significant challenges and gaps in the literature. Existing methods often struggle with stability and convergence issues, particularly in multi-dimensional settings. Furthermore, systematic research is needed to optimize B-spline collocation methods for various kinds of heat equations in CFD applications. The purpose of this work is to design an improved B-spline collocation method for the CFD heat equation in order to overcome these issues. The aims of this work are to close the mentioned research gaps by creating a more effective B-spline collocation method and analyzing its accuracy, stability, and computing efficiency in CFD problems. These methodological techniques ensure a comprehensive introduction that sets a clear direction for the research, highlights its relevance, and situates it within the existing body of knowledge.

Definition: CFD is the study of fluid flow problems using numerical methods and algorithms. It has many applications, such as in the study of combustion, aerodynamics, hydrodynamics, and most famously, heat transfer problems.

- Applications of heat transfer: Heat transfer challenges arise in a variety of industries, including industrial, aerospace, automotive, and energy. Conduction, convection, and radiation are examples of heat transport phenomena that may be simulated and analyzed using CFD in a variety of complicated geometries and circumstances [26].

- The relevance of numerical schemes: These methods are required for accurate prediction and analysis of complex fluid flow and thermal energy transfer since these problems frequently lack analytical answers. Numerical approaches provide computer solutions to discretize governing equations and gain insights into fluid behavior and thermal characteristics that may not be achievable through experimental methods alone. Numerical approaches are important in the solution of large-scale piratical engineering problems due to their efficiency and flexibility.
- History and evolution: The development of numerical methods for resolving fluid flow equations in the early to mid-20th century is credited with giving rise to CFD. The FDM, finite element method (FEM), and finite volume method play important roles in the growth of CFD.

4.2 FUNDAMENTALS OF HEAT TRANSFER

Conduction is the process by which heat is transferred through a substance or directly between two materials. It is governed by temperature gradients and is described by Fourier's law, which asserts that the rate of heat transfer corresponds to the negative temperature gradient.

Convection is the transmission of heat between a flowing fluid and a solid surface. It can occur spontaneously due to buoyancy considerations in fluids or be generated by a solid surface. Natural convection is explained by empirical correlations, whereas forced convection is explained by Newton's law of cooling.

Intensity transmission by electromagnetic waves is called radiation, and it occurs without the requirement for a medium. It is administered by Stefan-Boltzmann's regulation, which makes sense of the association between the intensity move rate and the temperature's fourth power.

4.2.1 Boundary Conditions (BCs) and Initial Conditions (ICs)

- BCs: This encompasses determined temperatures, heat motions, or convective BCs. They characterize the circumstances at the border of the system and influence the way of behavior of heat transfer.
- ICs: At the beginning of the simulation, the initial conditions (ICs) specify the system's temperature distribution. They may require defining initial temperature gradients and are essential for solving transient heat transfer problems. These equations and principles controlling heat transport issues may be solved numerically using methods like the B-spline collocation approach, which we shall discuss in the following sections.

4.2.2 Outline of the B-Spline Collocation Strategy

Definition: B-splines are mathematical functions that are employed in data interpolation and approximation. These are piecewise polynomial functions defined on knots on local domains. The preliminary work on the B-spline was first introduced by Schoenberg [5] in 1946, with the assumption that the data was a smooth piecewise polynomial approximation of equidistant data. Cox-de Boor then continued Schoenberg's work, and in the early 1970s, he devised a recurrence relation that is used in the creation of B-spline. Suppose the region $[a, b]$ be the region of interest, and let $R = (z_i)_{i=1}^{\infty}$ be a sequence of non-decreasing real numbers such that $a = z_0$ and $b = z_N$. These z_0 are known as the knot vectors, and the sequence R is called the knot sequence. The ith knot span refers to the half-open range $[z_i, z_{i+1})$.

4.2.2.1 Properties

B-spline basis functions have many important properties that make them very useful in the fields of computer graphics, curve fitting, and numerical analysis. These are listed as follows:

- Local support: B-splines have finite support over small intervals, making them computationally efficient.
- Smoothness: B-splines possess continuity properties, ensuring smooth transitions between adjacent segments.
- Degree of freedom control: The sequence of B-splines impacts their ability to depict complicated geometries and functions.
- Hierarchical structure: B-splines can be hierarchically refined, enabling adaptive mesh refinement and local refinement.

4.2.3 Cox-de Boor Recursion Formula

Using the Cox-de Boor recursion formula, the ith B-spline of degree q (order $(q + 1)$) may be represented by $\psi_{i,q}(z)$ as follows: [27]

$$\Psi_{i,q}(z) = \frac{z - z_i}{z_{i+q} - z_i}\Psi_{i,q-1}(z) + \frac{z_{i+1+q} - z}{z_{i+1+q} - z_{i+1}}\Psi_{i+1,q-1}(z),\ q \geq 1 \tag{4.1}$$

4.2.3.1 B-Spline of Order 1 and Degree Zero

The B-spline of degree zero for increasing the sequence of knot $(z_i)_{i=1}^{\infty}$ is expressed as

$$\Psi_{i,0}(z) = \begin{cases} 1, & z_i \leq z < z_{i+1} \\ 0, & otherwise \end{cases} \tag{4.2}$$

The nature and behavior of the B-spline of degree zero are similar to the step function. The only position where a zero-degree B-spline does not equal zero is $[z_i, z_{i+1})$.

4.2.3.2 B-Spline of Order 2 and Degree One

Applying $q = 1$ in equation (4.1) and implementing equation (4.2), we get

$$\Psi_{i,1}(z) = \frac{z - z_i}{z_{i+1} - z_i}\Psi_{i,0}(z) + \frac{z_{i+2} - z}{z_{i+2} - z_{i+1}}\Psi_{i+1,0}(z)$$

$$\frac{z - z_i}{z_{i+1} - z_i}\begin{cases} 1, & z_i \leq z < z_{i+1} \\ 0, & otherwise \end{cases} + \frac{z_{i+2} - z}{z_{i+2} - z_{i+1}}\begin{cases} 1, & z_{i+1} \leq z < z_{i+2} \\ 0, & otherwise \end{cases}$$

$$\Psi_{i,1}(z) = \frac{1}{h}\begin{cases} z - z_i, & z_i \le z < z_{i+1} \\ z_{i+2} - z, & z_{i+1} \le z < z_{i+2} \\ 0, & otherwise. \end{cases} \tag{4.3}$$

Equation (4.3) is the required B-spline of degree 1. Here, $[z_i, z_{i+1})$ and $[z_{i+1}, z_{i+2})$ are the two-knot spans of equation (4.3). A triangle function which also known as the tent function or HAT function is analogous to the first-degree B-spline for two knot spans $[z_i, z_{i+1})$ and $[z_{i+1}, z_{i+2})$.

4.2.3.3 B-Spline of Order 3 and Degree Two (Quadratic B-Spline)

The quadratic B-spline is obtained by substituting $q = 2$ in the Cox and de Boor relation, which is represented in equation (4.1), and using equation (4.3) as follows:

$$\Psi_{i,2}(z) = \frac{z - z_i}{z_{i+2} - z_i}\Psi_{i,1}(z) + \frac{z_{i+3} - z}{z_{i+3} - z_{i+1}}\Psi_{i+2,0}(z)$$

$$= \frac{1}{2h^2}\begin{cases} (z - z_i)^2, & z_i \le z < z_{i+1} \\ (z - z_i)(z_{i+2} - z) + (z_{i+3} - z)(z - z_{i+1}), & z_{i+1} \le z < z_{i+2} \\ (z_{i+3} - z)^2, & z_{i+2} \le z < z_{i+3} \\ 0, & otherwise. \end{cases} \tag{4.4}$$

Similarly, using the recurrence formula of Cox-de Boor equation (4.1), higher degrees of B-spline as cubic, quartic, quintic, sextic, septic, octic, nonic, and decatic B-spline can be generated.

4.3 BENEFITS OF THE B-SPLINE COLLOCATION TECHNIQUE IN CFD

Accuracy: High-order accuracy provided by the B-spline collocation method makes it possible to properly reflect processes like thermal energy transfer and fluid movement.

Flexibility: B-splines allow for flexible grid generation and adaptation, facilitating efficient resolution of flow features and boundary layers.

Local refinement: B-splines enable local refinement of the computational grid, focusing computational resources where they are most needed.

Preservation characteristics: In mathematical reenactments, B-splines protect protection properties, guaranteeing the protection of mass, force, and energy.

4.3.1 Comparative Analysis with Other Numerical Techniques

Finite difference scheme: The B-spline collocation approach yields better accuracy than FDM while dealing with complex BCs and irregular geometries. It gives more flexibility in grid creation than FDM because it employs a spline-based scheme rather than a grid-based one.

Finite element method (FEM): The B-spline collocation approach and FEM are similar in terms of piecewise polynomial approximation. But the B-spline collocation scheme yields more accuracy and better solutions due to the properties of B-spline and collocation methodology. FEM needs more work than the B-spline collocation method when managing complex geometries and grid generation. The advantage of the B-spline collocation approach is that it can be solved numerically to solve fluid flow and heat transfer issues in CFD.

4.3.2 Discretization Techniques

Collocation points: These points are located in the computational domain where the approximate solutions are evaluated. In the B-spline collocation method, the collocation points are selected so that they coincide with the knots of B-spline functions.

Benefits: Collocation points enable the direct application of differential equations at discrete places, therefore streamlining the numerical implementation and guaranteeing correctness.

4.3.3 Discretization of Time

Time-stepping techniques: It is the process of dividing the temporal domain into discrete intervals and evaluating the solutions at each time level.

Explicit techniques: In this process, the solution of the governing equations is obtained from the previous time step. This takes place in the numerical method like forward Euler or the Runge-Kutta method.

Implicit techniques: At each time step, the methods, like the Crank-Nicolson or backward Euler method, convert the governing equation to a set of equations. They produce more stable schemes.

Advantages: To research time-dependent heat transfer systems, time-stepping techniques are essential because they enable the modeling of transitory events and the development of the solution over time.

4.3.4 Error Analysis

Absolute error is the separation between an approximate and exact solution. In numerical analysis, it gives the precision of an error without taking into account its sign. It is defined as follows:

$$\text{Absolute error} = \left|\text{Exact value} - \text{Approximate value}\right|$$

Notes:

- Analytic value: The value being approximated is called the analytic or exact value, which might be theoretical, analytical, or real.
- Approximate value: An approximate value is a value that can be evaluated by a numerical method, measurement, or estimation.

- Magnitude: The magnitude of the absolute error is always positive, as it is the difference between the exact and approximate value.
- Significance: The efficiency of the scheme is estimated by the absolute error. It is easy to understand and easy to compute.

4.3.5 L_2 Error Norm

It is the square root of the integral of the square of the absolute error across the domain of interest. The L_2 error norm is expressed as follows:

$$L_2 = \sqrt{h\sum_{i=0}^{N}\left|u_{exact} - u_{num}\right|^2} \tag{4.5}$$

where, u_{exact} is the exact solution and u_{num} is the numerical solution.

Significance:

- Finite element analysis (FEA): A piecewise polynomial function is used in FEA to estimate the solution to a differential equation. One may evaluate the degree to which the finite element solution approximates the exact solution using the L_2 error norm.
- Finite difference method (FDM): In the solution of a PDE, the error norm L_2 is used to calculate the error at discrete node points. It is very useful in calculating the accuracy of the scheme.
- Convergence analysis: The rates of convergence of numerical techniques are examined using the L_2 error norm. The order of accuracy of the numerical method can be examined by L_2 error norm.

4.3.6 L_∞ Error Norm

It is the maximum absolute error between the two functions for every point in the domain. It can be represented as follows:

$$L_\infty = \max_{0\le i\le N}\left|u_{exact} - u_{num}\right| \tag{4.6}$$

Significance:

- Uniform convergence: The idea of uniform convergence is closely associated with the L_∞ norm. A series of functions converges uniformly to a limiting function, if the L_∞ norm of the difference between the functions and the limit approaches 0. This kind of convergence is more stable.
- Error bound approximation: The L_∞ norm provides a simple and comprehensible approximation of the largest error using polynomials, splines, or other basis functions. This is especially helpful in applications like computer graphics, where minimizing the largest difference between an estimated curve and the exact curve is necessary to achieve the visual quality.

4.3.7 Application of Cubic B-Spline Collocation in the Heat Equation

The B-spline collocation method is used to discretize the heat conduction equation. This involves expressing the differential equations for B-spline basis functions and collocation points. For the purpose of solving PDEs, such as the heat equation, the B-spline of order three is a useful numerical method.

The 1D heat equation is represented as follows:

$$u_t = \alpha u_{zz} \tag{4.7}$$

where α represents thermal diffusivity, t is time, x is the spatial coordinate, and $u(x, t)$ is the temperature distribution.

To approximate the solution of heat equation with the help of B-spline collocation technique, we began by expressing the solution as a linear combination of B-spline basis functions. This transforms the problem into a system of algebraic equations.

Let the solution $u(\mathrm{z}, t)$ be denoted as:

$$u(\mathrm{z},t) = \sum_{i=-1}^{N+1} c_i(t)\Psi_i(\mathrm{z}),\ i = 1(1)N \tag{4.8}$$

where $\Psi_i(\mathrm{z})$ is the basis of B-spline function and $c_i(t)$ is the coefficient to be determined in temporal domain.

4.3.8 Algorithm

Step 1: Discretize the spatial domain

Divide the spatial domain $[c, d]$ by the nodes z_i into N sub-intervals $[\mathrm{z}_i, \mathrm{z}_{i+1}]$, $i = 0(1)N-1$ with the spatial size $h = \frac{\mathrm{z}_{i+1}-\mathrm{z}_i}{N}$ and temporal space $t_j = j\Delta t, j = 0, 1\ldots, M$.

Step 2: Construct cubic B-splines

Construct the cubic B-spline basis functions $\Psi_i(\mathrm{z})$, $i = -1, 0, 1, \ldots, N+1$.

Step 3: Approximate the solution

Represent the approximate solution $u(\mathrm{z}, t)$ as a linear combination of the B-spline basis functions, i.e., $u(\mathrm{z}, t) = \sum_{i=-1}^{N+1} c_i(t)\Psi_i(\mathrm{z})$, where $c_i(t)$ are the time-dependent coefficients to be determined.

Step 4: Collocation points

Select the collocation points z_i within each sub-interval $[\mathrm{z}_i, \mathrm{z}_{i+1}]$.

Step 5: Derive the collocation equations

The heat equation is transformed into a linear set of equations at each collocation point z_i, which is solved by the Thomas method.

Step 6: Incorporate initial and boundary conditions

Use the initial time temperature distribution $u(\mathrm{z}, 0) = f(\mathrm{z})$ to determine the initial coefficients c_i^0.

Apply the boundary conditions (Dirichlet or Neumann) to adjust the equations. For Dirichlet conditions, $u(c, t) = g_1(t)$ and $u(d, t) = g_2(t)$.

Step 7: Time integration

Use an appropriate time-stepping method (Runge-Kutta and Crank-Nicolson) to solve the system of ordinary differential equations for $c_i^n(t), n = 1(1)M$.

Step 8: Analyze the absolute error

Exact solution: The analytical solution to the heat equation is $u_{exact}(z,t)$.

Approximate solution: $u_{approximate}(z,t)$ is the approximate solution obtained using the cubic B-spline collocation method.

Absolute error: $E(z,t) = | u_{exact}(z,t) - u_{approximate}(z,t) |$

Example:

Consider the exact solution of heat equation is $u(z,t) = \sqrt{\left(\frac{20}{20+t}\right)} e^{\left[\frac{-(z-2-\alpha t)^2}{4\beta(t+20)}\right]}$, $y \in [0,1]$ and $t \geq 0$,

with the initial condition $u(z,0) = e^{\left[\frac{-(z-2)^2}{80\beta}\right]}$

and the boundary condition are $u(0,t) = \sqrt{\left(\frac{20}{20+t}\right)} e^{\left[\frac{-(a-2-\alpha t)^2}{4\beta(t+20)}\right]}$, $u(1, 0) = \sqrt{\left(\frac{20}{20+t}\right)} e^{\left[\frac{-(b-2-\alpha t)^2}{4\beta(t+20)}\right]}$.

For this, the parameters are $\alpha = 0.8$, $\beta = 0.1$.

4.3.9 MATLAB® Program

Program for the heat problem:

```
Clc;
Clear;
%% initialize spatial size%%
a = ;
b = ;
h = ;
x = x0: h: x1;
size = x;
M = length(x);
%% initialize temporal size%%
t0 = ;
t1 = ;
Δt = ;
t = t0: k: t1;
N = length(t);
```

Declare the parameters according to the problems

```
%% initial condition, boundary condition and analytical
solution%%
for j=1:length(t)
  for i=1:length(x)
   exact(i,j)=;
  end
end

for i=1:length(x)
   ic(i)=;
```

Declare the analytical solution and initial and boundary conditions at each time and space level according to the problems.

```
    icpt(i)=;
    icpx(i)=;
end
for i=1:length(t)
    bca(i)=;
    bcb(i)=;
end
%% B-spline basis function %%
for i=1: length(x)
  u(i,i:i+2)  = [1/6 4/6 1/6];
  ux(i,i:i+2)  = [-1/2h 0 1/2h];
  uxx(i,i:i+2)= [1/h^2 -2/h^2 1/h^2];
end
%% Approximate the solution at t = t0 %%
A=[ux(1,:);u;ux(end,:)];
B=[icpx(1) ic icpx(end)]';
Q(:,1)=matA0\matB0;
sol(:,1)=u*Q(:,1);
```

1. Declare the approximate solution and its derivatives.
2. Approximate solution at $t = t_0$

4.3.9.1 Remarks

- From Table 4.1 and Table 4.2 it is evident that the absolute error and L_2 and L_∞ are better than the results available in the literatures.
- The analytical solution and the numerical solution are very close to each other, which is evident from Figures 4.1 and 4.2.

4.3.10 Multi-Dimensional Heat Transfer Problems

Spatial discretization in multiple dimensions: The B-spline collocation method can extend in higher-dimensional B-splines and grid structure to handle multi-dimensional domains.

Solution strategy: It can develop the algorithms and solvers which are efficiently capable for solving coupled heat transfer equations in multiple dimensions of heat transfer (conduction, convection, and radiation) and complex geometries conditions.

Applications: It explore applications of multi-dimensional heat transfer modeling in engineering systems such as heat exchangers, electronics cooling, and building energy systems.

4.3.11 Turbulent Heat Transfer Modeling Using B-Spline Collocation

Turbulence modeling: In order to simulate turbulent heat transfer phenomena, turbulence models such as Reynolds-averaged Navier-Stokes (RANS) or large eddy simulation (LES) can be applied into the B-spline collocation framework [28].

Turbulence modeling strategies: Turbulence closure models can implement wall functions, and subgrid-scale models within the B-spline collocation method to accurately capture the effects of turbulence on heat problem.

TABLE 4.1
Error Analysis (Absolute Error and Error Norms L_2, L_∞) for $h = 0.01, \Delta t = 0.001$ in the Range [0, 1] with the Literatures

	x / t	$\frac{1}{10}$	$\frac{3}{10}$	$\frac{5}{10}$	$\frac{7}{10}$	$\frac{9}{10}$	L_2	L_∞
Current method	1	3.2498×10^{-10}	1.2141×10^{-10}	2.2952×10^{-9}	3.1978×10^{-9}	2.5078×10^{-9}	2.1331×10^{-9}	3.7818×10^{-10}
	2	6.9019×10^{-13}	2.0611×10^{-10}	6.8607×10^{-10}	1.2642×10^{-9}	1.1218×10^{-9}	8.0490×10^{-10}	1.7092×10^{-10}
	3	1.5542×10^{-10}	3.8345×10^{-10}	4.6505×10^{-10}	3.8847×10^{-10}	1.8705×10^{-10}	3.3798×10^{-10}	2.5359×10^{-11}
	4	1.717×10^{-10}	5.1571×10^{-10}	8.2443×10^{-10}	1.0087×10^{-9}	7.2409×10^{-10}	7.0291×10^{-10}	1.0684×10^{-10}
	5	1.2889×10^{-10}	4.1817×10^{-10}	7.2610×10^{-10}	9.6107×10^{-10}	7.2748×10^{-10}	6.5153×10^{-10}	1.0827×10^{-10}
[14]	1	3.30×10^{-10}	1.218×10^{-9}	2.30×10^{-9}	3.20×10^{-9}	2.51×10^{-9}	–	–
	2	2.99×10^{-12}	2.10×10^{-10}	6.90×10^{-10}	1.27×10^{-9}	1.12×10^{-9}	–	–
	5	1.29×10^{-10}	4.18×10^{-10}	5.26×10^{-10}	6.54×10^{-10}	7.27×10^{-10}	–	–
[15]	1	3.40×10^{-5}	3.10×10^{-5}	2.70×10^{-5}	2.50×10^{-5}	2.00×10^{-5}	–	–
	5	9.06×10^{-10}	1.84×10^{-9}	1.33×10^{-9}	3.83×10^{-10}	1.31×10^{-9}	–	–
[16]	1	9.96×10^{-9}	2.70×10^{-8}	3.78×10^{-8}	3.99×10^{-8}	2.55×10^{-8}	–	–
	2	8.45×10^{-9}	2.71×10^{-8}	4.59×10^{-8}	5.89×10^{-8}	4.36×10^{-8}	–	–
[17]	1	2.44×10^{-8}	4.57×10^{-8}	2.73×10^{-8}	2.04×10^{-8}	4.34×10^{-8}	–	–
	2	3.50×10^{-9}	7.57×10^{-9}	5.46×10^{-9}	2.71×10^{-9}	7.24×10^{-9}	–	–
	5	8.77×10^{-10}	1.68×10^{-9}	1.02×10^{-9}	7.68×10^{-10}	1.56×10^{-9}	–	–
[18]	1	8.71×10^{-9}	2.05×10^{-8}	3.71×10^{-8}	3.74×10^{-8}	1.65×10^{-8}	–	–
	2	7.71×10^{-9}	3.05×10^{-8}	4.71×10^{-8}	3.74×10^{-8}	1.65×10^{-8}	–	–
	5	1.73×10^{-9}	3.91×10^{-10}	3.59×10^{-9}	6.94×10^{-9}	9.13×10^{-9}	–	–

TABLE 4.2
Error Analysis with the Parameters $h = 0.05, \Delta t = 0.001, \alpha = 0.6,$ ***nd*** $\beta = 0.01$ ***in*** $[0,1]$

t	$x = 5/100$	$x = 25/100$	$x = 45/100$	$x = 65/100$	$x = 85/100$	L_2	L_∞
1	4.2413×10^{-4}	1.362×10^{-3}	3.976×10^{-3}	1.0554×10^{-2}	2.547×10^{-2}	5.1594×10^{-8}	1.3330×10^{-7}
2	1.2089×10^{-5}	4.8362×10^{-5}	1.7665×10^{-4}	5.8917×10^{-4}	1.79×10^{-3}	3.6004×10^{-8}	6.5441×10^{-8}
3	2.1445×10^{-7}	1.0485×10^{-6}	4.6996×10^{-6}	1.9308×10^{-5}	7.2716×10^{-5}	7.3073×10^{-9}	1.7809×10^{-8}
4	2.5124×10^{-9}	1.4767×10^{-8}	7.9840×10^{-8}	3.9711×10^{-7}	1.8171×10^{-6}	5.2430×10^{-10}	1.4162×10^{-9}
5	2.04314×10^{-11}	1.42260×10^{-10}	9.1412×10^{-10}	5.4211×10^{-9}	2.9673×10^{-8}	1.9086×10^{-11}	5.4828×10^{-11}

Assessment and authentication: To determine the accuracy of the numerical predictions, validate the turbulent heat transfer simulation using experimental data and benchmark cases.

4.3.12 Parallelization and Optimization of Large-Scale Simulations

Parallel computing: Applying parallelization approaches like domain decomposition or message passing interface (MPI) to scale out computational workload over multiple processors or compute nodes.

Load balancing: Develop load-balancing strategies that evenly distribute computing tasks among several processes, maximizing resource utilization and minimizing computational overhead.

Optimization techniques: Investigate optimization algorithms and strategies like gradient-based optimization and genetic algorithms to optimize simulation parameters, grid resolution, and numerical schemes for large-scale simulations.

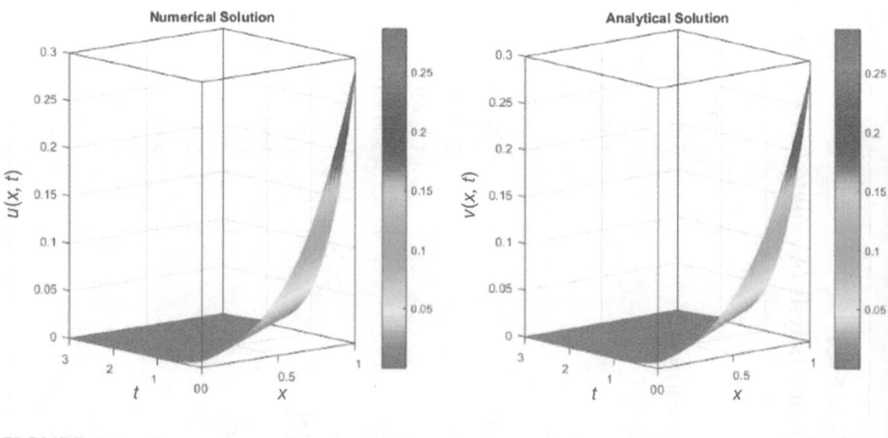

FIGURE 4.1 Comparison of approximate and exact solution with the parameters $h = 0.05$, $\Delta t = 0.001$, $\alpha = 0.6$, $\beta = 0.01$, and $t = 3$ in the interval $[0,1]$.

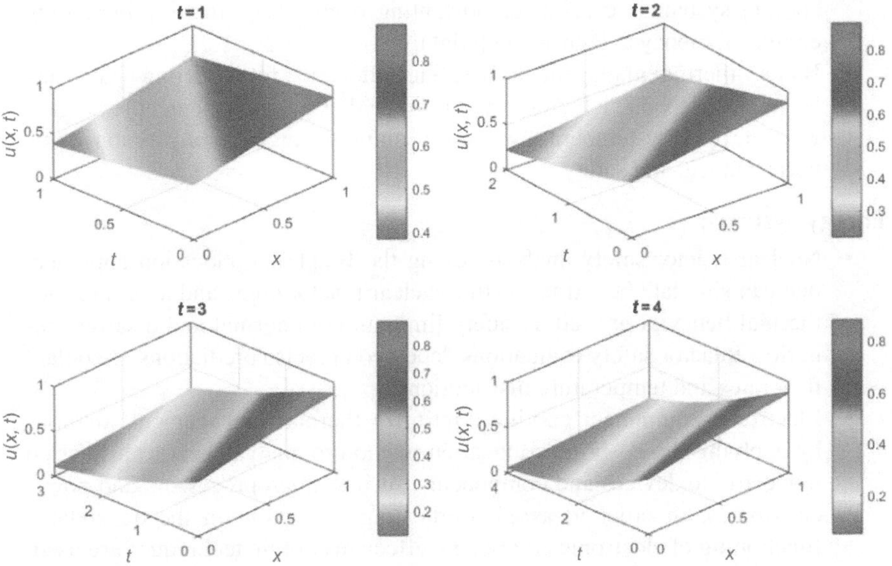

FIGURE 4.2 Approximate solution with the parameters at $h = 0.01$, $\Delta t = 0.001$, $\alpha = 0.8$, $\beta = 0.1$, and $y \in [0,1]$ with different times $t = 1(1)4$.

The capacity for heat transfer modeling by addressing these challenging subjects can be increased by B-spline collocation techniques. This will allow for precise and effective simulations of turbulent, large-scale, multi-dimensional heat transfer phenomena in a variety of engineering and scientific applications.

4.3.13 Applications and Investigation

Various applications are explained as follows.

Aerospace Engineering:

- Heat transfer in turbomachinery: To model heat transmission in gas turbine engines, including internal cooling passageways, blade surfaces, and combustor liners, use the B-spline collocation approach. Precise estimation of heat transfer rates and temperature distributions is essential for enhancing engine longevity and performance.
- Thermal management of hypersonic vehicles: Use B-spline collocation to study the processes of heat transport on the surfaces of hypersonic vehicles during high-speed flight. To ensure structural integrity and build thermal protection systems, it is crucial to comprehend temperature distributions and thermal loads.

Automotive Engineering:

- Cooling system design: Utilize the B-spline collocation technique to investigate heat transfer processes in automotive cooling systems, such as radiators, coolant channels, and engine parts. Improving the performance of the

cooling system is crucial for preventing overheating, thereby increasing engine efficiency and ensuring reliability.
- Battery thermal management: B-spline collocation for modeling heat transfer in electric vehicle batteries to optimize thermal management strategies. Keeping battery temperatures within optimal ranges improves longevity, performance, and safety.

Energy Systems:

- Nuclear reactor safety analysis: Using the B-spline collocation approach, one can simulate heat transport in nuclear reactor cores and assess reactor thermal behavior as well as safety limits in both normal and disaster scenarios. Reactor safety evaluations depend on precise predictions of coolant flow rates and temperature distributions.
- Electronic component cooling: Optimize thermal management solutions by applying the B-spline collocation method to analyze heat transmission in electronic devices and components, such as microprocessors and power electronics. In order to avoid overheating and maintain the dependable functioning of electronic equipment, effective cooling techniques are used.
- Solar energy systems: Examine heat transmission in thermal energy storage systems and solar collectors by employing B-spline collocation to maximize system performance and efficiency. Gaining knowledge of heat transfer mechanisms improves the use of solar energy for heating and electricity generation.

Building and HVAC (Heating, Ventilation, and Air Conditioning) Systems

- Building energy performance: Use the B-spline collocation approach to assess heat transfer in HVAC systems, building envelopes, and energy-efficient building designs in order to maximize thermal comfort and energy efficiency. Accurate thermal modeling is used to reduce energy consumption and operating costs while preserving interior comfort [29].

Through presenting these real-world examples and case studies, we show how the B-spline collocation method can be applied practically and effectively to address heat transfer issues in a variety of industries, resulting in better system performance, better engineering designs, and long-lasting technological advancements.

4.3.14 Future Directions and Challenges

Various directions of CFD analysis and challenges faced are described briefly as follows.

Emerging Trends in CFD and Heat Transfer Research:

- Multi-physics and multi-scale modeling: Integrating CFD with other physics, such as electromagnets, solid mechanics, and chemistry, to simulate complex multi-physics phenomena and multi-scale processes.
- High-performance computing (HPC): Continued advancements in parallel computing architectures, algorithms, and software tools to enable faster, more efficient simulations of larger and more complex systems.

Open Challenges and Areas for Further Improvement:

- Turbulence modeling: Improving turbulence modeling approaches, including the development of hybrid RANS-LES models and scale-resolving simulations, to capture turbulent flow and heat transfer phenomena accurately.
- Uncertainty quantification: Enhancing techniques for uncertainty quantification and sensitivity analysis to assess and mitigate uncertainties in CFD and heat transfer simulations, particularly in engineering design and risk assessment applications.
- Coupled physics simulations: Addressing challenges associated with coupled physics simulations, such as fluid-structure interaction, conjugate heat transfer, and multi-phase flow, to simulate more realistic and comprehensive engineering systems.

Potential Applications and Advancements in B-Spline Collocation Methods:

- Adaptive mesh refinement: Advancing adaptive mesh refinement techniques within the B-spline collocation framework to dynamically adapt the computational grid based on solution features and physics, improving simulation accuracy and efficiency.
- Uncertainty quantification and optimization: Integrating uncertainty quantification and optimization techniques with B-spline collocation methods to enable robust design optimization and decision-making under uncertainty in engineering applications.
- Parallelization and scalability: We are further optimizing parallelization strategies and the scalability of B-spline collocation methods to leverage HPC resources for large-scale simulations of complex heat transfer problems.

By addressing these future directions and challenges, researchers and practitioners in the fields of CFD and heat transfer can advance the state of the art, develop innovative solutions to real-world problems, and contribute to scientific and technological progress in diverse industries and applications.

4.4 THE APPLICATION OF B-SPLINE COLLOCATION FOR HEAT TRANSPORT IN CFD AND ITS PRACTICAL IMPACTS

In CFD, B-spline collocation techniques are useful for solving PDEs, such as those governing heat transfer. Here are a few instances of B-spline collocation for CFD heat transfer in practical problems.

i. Enhanced precision and movement:
    - Better continuity: When compared to traditional finite element approaches, B-splines produce smoother solutions by ensuring better continuity at element interfaces. Smooth temperature gradients are essential when dealing with heat transfer issues.
    - Reduction of errors: B-spline basis functions generate more accurate and authentic simulations, which prevent the numerical oscillations and errors.

ii. Efficient handling of complex geometries:
   - Flexible mesh generation: B-splines don't require complicated meshing techniques to readily conform to complex boundaries and geometries. Due to their adaptability, they may be used in simulations with complex designs, as those seen in automotive or aerospace applications.
   - Local improvement: B-spline collocation allows for local refinement without requiring changes to the entire mesh. Therefore, by focusing computational resources on regions or areas with significant temperature gradients or critical heat transmission zones, efficiency and accuracy may be maximized.

iii. Diminished computational cost:
   - Construction of spare matrix: Generally, B-spline techniques are used to construct sparse matrices, which are easier to solve than dense matrices.
   - Parallelization: B-spline collocation methods are well suited for parallel computing. Their design enables effective implementation on simultaneous processors, substantially reducing processing time.

iv. Enhanced stability:
   - Robustness in high gradients: B-splines provide consistency whenever dealing with excessive temperature gradients, which are common in heat transfer problems. By virtue of its versatility, the numerical solution is assured to be accurate and stable throughout a wide range of operating circumstances.

v. Applicability to non-linear problems:
   - Non-linear heat transfer: B-spline collocation outperforms certain standard strategies for dealing with non-linear heat transport scenarios, such as radiation or phase shift. This expands the range of technical obstacles that the technique may be utilized to solve in engineering problems.

4.4.1 Enhancing Sustainable Performance in Manufacturing with B-Splines for Heat Transfer in CFD

B-splines are a powerful technique for improving the precision and efficacy of simulations in CFD, particularly in heat transfer studies. Sustainable performance in manufacturing and operational operations may be considerably enhanced by mixing B-splines with artificial intelligence (AI) and information communication technologies (ICT), as follows:

i. Increased simulation accuracy:
   - Accuracy in complex modeling geometries: Heat transfer models are more accurate when complex geometries are represented by continuous B-spline basis functions. This precision helps in the creation of products and processes that optimize thermal performance while consuming less energy and waste. Decision-making may therefore be based on more reliable data, supporting long-term and efficient operations.
   - Better mesh quality: B-spline enables the creation of superior meshes that lower numerical errors and increase simulation stability.

ii. Integration with AI for optimization
   - Data-driven insights: AI systems can analyze the huge amounts of data generated by CFD simulations to identify trends and enhance heat transfer procedures. For example, machine learning algorithms are able to forecast the cooling system designs that will save the most money on operating costs while consuming the least amount of energy.
   - Adaptive control system: By integrating AI with a B-spline-based CFD model, AI may be used to construct a real-time adaptive control system that responds to shifting operating conditions. This dynamic adjustment improves sustainability and leads to more efficient heat management.

iii. Leveraging ICT for enhanced communication and collaboration:
   - Real-time control and monitoring: Advanced ICT systems make it possible to watch heat transfer processes in real time. Manufacturers may monitor temperatures and heat flow in real time and make quick modifications to optimize performance by merging B-spline CFD models with Internet of Things sensors.
   - Collaborative platforms: The collaborations between engineers, designers, and operators are facilitated by ICT by providing them a single platform for accessing and evaluating the outcomes of CFD simulations. Together, we can ensure that all parties are pursuing the same objective, which is to increase sustainability.

iv. Practice for sustainable manufacturing:
   - Energy efficiency: The system can be generated by the utilization of the B-spline collocation technique, which can use less energy for heating or cooling for heat transfer simulations. This minimizes greenhouse emissions and advances the industrial sector's overall sustainability aims.

4.4.2 B-Spline for Environmentally Sustainable Production

B-splines are the mathematical tools that are used in CFD to produce smooth curves and surfaces. They have a number of properties that make them useful for modeling complex systems. The following are some ways that the use of B-splines in sustainable manufacturing processes can lead to socio-environmental-friendly production:

- Precision and accuracy: When expressing boundaries and geometries, B-splines offer a great degree of precision. With more modeling of thermal characteristics and heat flow made possible by this accuracy, heat transfer models can forecast temperature distributions and thermal stresses with greater accuracy. As a result, energy consumption is decreased during the design of cooling or heating systems.
- Smoothness and continuity: Smooth transitions between various model components are guaranteed by B-splines. This smoothness removes artificial discontinuities that might skew the findings of heat transfer simulations, which is essential for accuracy. Manufacturers may create procedures that maximize energy efficiency and save waste by guaranteeing more accurate simulations.

- Adaptability and flexibility: B-splines are easily adaptable to a wide range of BCs and complex geometries. With the typical construction of goods consisting of multiple materials and complex geometries, this adaptability is essential for sustainable production.
- Complex geometries and a variety of BCs can be readily adapted to by B-splines. This adaptability is crucial for sustainable production, as goods are typically constructed of many materials and have complex geometries. Precise modeling of this complex geometry enhances manufacturing process management by reducing material waste and increasing energy efficiency.
- Better energy management: B-spline improves in the precise control of thermal processes, which enables better energy management. By optimizing their heating and cooling cycles, manufacturers may reduce energy consumption, greenhouse emissions, and operating expenses.

4.4.2.1 Socio-Environmental Benefits

- Reduced energy consumption: Efficient heat transfer modeling leads to less energy use, decreasing the carbon footprint of manufacturing processes.
- Resource conservation: Minimizing material waste and improving product quality conserve natural resources.
- Cost savings: Energy and material efficiency lead to significant cost reductions, making sustainable practices economically viable.

4.4.3 Implications of the Study

Few implications of the work is explained with following descriptions.

4.4.3.1 Implications for Academicians

The study of B-spline collocation in CFD heat transfer has significant significance for academics. The first benefit is that it provides a solid basis for developing and refining numerical methods for heat transfer measurement. B-spline are flexible and precise, which is important for advancing theoretical understanding and computational techniques since it allows researchers to more accurately approximate complex geometries and BCs.

Furthermore, this work bridges disciplinary boundaries by addressing gaps in the domains of computational mathematics, thermal science, and fluid mechanics. Scholars may investigate the potential for further optimization of B-spline algorithms, their applicability in resolving diverse heat transfer problems, and their potential to facilitate the development of computational instruments that exhibit greater scalability and efficacy.

4.4.3.2 Implications for Practitioners

The consequences of this work are equally significant for practitioners in the fields of engineering and applied sciences. More precise and dependable simulations, which are essential for developing and optimizing thermal systems across a range of sectors, including aerospace, automotive, and energy, can result from the use of B-spline collocation techniques in CFD.

The increased accuracy and computational efficiency of B-splines may be used by practitioners to better the performance and design of electronic cooling systems, HVAC systems, and heat exchangers. This may lead to lower costs, more energy efficiency, and enhanced thermal management system dependability and safety.

Moreover, the validation and verification of CFD models may be expedited by the application of B-spline collocation techniques. This decreases the time and resources needed for experimental testing and increases the confidence in simulation results, which helps with decision-making.

Conclusively, the investigation of B-spline collocation in heat transfer for CFD not only accelerates academic research but also yields realistic advantages that can be used immediately in engineering practice. It is a major advancement in the ongoing search for computer techniques in heat transfer analysis that are more precise and effective.

4.5 CONCLUSION

The B-spline collocation method offers several advantages that make it well suited for advancing CFD in heat transfer problems. These include the following:

- High-order accuracy and flexibility in representing complex geometries and functions.
- Efficient utilization of computational resources through local refinement and hierarchical structure.
- Ability to handle various heat transfer phenomena with accuracy and stability, including conduction, convection, and radiation.

The B-spline collocation technique is an effective technique in heat transfer problems. It also includes the larger area of computational technique in scientific and technical research influencing the course of future innovation and discovery. B-splines collocation has the potential to increase accuracy, computational efficiency, mesh generation flexibility, numerical stability, application to multi-physics issues, and scalability to large-scale problems when used in CFD heat transfer simulations.

B-spline collocation techniques are useful for advanced engineering analysis and design optimization because they improve the precision, effectiveness, and robustness of CFD simulations in heat transfer. Additionally, it results in more socially and environmentally conscious production techniques.

REFERENCES

1. Kutluay, Selçuk, Nuri Murat Yağmurlu, and Ali Sercan Karakaş. "An effective numerical approach based on cubic Hermite B-spline collocation method for solving the 1D heat conduction equation." *New Trends in Mathematical Sciences* 10, no. 4 (2022): 20–31.
2. Tassaddiq, Asifa, Muhammad Yaseen, Aatika Yousaf, and Rekha Srivastava. "A cubic B-spline collocation method with new approximation for the numerical treatment of the heat equation with classical and non-classical boundary conditions." *Physica Scripta* 96, no. 4 (2021): 045212.

3. Rabah, Ahlem Ben, Shaher Momani, and Omar Abu Arqub. "The B-spline collocation method for solving conformable initial value problems of non-singular and singular types." *Alexandria Engineering Journal* 61, no. 2 (2022): 963–974.
4. Annasabi, Zouheir, F. Erchiqui, and Mhamed Souli. "Application of the B-spline method to solve nonlinear problem of heat conduction with radiation boundary conditions using Kirchhoff transformation." *Journal of Heat Transfer* 144, no. 5 (2022): 051401.
5. Schoenberg, I. J. "On trigonometric spline interpolation." *Journal of Mathematics and Mechanics* 13, no. 5 (1964): 795–825.
6. Jena, Saumya Ranjan, and Archana Senapati. "One-dimensional heat and advection-diffusion equation based on improvised cubic B-spline collocation, finite element method and Crank-Nicolson technique." *International Communications in Heat and Mass Transfer* 147 (2023): 106958.
7. Shafiq, Madiha, Muhammad Abbas, Emad K. El-Shewy, Mahmoud AE Abdelrahman, Noura F. Abdo, and Ali A. El-Rahman. "Numerical investigation of the fractional diffusion wave equation with the Mittag-Leffler function." *Fractal and Fractional* 8, no. 1 (2023): 18.
8. Hamad, Q, and Younis A Sabawi. "Numerical solution of the heat equation by cubic B-spline collocation method." *Eurasian Journal of Science and Engineering* 9, no. 2 (2023): 252–261.
9. Raslan, K. R., and Khalid K. Ali. "A new structure formulation for cubic B-spline collocation method in three and four-dimensions." *Nonlinear Engineering* 9, no. 1 (2020): 432–448.
10. Mittal, R. C., and Rajni Rohila. "The numerical study of advection–diffusion equations by the fourth-order cubic B-spline collocation method." *Mathematical Sciences* 14 (2020): 409–423.
11. Dwivedi, Kushal Dhar, Subir Das, and Dumitru Baleanu. "Numerical solution of highly non-linear fractional order reaction advection diffusion equation using the cubic B-spline collocation method." *International Journal of Nonlinear Sciences and Numerical Simulation* 23, nos. 7–8 (2022): 1157–1172.
12. Umer, Aqsa, Muhammad Abbas, Madiha Shafiq, Farah Aini Abdullah, Manuel De la Sen, and Thabet Abdeljawad. "Numerical solutions of Atangana-Baleanu time-fractional advection diffusion equation via an extended cubic B-spline technique." *Alexandria Engineering Journal* 74 (2023): 285–300.
13. Shafiq, Madiha, Muhammad Abbas, Homan Emadifar, Ahmed SM Alzaidi, Tahir Nazir, and Farah Aini Abdullah. "Numerical investigation of the fractional diffusion wave equation with exponential kernel via cubic B-spline approach." *Plos One* 18, no. 12 (2023): e0295525.
14. Jena, Saumya Ranjan, and Guesh Simretab Gebremedhin. "Computational technique for heat and advection–diffusion equations." *Soft Computing* 25, no. 16 (2021): 11139–11150.
15. Dehghan, Mehdi. "Weighted finite difference techniques for the one-dimensional advection–diffusion equation." *Applied Mathematics and Computation* 147, no. 2 (2004): 307–319.
16. Mittal, R. C., and R. K. Jain. "Redefined cubic B-splines collocation method for solving convection–diffusion equations." *Applied Mathematical Modelling* 36, no. 11 (2012): 5555–5573.
17. Nazir, Tahir, Muhammad Abbas, Ahmad Izani Md Ismail, Ahmad Abd Majid, and Abdur Rashid. "The numerical solution of advection–diffusion problems using new cubic trigonometric B-splines approach." *Applied Mathematical Modelling* 40, nos. 7–8 (2016): 4586–4611.
18. Mohammadi, Reza. "Exponential B-spline solution of convection-diffusion equations." *Applied Mathematics* 4, no. 6 (2013): 933–944.

19. Gnedin, Nickolay Y., Vadim A. Semenov, and Andrey V. Kravtsov. "Enforcing the Courant–Friedrichs–Lewy condition in explicitly conservative local time stepping schemes." *Journal of Computational Physics* 359 (2018): 93–105.
20. Khan, Beenish, Muhammad Abbas, Ahmed SM Alzaidi, Farah Aini Abdullah, and Muhammad Bilal Riaz. "Numerical solutions of advection diffusion equations involving Atangana–Baleanu time fractional derivative via cubic B-spline approximations." *Results in Physics* 42 (2022): 105941.
21. Yaseen, Muhammad, Muhammad Abbas, and Muhammad Bilal Riaz. "A collocation method based on cubic trigonometric B-splines for the numerical simulation of the time-fractional diffusion equation." *Advances in Difference Equations* 2021 (2021): 1–19.
22. Akram, Tayyaba, Muhammad Abbas, Azhar Iqbal, Dumitru Baleanu, and Jihad H. Asad. "Novel numerical approach based on modified extended cubic B-spline functions for solving non-linear time-fractional telegraph equation." *Symmetry* 12, no. 7 (2020): 1154.
23. Reutskiy, Sergiy, Yuhui Zhang, Ji Lin, Jun Lu, Haifeng Xu, and Yongjun He. "A novel B-spline method to analyze convection-diffusion-reaction problems in anisotropic inhomogeneous medium." *Engineering Analysis with Boundary Elements* 118 (2020): 216–224.
24. Hidayat, Mas Irfan Purbawanto, Bambang Ariwahjoedi, and Setyamartana Parman. "A new meshless local B-spline basis functions-FD method for two-dimensional heat conduction problems." *International Journal of Numerical Methods for Heat & Fluid Flow* 25, no. 2 (2015): 225–251.
25. Jena, Saumya Ranjan, and Archana Senapati. "Explicit and implicit numerical investigations of one-dimensional heat equation based on spline collocation and Thomas algorithm." *Soft Computing* (2024): 1–22, Volume 28, pages 12227–12248. https://link.springer.com/article/10.1007/s00500-024-09925-3.
26. Das, Shashikant, and Sudhakar Subudhi. "Studying the physiology of the human body using a mathematical model of transient energy balance." *Sādhanā* 48, no. 4 (2023): 245.
27. De Boor, Carl, and Boor Carl De. *A Practical Guide to Splines* 27, no. 1–392 (1978): 10. https://www.stat.cmu.edu/~brian/valerie/617-2022/week07/spline%20references/pdfcookie.com_a-practical-guide-to-splines.pdf
28. Morinishi, Youhei, Shinji Tamano, and Koichi Nakabayashi. "A DNS algorithm using B-spline collocation method for compressible turbulent channel flow." *Computers & Fluids* 32, no. 5 (2003): 751–776.
29. Afroz, Zakia, G. M. Shafiullah, Tania Urmee, and Gary Higgins. "Modeling techniques used in building HVAC control systems: A review." *Renewable and Sustainable Energy Reviews* 83 (2018): 64–84.

5 IoT-Enhanced Statistical Methods for Process Optimization in Biotechnology and Pharmaceutical Manufacturing

Polaki Suman, B. Rabi Prasad, Ghanishtha Prusty, Ch. Mohan Babu, and Islauddin Khan

5.1 INTRODUCTION

Process optimization systematically enhances manufacturing parameters to achieve the most efficient and effective production outcomes. It employs statistical, mathematical, and computational techniques to identify the optimal process conditions that maximize desired outputs while minimizing undesired ones [Fytopoulos & Pardalos (2023)]. Process optimization is essential for ensuring production processes' quality, efficiency, and reliability in the biotechnology and pharmaceutical industries, where precision and consistency are critical [Fytopoulos et al. (2022)]. The importance of process optimization in these sectors cannot be overstated. By fine-tuning process parameters, companies can increase product yield, reduce waste, and minimize resource consumption—critical factors in industries where raw materials and energy costs significantly impact production expenses [Faisal et al. (2024)]. Moreover, optimization plays a crucial role in maintaining the high standards of product quality required in these industries, ensuring uniformity by reducing production variability, which is vital for meeting regulatory standards and safeguarding patient safety [Nunavath et al. (2024)].

Optimization helps in cost reduction to run processes; companies can achieve significant savings through reduced use of expensive reagents, lower energy consumption, or minimized downtime [Naugolnova (2023)]. These savings can be reinvested in research and development or passed on to consumers, enhancing a company's competitive edge. As sustainability becomes increasingly prioritized in the biotechnology and pharmaceutical sectors, process optimization contributes by minimizing waste generation and reducing environmental impact. Optimized processes align with the growing emphasis on sustainable practices and ensure compliance

DOI: 10.1201/9781003503323-5

with stringent regulatory requirements set by agencies like the U.S. Food and Drug Administration (FDA) and the European Medicines Agency (EMA). Companies that successfully optimize their processes in a competitive market can achieve faster production times, higher-quality products, and lower costs, gaining a significant advantage over competitors [Rijamampianina et al. (2003)].

Complex and sensitive processes characterize the biotechnology and pharmaceutical sectors. Products in these industries often involve living organisms or highly reactive chemical compounds, requiring precise control over the manufacturing environment. In this context, process optimization is essential for improving efficiency, reducing costs, and ensuring that the final product is safe, effective, and of the highest quality [Fytopoulos & Pardalos (2023)]. For instance, even slight variations in temperature, pH, or nutrient concentrations can significantly impact yield and quality in biologics production. Similarly, optimizing synthesis routes and purification steps in pharmaceutical manufacturing can lead to higher product purity and fewer side effects [Fytopoulos et al. (2022)].

This chapter aims to explore and apply advanced statistical methods for process optimization in biotechnology and pharmaceutical manufacturing [(Cooksey, 2020)]. The goal is to provide a comprehensive guide to these methods, enhancing production efficiency, product quality improvement, and ensuring regulatory compliance. By integrating traditional statistical approaches with modern techniques like machine learning (ML) and genetic algorithms (GAs), this chapter demonstrates tools that effectively address the complexities and challenges inherent in highly regulated industries.

This chapter covers a range of statistical techniques, from foundational descriptive statistics and hypothesis testing to more advanced methodologies like design of experiments (DoE), response surface methodology (RSM), and multivariate analysis (MVA). It also delves into cutting-edge optimization strategies, including ML, the Internet of Things (IoT), and GAs. The chapter is designed to be practical and relevant, with real-world case studies illustrating these techniques' application in optimizing processes within the biotechnology and pharmaceutical sectors. Additionally, the scope includes considerations of regulatory compliance, data integrity, and the impact of emerging technologies like Industry 4.0 on process optimization. The amalgamation of statistical methods with innovative technologies, for instance, the IoT, is revolutionizing process optimization in biotechnology and pharmaceutical paramount [Jaiswal et al. (2023)]. IoT ensures realistic data collection, monitoring, and analysis across various stages of production, creating an interconnected system that enhances decision-making and process control.

The newness of this chapter stems from its holistic approach to process optimization, combining traditional statistical methods with pioneering computational techniques. This dual approach provides a robust framework for improving process efficiency and product quality while opening new avenues for innovation in biomanufacturing [Muduli et al. (2022)]. Including ML and GAs as optimization tools in this context is particularly innovative, offering a forward-looking perspective on how these technologies can be harnessed to achieve the evolving demands of the biotechnology and pharmaceutical industries. Furthermore, the chapter addresses the challenges of maintaining regulatory compliance in a rapidly changing technological landscape, making it a valuable resource for practitioners seeking to stay ahead in the field.

5.2 FUNDAMENTALS OF STATISTICAL METHODS

Statistical methods are essential for optimizing processes in biotechnology and pharmaceutical manufacturing, as they offer a structured approach to data analysis, enabling informed decision-making and process improvements, as shown in Table 5.1. Descriptive statistics, a fundamental component of statistical methods, play a crucial role in summarizing and characterizing key features of datasets, which are critical for understanding data patterns, central tendencies, and variability. These factors are vital for optimizing manufacturing processes. Measures of central tendency, such as the mean, median, and mode, help identify typical values within a dataset. For example, in a dataset like [2, 3, 5, 7, 11], the mean is 5.6, the median is 5, and the dataset lacks a mode. Understanding these central tendencies provides insight into what constitutes a 'typical' value in the data, which is crucial for process optimization.

In addition to central tendency, measures of dispersion, such as the range and standard deviation, provide insights into the spread or variability of the data. For the same dataset, the range is 9, and the standard deviation is approximately 3.36. These dispersion metrics are critical for understanding the extent of variability within the process, which can significantly impact the quality and consistency of manufacturing outputs. High variability may indicate issues that need addressing to achieve optimal and consistent results.

Graphical summaries, including histograms, box plots, and scatter plots, are also vital in statistical analysis. Graphical representation of data makes it easier to identify patterns, trends, and outliers that might not be immediately apparent through numerical analysis alone. In biotechnology and pharmaceutical manufacturing, these graphical methods are extensively applied to monitor production quality, assess process stability, and identify areas for improvement. These tools facilitate a deeper

TABLE 5.1
Common Statistical Methods and Their Applications in Biotechnology and Pharmaceutical Manufacturing

Statistical Method	Application Area	Purpose	Example Use Case
Descriptive Statistics	Data Analysis and Process Monitoring	Summarize dataset characteristics	Monitoring consistency in product quality
Hypothesis Testing	Quality Control and Process Validation	Test for evidence to reject a null hypothesis	Comparing yields of new vs. existing processes
Multivariate Analysis	Process Monitoring and Quality Control	Analyse multiple variables to identify patterns	Identifying variables affecting product stability
Statistical Process Control	Continuous Process Monitoring	Real-time process monitoring and control	Ensuring tablet weight consistency
Regression Analysis	Process Optimization and Predictive Modelling	Model relationships between parameters and outcomes	Predicting temperature impact on enzyme activity
ANOVA	Process Comparison and Optimization	Compare group means for significant differences	Comparing effectiveness of fermentation media

understanding of process behaviour by visually summarizing data, enabling more effective process control and optimization strategies.

5.3 DoE IN IoT-ENABLED BIOTECHNOLOGY AND PHARMACEUTICAL PROCESSES

The DoE is crucial in optimizing biotechnology and pharmaceutical manufacturing processes. It involves a systematic approach to designing, executing, analysing, and interpreting controlled experiments to evaluate factors influencing a particular outcome [(Lamidi et al., 2024)].

5.3.1 Principles of Experimental Design

The core principles of DoE include randomization, replication, and blocking. Randomization ensures unbiased results by randomly assigning treatments to experimental units, minimizing confusing effects from affecting the outcome, as demonstrated in studies evaluating the impact of different culture media on cell growth [Kaltenbach (2021)]. Replication involves repeating experiments to estimate data variability and increase result reliability. For instance, multiple trials in drug formulation testing reduce uncertainty and provide a more accurate estimate of the drug's effectiveness [(Beg et al., 2019; Jankovic et al. 2021)]. Blocking groups similar experimental units to control known sources of variability, enhancing the detection of treatment effects, a method commonly used in agricultural experiments where soil type variability needs to be controlled.

5.3.2 Standard Experimental Designs

There are several standard experimental designs in DoE. However, RSM, fractional factorial design, and complete factorial design are widely accepted designs. Complete factorial design examines all possible combinations of factors across various levels, offering a comprehensive understanding of factor effects and interactions. However, the required number of experiments can become substantial as the number of factors increases [Adams et al. (2023)]. Fractional factorial design reduces the experimental workload by focusing on a subset of possible combinations, yielding valuable insights into primary effects and some interactions. RSM employs statistical techniques to model and optimize processes influenced by multiple variables, identifying optimal process conditions by exploring the response surface with polynomial models [Krishnamurthy et al. (2024)].

5.3.3 Standard Experimental Designs for IoT Systems

IoT-enabled systems make it easier to collect and analyse the large volume of data generated by full factorial design. A subset of full fractional design is equally useful for reducing the number of experimental runs without reducing too much information. IoT devices continuously update the model with real-time data for dynamic optimization by RSM.

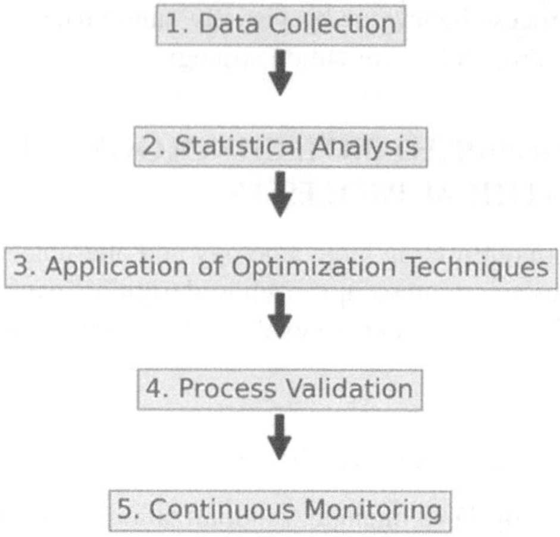

FIGURE 5.1 Process optimization workflow.

5.3.4 Applications of DoE in Biotechnology and Pharmaceutical Manufacturing

DoE is applied in several critical areas, including process optimization, formulation development, and analytical method development. Process optimization uses DoE to identify optimal conditions for maximizing bioproduct yield, purity, and potency, such as optimizing fermentation conditions for producing recombinant proteins by studying variables like temperature, pH, and nutrient concentrations [Du et al. (2022)]. Formulation development enables DoE to create stable and effective drug formulations by evaluating the effects of excipients, processing methods, and storage conditions, as seen in optimizing tablet formulations to ensure stability and bioavailability [Dhoot et al. (2019); Shi et al., 2021]. Analytical method development enhances the accuracy and precision of analytical methods by studying variables such as solvent composition, temperature, and detector settings, demonstrated in the optimization of high-performance liquid chromatography (HPLC) methods for impurity separation in drug substances [Patil et al. (2019)] (Figure 5.1).

5.4 RESPONSE SURFACE METHODOLOGY (RSM)

Various techniques, such as statistical and mathematical theories, together form the RSM, which helps to enhance and optimize processes. Especially, when there is an impact on process performance attributes for multiple input variables. RSM is highly effective in biotechnology and pharmaceutical manufacturing, where the complexity of processes requires more sophisticated approaches than traditional trial-and-error methods. To examine the relationship between one or more response variables as well as several explanatory variables, experiments are designed. The RSM fits a polynomial equation to the experimental data. This approach allows for a deeper understanding of the interaction between variables and the identification of optimal conditions for desired responses [Kumar & Reji (2022)].

5.4.1 Modelling and Optimization Techniques in RSM

The RSM process involves several key steps: identifying the critical factors influencing the response, designing the experiments, executing them, fitting a model to the data, and applying this model for optimization. The first step, identifying key factors, involves determining the significant variables affecting the response. These could include process parameters such as temperature, pH, agitation speed, and nutrient concentrations in a bioprocess or formulation-related variables like the type and concentration of excipients in pharmaceutical manufacturing (Sampathkumar & Kerwin 2024). Designing experiments in RSM commonly utilizes designs like Box-Behnken design (BBD), central composite design (CCD), and Doehlert design. These approaches enable efficient exploration of the factor space and facilitate the fitting of quadratic models, which are essential for capturing the curvature of the response surface. The next step, executing the experiments, involves collecting data on the response variables, where accurate and consistent execution is crucial for reliable data. Following data collection, fitting a model involves applying a polynomial, often a second-order (quadratic) model, to the data, represented by the formula,

$$Y = \beta_0 + \sum_{i=1}^{k} \beta_i X_i + \sum_{i=1}^{k} \beta_{ii} X_i^2 + \sum_{i<j}^{k} \beta_{ij} X_i X_j + \epsilon$$

In this context, Y represents the response variable, X_i stands for the input factor, β refers to the coefficients, and ϵ symbolizes the error term. Finally, optimizing with the model involves exploring the response surface using visualization tools like contour and surface plots to identify conditions that maximize or minimize the response through optimization methods like gradient ascent or descent.

5.4.1.1 Case Studies in Pharmaceutical Formulations and Bioprocessing

RSM has broad applications in optimizing complex processes within pharmaceutical formulations and bioprocessing. For instance, in the optimization of tablet formulation, RSM is used to determine the ideal composition of tablets to achieve the desired disintegration time and dissolution profile. Researchers may use a central composite design to measure outcomes like disintegration time and dissolution rate, and researchers may vary the concentrations of excipients such as a binder, a disintegrative, and a lubricant. A quadratic model is applied to the data, and response surfaces are generated, enabling researchers to determine the optimal combination of excipient concentrations that minimize disintegration time and maximize the dissolution rate. This ensures quick and complete drug release [Jange et al. (2023)]. In bioprocess optimization for protein production, RSM is applied to optimize conditions for recombinant protein production in a bioreactor, considering factors like temperature, pH, agitation speed, and nutrient concentration. A BBD helps vary these factors to maximize protein yield and purity. Once a quadratic model is fitted to the experimental data, response surfaces are examined, and contour and surface plots are utilized to visualize the interaction effects between factors. This helps in optimizing the process for large-scale

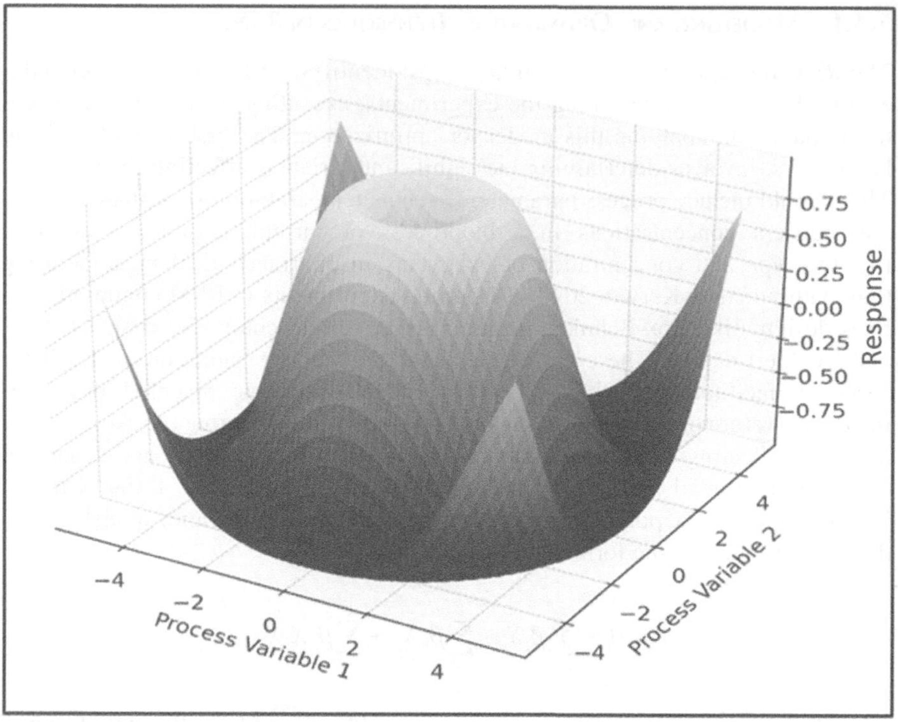

FIGURE 5.2 Response surface plot.

production [Rani & Vimolmangkang (2024)]. Similarly, enhancing biodegradation efficiency involves using RSM to optimize conditions for biodegradation in wastewater treatment. Factors such as microbial inoculum size, nutrient concentration, and aeration rate are varied to maximize pollutant degradation efficiency. Using a central composite design, researchers conduct experiments and fit a quadratic model to the biodegradation efficiency data, leading to the identification of optimal conditions for effective industrial wastewater treatment [Sravan et al. (2024)] (Figure 5.2).

5.4.2 Multivariate Analysis in Biotechnology and Pharmaceutical Manufacturing

MVA is a collection of statistical methods that examine multiple variables concurrently to comprehend their interrelationships and their collective influence on a specific outcome [S.A. Ioannis (2006)]. MVA is a remarkable tool in biotechnology and pharmaceutical manufacturing, where processes frequently involve complex relationships between multiple variables. This analytical approach enables a thorough understanding of the complex relationships within data, making it invaluable for quality control, process monitoring, and optimization [(Conner et al., 2020)].

5.4.3 Principal Component Analysis

This is a broadly used dimensional reduction method for datasets by making a new set of uncorrelated variables from the original variables. These uncorrelated variables are called principal components, which are ranked based on the amount of variance they capture from the dataset [Panda et al. (2024)]. In biotechnological processes, where parameters like temperature, pH, oxygen levels, and nutrient concentrations are monitored, principal component analysis (PCA) can streamline data interpretation by reducing the number of variables, effectively merging these measurements into principal components that emphasize the most crucial variations. The steps involved in PCA include mean and standardization, the calculation of the covariance matrix from the standardized data, eigenvalue decomposition to derive eigenvalues and eigenvectors from the covariance matrix, and the selection of principal components based on eigenvalues to form a simplified dataset that retains the key characteristics of the original data.

5.4.4 Partial Least Squares (PLS) Regression

It is a powerful multivariate technique that models the relationship between a set of independent variables (predictors) and dependent variables (responses). Partial least squares (PLS) is particularly effective in dealing with highly collinear and noisy data, which is common in biotechnological and pharmaceutical processes. PLS is designed to maximize the model's predictive power by considering predictors and responses simultaneously. In pharmaceutical manufacturing, PLS can relate process parameters such as mixing time, temperature, and ingredient concentrations to product quality attributes like potency and stability, aiding in process optimization. The steps in PLS include standardizing predictors and responses, extracting latent variables that capture the covariance between predictors and responses, fitting a linear model between these latent variables, and using the fitted model to predict responses based on new predictor data.

5.4.5 Applications in Quality Control and Process Monitoring

PCA and PLS are instrumental in maintaining product quality and ensuring effective process monitoring. In quality control, PCA helps identify significant sources of variation, such as how changes in temperature and pH affect product yield and purity. The detection of deviation from normal operating conditions is calculated by monitoring these principal components. PLS models, on the other hand, can predict quality attributes like dissolution rates and tablet hardness based on process parameters, enabling real-time quality control. For process monitoring, PCA and PLS are essential tools for reducing data dimensionality and highlighting critical variables. PCA can create control charts for principal components, signalling when a process deviates from expected patterns, while PLS is used to develop predictive models and soft sensors that estimate unmeasured process variables, such as biomass concentration in fermentation processes, based on easily measured variables like pH and oxygen uptake rate. These models provide real-time insights, facilitating better control and optimization of manufacturing processes.

5.4.6 Statistical Process Control (SPC)

To monitor the manufacturing process, statistical process control (SPC) is an essential technique. It operates at optimal levels and ensures consistently producing high-quality products. By utilizing statistical tools, SPC helps detect and correct manufacturing variations, thus enhancing product quality and process efficiency. In the pharmaceutical industry, implementing SPC involves using control charts for variables and attributes, conducting process capability analysis, and applying these tools to monitor critical quality attributes (CQAs).

5.4.6.1 Control Charts for Variables and Attributes

These are fundamental graphical tools in SPC, used to track the stability of a process over time and identify variations that may indicate potential issues. Control charts can be primarily classified into two types: variables and attributes.

a. **Control charts for variables**: Control charts focus on continuous data, such as weight measurements, temperature, pH, or concentration. R-chart and X-bar chart are the commonly used control charts. The X-bar chart tracks the process mean over time by plotting the average of sample data points at regular intervals. Control limits, including the lower control limit (LCL) and upper control limit (UCL), are calculated based on the process mean and standard deviation. Furthermore, the R-chart monitors the range within a sample, reflecting the process's variability by plotting the difference between the highest and lowest values in each sample. This type of chart helps identify shifts in process variability, with control limits determined by the average range and the inherent variability of the process.
b. **Control charts for attributes**: These charts, in contrast, are designed to monitor discrete data, such as the count of defective items or the occurrence of specific events. C-chart and p-chart are the two main control charts for attributes. The p-chart monitors the proportion of defective items in a sample by plotting the fraction of nonconforming units. The control limits for the p-chart are determined by the average proportion of defects and the sample size. Meanwhile, the c-chart tracks the number of defects per unit, plotting the number of defects observed in each sample. Control limits for the c-chart are typically calculated based on the average number of defects, often following a Poisson distribution.

5.4.6.2 Process Capability Analysis (PCA)

It is a critical component of SPC, used to evaluate how well a process can produce products that meet predetermined quality standards. This analysis involves comparing the output of a process to its specification limits to assess its performance. Two key metrics in process capability analysis (PCA) are the process capability index and the process capability (Cp) index adjusted for the process mean (Cpk). The Cp index evaluates a process's potential capability by comparing the range between the upper specification limit (USL) and lower specification limit (LSL) to the process variability. A Cp value exceeding 1 indicates that the process has the potential to meet the specifications consistently. In contrast, the Cpk index assesses a process's capability by accounting for the process mean and deviation from the target value. When the Cpk value is greater than 1, then it signifies that the process is both capable and well centred within the specification limits.

5.4.6.3 Implementing SPC in Pharmaceutical Manufacturing

Implementing SPC in pharmaceutical manufacturing involves several crucial steps to ensure effective process control and quality improvement. It is essential to identify CQAs, such as potency, purity, dissolution rate, and stability, which significantly impact product quality and patient safety. Appropriate control charts must be selected based on the type of data (continuous or discrete) and the CQAs being monitored. For instance, X-bar and R-charts are commonly used for continuous data like tablet weight, while p-charts are more suited for discrete data, such as the proportion of defective tablets [Sridevi (2024)]. Establishing control limits, monitoring the process, investigating issues, and conducting regular process capability analyses are necessary to maintain control and drive continuous improvement. Regularly reviewing and updating control limits, refining process parameters, and implementing best practices are critical for enhancing performance and ensuring product quality over time. Companies can significantly improve process stability, product quality, and overall efficiency by integrating SPC into pharmaceutical manufacturing. This leads to better compliance with regulatory standards and greater competitiveness in the market.

5.5 ADVANCED OPTIMIZATION TECHNIQUES IN BIOMANUFACTURING

Advanced optimization techniques have become indispensable tools in pursuing enhanced efficiency and quality within biomanufacturing. These methods leverage cutting-edge approaches, such as ML and GAs, and integrate them with traditional statistical techniques to optimize complex processes. This section delves into these advanced methods and provides case studies that illustrate their practical application in biomanufacturing, demonstrating their significant impact on process improvement.

5.5.1 Machine Learning Methods

These techniques are algorithms designed to identify patterns and relationships within large datasets, enabling predictions or decisions without explicit programming. In biomanufacturing, ML techniques are employed for various purposes, including process optimization, quality prediction, and anomaly detection [Wu et al. (2024); Polaki et al. (2024)]. The most widely recognized ML techniques are neural networks and GAs, each offering distinct advantages for optimizing manufacturing processes.

a. **Neural networks**: Neural networks, a subset of ML algorithms, are inspired by the structure and function of the human brain. These networks are made up of multiple layers of interconnected layers of neurons that process input data to generate predictive outputs. Neural networks are particularly effective at modelling complex, nonlinear relationships between process variables and outcomes. For example, in a biomanufacturing process, a neural network can be trained on historical data, including process parameters such as temperature, pH, and nutrient levels, along with corresponding product quality metrics like yield and purity. Once trained, the network can predict product quality based on current conditions, allowing operators to adjust parameters in real time to achieve optimal results.

b. **Genetic algorithms (GAs)**: GAs are optimization methods modelled after the concepts of natural selection and genetics. They work by evolving a population of possible solutions over multiple iterations, gradually improving towards the best solution for challenging problems. Each solution is evaluated using a fitness function, and the most successful candidates are combined and mutated to generate new possibilities. GAs can optimize process parameters in biomanufacturing by navigating a broad search space to identify optimal conditions. For instance, a GA might optimize fermentation conditions by adjusting factors such as agitation speed, oxygen flow rate, and nutrient concentration to maximize product yield.
c. **Integration with traditional statistical methods**: Integrating advanced optimization techniques with traditional statistical methods enhances their effectiveness and interpretability. Conventional approaches like DoE and SPC provide a structured framework for data collection and initial process understanding, which can augment the predictive and optimization capabilities of ML and GAs.
d. **Integration examples**:
    i. **DoE and neural networks**: Conducting a DoE helps identify the most influential process parameters and their interactions. The data from these experiments can be used to train a neural network model, which can then predict process outcomes for new parameter settings, facilitating continuous improvement and optimization.
    ii. **SPC and GAs**: SPC monitors the process and identifies deviations from control limits. When a deviation is detected, a GA is employed to explore and identify optimal parameter adjustments that restore the process to control while enhancing performance.
    iii. **RSM and ML**: RSM provides a polynomial model to explore the response surface. This model can be augmented with ML techniques to handle complex, nonlinear relationships that traditional RSM might not effectively capture, allowing for more accurate predictions and optimizations.

5.5.2 Case Studies on Advanced Optimization in Biomanufacturing

a. **Optimizing protein production in fermentation**: A biotechnology company aimed to optimize recombinant protein production in fermentation, involving multiple variables such as temperature, pH, agitation speed, and nutrient concentration. Traditional optimization methods struggled with the nonlinear interactions between these variables. The company could model the relationship between process variables and protein yield by conducting a DoE to gather initial data and train a neural network using this data. A GA was then applied to explore the search space and identify the optimal combination of variables, resulting in a 15% increase in protein yield compared to traditional methods.

b. **Enhancing bioreactor performance with integrated SPC and ML**: A pharmaceutical manufacturer sought to improve the performance of a bioreactor for producing monoclonal antibodies, where variations affected product quality and yield. The company implemented SPC to monitor critical process parameters and detect deviations. When deviations occurred, a GA was used to identify optimal parameter adjustments, and the SPC data was integrated with a neural network model to predict the impact of these adjustments on product quality and yield. This approach led to a 10% improvement in product yield and reduced quality variations.

c. **Quality prediction in tablet manufacturing**: A pharmaceutical company aimed to predict and control the quality of tablets produced by a high-speed press, with variables including compression force, tablet weight, and granule moisture content. The company collected historical process data and conducted a DoE to understand the relationships between variables. Based on these variables, the tablet quality attributes can be predicted using a neural network. Then, they trained to predict tablet quality attributes, such as the dissociation rate of the tablet and hardness. An SPC system was implemented to monitor real-time process data and detect deviations. Using the neural network predictions, the company could proactively adjust process parameters, resulting in a 20% reduction in quality defects and improved product consistency.

5.5.3 Case Studies on IoT and Statistical Methods in Action

a. **Process optimization**: Utilizing IoT in the fermentation process allows for real-time monitoring of key parameters like pH, temperature, and oxygen levels through IoT sensors. By applying DoE and MVA to the data collected from these sensors, the process efficiency can be significantly improved, leading to a 20% increase in yield [Swarbrick, B. (2018)]. Furthermore, the implantation of these advanced analytical techniques led to a reduction in process variability and a decrease in overall production costs.

b. **Pharmaceutical manufacturing**: Integrated IoT sensors are been used into tablet manufacturing process to monitor critical parameters such as granule, size, moisture content, and tablet hardness [Anandavel, S.V. (2021)]. By multivariant statistical process control (MSPC) techniques on the real-time IoT data, the product quality and regulatory standards are maintained [Balakrishna Kurup H. (2019)].

c. **Statistical method for IoT-enabled process optimization**: Descriptive statistics like mean, median, standard deviation, and range are very much essential for collecting the large datasets from IoT sensors [Krishnamurthi et al. (2020)]. Real-time dashboards can visualize these metrics by various tools: control charts, histograms, and scatter plots. Basic understanding of distribution to process data is crucial for optimizing variations in a product quality. Commonly used distributions include Poisson and exponential models. Hypothesis testing such as t-tests or ANOVA helps in comparing different conditions.

5.6 REGULATORY CONSIDERATIONS AND FUTURE DIRECTIONS IN BIOTECHNOLOGY AND PHARMACEUTICAL MANUFACTURING

Regulatory compliance is fundamental in biotechnology and pharmaceutical manufacturing, ensuring that products are safe, effective, and meet the highest quality standards. Regulatory agencies like the EMA and the FDA set rigorous protocols that pharmaceutical companies must follow to introduce their products to the market. Compliance with these regulations is mandatory by law and plays a vital role in upholding public confidence while ensuring the safety and effectiveness of medications [Gupta et al. (2022)].

5.6.1 U.S. Regulatory Framework and FDA Guidelines

In the United States, the FDA regulates the manufacture, distribution, and marketing of pharmaceutical products. Compliance with FDA regulations is mandatory for any company selling products in the United States. Among the essential FDA guidelines is good manufacturing practices (GMP), which outlines the criteria for methods, facilities, and controls in the manufacturing, processing, and packaging of pharmaceutical products. GMP guarantees that products are consistently manufactured and monitored in compliance with quality standards to protect consumer health. Furthermore, FDA guidelines on process validation require manufacturers to demonstrate that their processes consistently produce products that meet predetermined specifications and quality attributes. This involves rigorous testing and meticulous manufacturing process documentation to ensure every step meets the necessary standards. Another critical regulation is the FDA's 21 CFR Part 11, which governs electronic records and signatures, setting criteria for these records to be considered trustworthy, reliable, and equivalent to paper records. Compliance with this regulation ensures the integrity of electronic records throughout the manufacturing process.

5.6.2 European Regulatory Framework: EMA Guidelines

In Europe, the EMA is an essential unit in managing the scientific assessment, regulation, and safety surveillance of medicines across the European Union (EU). Like the FDA, the EMA enforces GMP guidelines to ensure consistent product quality, as outlined in EudraLex, Volume 4, the primary source of GMP regulations within the EU. The International Council for Harmonisation (ICH) also provides harmonized guidelines adopted by both the FDA and the EMA, covering various aspects of pharmaceutical development. These include guidelines on pharmaceutical development (Q8), quality risk management (Q9), and pharmaceutical quality systems (Q10), which are essential for ensuring that products meet necessary standards for safety, efficacy, and quality. The EMA's marketing authorization process thoroughly reviews a medicinal product's quality, safety, and efficacy before it can be marketed in the EU. This process includes assessing clinical trial data, manufacturing processes, and product labelling to ensure that all aspects of the product meet stringent regulatory standards.

5.6.3 Ensuring Data Integrity and Reproducibility

Data integrity and reproducibility are cornerstones of regulatory compliance in pharmaceutical manufacturing. Data integrity refers to data's completeness, consistency, and accuracy throughout its lifecycle. The ALCOA+ principles serve as a cornerstone for ensuring data integrity, emphasizing the importance of data being traceable, readable, recorded in real time, authentic, precise, comprehensive, harmonized, long lasting, and easily retrievable. As electronic records become increasingly prevalent, adherence to regulations like 21 CFR Part 11 is paramount for electronic data management systems. These systems must incorporate robust security measures, detailed audit logs, and reliable data backup protocols to safeguard data. Maintaining data integrity is critical for generating trustworthy records, which are essential for regulatory compliance and upholding product quality (Reddy & Henze 2023).

Reproducibility, the ability to obtain consistent results using the same method under identical conditions, is equally important in pharmaceutical manufacturing. Maintaining reproducibility requires the use of standard operating procedures (SOPs) to ensure consistency in manufacturing processes. SOPs provide detailed guidelines, ensuring procedures are uniformly executed, reducing variability, and improving reliability. Training and competency assessments are also crucial, ensuring that all personnel are adequately trained and capable of performing their tasks to the highest standards. Process validation is another critical strategy involving studies that confirm that the manufacturing process consistently produces products meeting predetermined specifications. Regular internal and external audits are necessary to maintain ongoing compliance with SOPs and regulatory standards, identify deviations from standard procedures, and enable continuous process improvement.

For example, in a biomanufacturing process for producing monoclonal antibodies, ensuring data integrity and reproducibility involves capturing real-time process data through electronic batch records (EBRs), establishing SOPs for each stage of production, and conducting process validation studies to confirm the consistency of antibody production. Regular internal audits ensure compliance with regulatory requirements and help identify opportunities for further process improvements. By adhering to these practices, manufacturers can maintain high standards of quality, ensuring their products are safe, effective, and reliable.

5.6.4 Future Directions and Trends in Biotechnology and Pharmaceutical Manufacturing

Process optimization in biotechnology and pharmaceutical manufacturing is rapidly evolving, driven by advancements in statistical methods, big data, analytics, and the transformative impact of Industry 4.0. This section explores the emerging trends and future directions shaping the landscape of sustainable manufacturing in these industries [Arden et al. (2021)].

a. **Emerging statistical methods**: The continuous development of statistical methods opens new opportunities for process optimization. Some key trends include the following.

b. **Bayesian statistics**: Bayesian methods, which incorporate prior knowledge and current data to make statistical inferences, are gaining traction. These methods are instrumental in scenarios with limited data or where integrating expert knowledge into the analysis is beneficial. Bayesian approaches can improve decision-making in clinical trials, quality control, and predictive modelling in biomanufacturing.
c. **Machine learning and artificial intelligence (AI)**: ML and AI are rapidly transforming the field of statistical analysis. These advanced techniques empower researchers to process vast datasets. Algorithms, like deep learning and reinforcement learning, are being employed to optimize the combination of ML/AI and statistical analysis. Sophisticated ML algorithms, including deep learning and reinforcement learning, are increasingly being used to optimize processes, forecast outcomes, and automate decision-making [(Mehta et al., 2024)].
d. **Nonparametric methods**: Nonparametric statistical methods are becoming increasingly important because they do not assume a specific distribution for the data. These flexible and robust methods are suitable for analysing data with unknown or irregular distributions. They are instrumental in quality control and process monitoring where traditional parametric assumptions may not hold.
e. **Hybrid models**: Combining different statistical techniques into hybrid models enhances their robustness and accuracy. For example, integrating neural networks with traditional regression models or using ensemble methods (such as random forests) can improve predictive performance and reliability in process optimization.
f. **Impact of Industry 4.0 on process optimization**: Industry 4.0, often known as the Fourth Industrial Revolution, involves integrating digital technologies, cyber-physical systems, and the IoT in manufacturing, significantly influencing process optimization in the biotechnology and pharmaceutical sectors [Khang et al. (2024)].
g. **Smart manufacturing**: Industry 4.0 revolutionizes manufacturing by empowering the creation of intelligent, automated, and interconnected systems capable of self-optimization. Leveraging the IoT, sensors, and actuators, these systems capture real-time shop floor data, which is then analysed using cutting-edge analytics and ML techniques. This synergy enables continuous process refinement, efficiency gains, and enhanced product quality [Deepak Kumar et al. (2023)].
h. **Digital twins**: A digital twin is a virtual representation of a physical manufacturing process or system that leverages real-time data to simulate, predict, and enhance performance. In biomanufacturing, digital twins can mimic intricate processes, enabling manufacturers to test different scenarios, pinpoint ideal conditions, and anticipate the impact of changes without affecting real-world production [Udugama et al. (2021)].
i. **Cyber-physical systems (CPS)**: CPS combines physical processes with digital technologies, enabling seamless interaction. In biomanufacturing, CPS allows for real-time process monitoring and control, enabling adaptive

and autonomous optimization. For instance, a CPS can dynamically adjust fermentation conditions based on sensor data to maintain ideal growth environments for microbial cultures [Pandey et al. (2023)].

j. **Advanced robotics and automation**: Industry 4.0 promotes the adoption of advanced robotics and automation, boosting process efficiency and minimizing human error. Automated systems perform repetitive and complex tasks with high precision and consistency. In pharmaceutical manufacturing, robotic systems are used for tasks like aseptic filling, packaging, and quality inspection, ensuring products meet high standards of quality and safety [Mule et al. (2023)].

k. **Blockchain for traceability**: Blockchain technology offers a secure and transparent method for tracking and verifying data throughout the manufacturing process. In biomanufacturing, it can ensure raw material traceability, monitor production workflows, and validate the integrity of quality control information, thus improving transparency, accountability, and regulatory compliance [Clauson et al. (2022)].

For example, a biopharmaceutical company adopting Industry 4.0 technologies to optimize its monoclonal antibody production process might install IoT sensors throughout the production line to collect real-time data on critical process parameters. The company can simulate different conditions and identify optimal settings by creating a digital twin of the production process. Advanced analytics and ML algorithms analyse the data to predict product yield and quality, while automated systems adjust process parameters in real time based on these insights, significantly improving process efficiency and product consistency.

5.7 CONCLUSION

This chapter has emphasized the critical role of statistical methods in optimizing biotechnology and pharmaceutical manufacturing processes. From descriptive statistics to advanced techniques like ML and GAs, these methods are essential for enhancing product quality, yield, and overall process efficiency. The integration of IoT further amplifies the transformative potential of statistical methods. The application of statistical tools such as DoE, RSM, and SPC was illustrated through examples that demonstrate their impact on production improvements. Regulatory compliance, particularly with standards set by the FDA and EMA, is vital for maintaining industry credibility. Data integrity and reproducibility are crucial for meeting these stringent standards and sustaining high product quality. Looking ahead, the integration of emerging statistical methods and Industry 4.0 technologies, including big data analytics and AI, presents significant opportunities for real-time process optimization and control. These advancements are essential for meeting evolving industry demands and ensuring competitiveness in a highly regulated environment. The commitment to continuous improvement, supported by statistical methods and technological innovation, is indispensable for the sustainable growth of the biotechnology and pharmaceutical sectors. Manufacturers can deliver high-quality, effective products that improve patient outcomes and contribute positively to public health by consistently refining processes and embracing new technologies. This approach will drive the industry's future success, ensuring its continued relevance and excellence.

REFERENCES

Adams, A. M., McCarthy, P. M., and Nelson, D. W.: Factorial design. In Translational Surgery. Elsevier (pp. 315–320) (2023).

Anandavel, S.V.: Analysis of manufacturing processes according to FMEA techniques and Implementation of IoT systems to prevent process failures (Doctoral dissertation, Politecnico di Torino). (2021).

Arden, N.S., Fisher, A.C., Tyner, K., Yu, L.X., Lee, S.L., and Kopcha, M.: Industry 4.0 for pharmaceutical manufacturing: Preparing for the smart factories of the future. Int. J. Pharm. 602, 120554 (2021). https://doi.org/10.1016/j.ijpharm.2021.120554

Balakrishna Kurup, H.: Prioritization of the product characteristics for implementation of process control in powertrain production. Master's thesis, Chalmers University of Technology, Gothenburg, Sweden (2019).

Beg, S., Swain, S., Rahman, M., Hasnain, M.S., and Imam, S.S.: Application of design of experiments (DoE) in pharmaceutical product and process optimization. In: Pharmaceutical Quality by Design. Elsevier, pp. 43–64 (2019).

Clauson, K.A., Crouch, R.D., Breeden, E.A., and Salata, N.: Blockchain in pharmaceutical research and the pharmaceutical value chain. In: Blockchain in Life Sciences. Springer Nature Singapore, pp. 25–52 (2022).

Conner, J., Wuchterl, D., Lopez, M., Minshall, B., Prusti, R., Boclair, D., Peterson, J., and Allen, C.: The biomanufacturing of biotechnology products. In: Biotechnology Entrepreneurship. Elsevier, pp. 447–480 (2020).

Cooksey, R.W.: Descriptive statistics for summarising data. In: Illustrating Statistical Procedures: Finding Meaning in Quantitative Data. Springer Singapore, pp. 61–139 (2020).

Deepak Kumar, S., Arun Manohar, G., Surya Teja, R., Ramana Rao, P.S.V., Mandal, A., Behera, A., and Srinivasa Rao, P.: Smart manufacturing technologies in industry 4.0: Challenges and opportunities. In: Intelligent Manufacturing Management Systems. Wiley, pp. 23–57 (2023). https://doi.org/10.1002/9781119836780.ch2

Dhoot, A.S., Fernandes, G.J., Naha, A., Rathnanand, M., and Kumar, L.: Design of experiments in pharmaceutical development. Pharm. Chem. J. 53(8), 730–735 (2019). https://doi.org/10.1007/s11094-019-02070-4

Du, Y.-H., Wang, M.-Y., Yang, L.-H., Tong, L.-L., Guo, D.-S., and Ji, X.-J.: Optimization and scale-up of fermentation processes driven by models. Bioengineering 9(9), 473 (2022). https://doi.org/10.3390/bioengineering9090473

Faisal, F., Rasul, M. G., Chowdhury, A. A., & Jahirul, M. I.: Optimisation of process parameters to maximise the oil yield from pyrolysis of mixed waste plastics. Sustainability 16(7), 2619 (2024). https://doi.org/10.3390/su16072619

Farid, S.S., Baron, M., Stamatis, C., Nie, W., and Coffman, J.: Benchmarking biopharmaceutical process development and manufacturing cost contributions to R&D. mAbs 12(1), (2020). https://doi.org/10.1080/19420862.2020.1754999

Fytopoulos, A., Ramachandran, R., and Pardalos, P. M.: (Eds.) (Optimization of pharmaceutical processes (2nd ed.). Springer Nature (2022).

Fytopoulos, A., & Pardalos, P. M.: Benefits of Application of Process Optimization in Pharmaceutical. In Trends in Biomathematics: Modeling Epidemiological, Neuronal, and Social Dynamics: Selected Works from the BIOMAT Consortium Lectures (2023).

Gupta, S., Baboota, S., Ali, J., and Dang, S.: Regulatory affairs for chemistry, manufacturing, and controls. In: Regulatory Affairs in the Pharmaceutical Industry. Elsevier, pp. 75–87 (2022).

Ioannis, S.A.: Multivariate analysis. In: Handbook of Food and Bioprocess Modeling Techniques. CRC Press, pp. 323–356 (2006).
Jaiswal, S.K., Mukti, S.K., and Rath, K.C.: E-waste control and its recycling to build sustainable society in the global context. In: Handbook of Research on Applications of AI, Digital Twin, and Internet of Things for Sustainable Development. IGI Global, pp. 200–222 (2023).
Jange, C.G., Wassgren, C.R., and Ambrose, K.: The significance of tablet internal structure on disintegration and dissolution of immediate-release formulas: A review. Powders 2(1), 99–123 (2023). https://doi.org/10.3390/powders2010008
Jankovic, A., Chaudhary, G., and Goia, F.: Designing the design of experiments (DOE) – an investigation on the influence of different factorial designs on the characterization of complex systems. Energy Build. 250, 111298 (2021). https://doi.org/10.1016/j.enbuild.2021.111298
Kaltenbach, H.-M.: Principles of experimental design. In: Statistical Design and Analysis of Biological Experiments. Springer International Publishing, pp. 1–13 (2021).
Khang, A., Rath, K. C., Kumar, S. S., Satapathy, A. K., and Agrawal, A.: Chapter 14: Robotic Process. In A. Khang, V. A. Hajimahmud, A. Misra, & E. Litvinova (Eds.), Machine Vision and Industrial Robotics in Manufacturing: Approaches, Technologies, and Applications. CRC Press, Taylor & Francis Group (pp. 252–267) (2024). https://doi.org/10.1201/9781003438137
Krishnamurthi, R., Kumar, A., Gopinathan, D., Nayyar, A., and Qureshi, B.: An overview of IoT sensor data processing, fusion, and analysis techniques. Sensors 20(21), 6076 (2020).
Krishnamurthy, M., Fuad, A., Ruiyong, Z., and Eerla, R.: Application of response surface methodology (RSM) in the statistical evaluation of biodiesel production from the neutral lipids of the Coelastrella-Nannochloropsis consortium. Environ. Res. 243, 117829 (2024). https://doi.org/10.1016/j.envres.2023.117829
Kumar, R., & Reji, M.: Response surface methodology (RSM): An overview to analyze multivariate data. Indian J. Microbiol. Res. 9(4), 241–248 (2022).
Lamidi, S., Olalere, R., Yekinni, A., and Adesina, K.: Design of Experiments (DOE): Applications and benefits in quality control and assurance. In: Quality Control and Quality Assurance – Techniques and Applications. IntechOpen (2024).
Mehta, A., Niaz, M., Adetoro, A., and Nwagwu, U.: Advancements in manufacturing technology for the biotechnology industry: The role of artificial intelligence and emerging trends. Int. J. Chem. Math. Phys. 8(2), 1–12 (2024). https://aipublications.com/uploads/issue_files/3IJCMP-JUN20241-Advancements.pdf
Muduli, K., Raut, R., Narkhede, B.E., and Shee, H.: Blockchain technology for enhancing supply chain performance and reducing the threats arising from the COVID-19 pandemic. Sustainability 14(6), 3290 (2022).
Mule, A., Puranik, A., and Tarkase, K.N.: Recent automation of Industry 4.0 in pharmaceutical manufacturing. Int. J. Pharm. Res. Appl. 8, 1939–1945 (2023).
Naugolnova, I.: Process Optimization in Chemical Engineering: Strategies for Cost Reduction and Efficiency. Chem. Eng. J. Adv. 15 (2023).
Nunavath, R.S., Singh, M.T., Jain, A., Chakma, M., Arivuselvam, R., and Azeeze, M.S.T.A.: Quality by design in pharmaceuticals: A review of its impact on regulatory compliance and product quality. Drug Research, 74(1), 18–23 (2024). https://doi.org/10.1055/a-2185-4916
Panda, S.K., Rath, K.C., & Mishra, S.: The new age of manufacturing from concept to creation: Innovative world of 3D printing. In A. Khang, K. C. Rath, & S. S. Kumar (Eds.), Evolutionary Manufacturing, Design and Operational Practices for Resource and Environmental Sustainability. CRC Press, Taylor & Francis Group, pp. 223–244 (2024). https://doi.org/10.1201/9781003438137-14

Pandey, K., Pandey, M., Kumar, V., Aggarwal, U., and Singhal, B.: Bioprocessing 4.0 in biomanufacturing: Paving the way for sustainable bioeconomy. Syst. Microbiol. Biomanuf. (2023). https://doi.org/10.1007/s43393-023-00206-y

Patil, M.M.S., Patil, D.R.R., Chalikwar, D.S.S., Surana, D.S.J., and Firke, D.S.D.: Analytical method development and validation: A review. Int. J. Pharm. Biol. Sci. Arch. 7(3), (2019). https://doi.org/10.32553/ijpba.v7i3.126

Polaki, S., Rabi, P.B., Pushpalatha, G., and Ghanistha, P.: Advanced conservation strategies for anticancer plants: Integrating technology and biodiversity preservation. In: J. A. Daniels (Ed.), Advances in Environmental Research. Nova Science Publisher Inc., Vol. 101, pp. 61–108 (2024).

Rani, D., and Vimolmangkang, S.: Production of recombinant proteins using plant cell suspension cultures and bioreactor engineering: A short review. In: Concepts and Strategies in Plant Sciences. Springer Nature Singapore, pp. 141–161 (2024).

Reddy, T.A., and Henze, G.P.: Probability concepts and probability distributions. In: Applied Data Analysis and Modeling for Energy Engineers and Scientists. Springer International Publishing, pp. 31–74 (2023).

Rijamampianina, R., Abratt, R., & February, Y.: A framework for concentric diversification through sustainable competitive advantage. Management Decision 41(4), 362–371 (2003).

Sampathkumar, K., and Kerwin, B.A.: Roadmap for drug product development and manufacturing of biologics. J. Pharm. Sci. 113(2), 314–331 (2024). https://doi.org/10.1016/j.xphs.2023.11.004

Shi, G., Lin, L., Liu, Y., Chen, G., Luo, Y., Wu, Y., and Li, H.: Pharmaceutical application of multivariate modelling techniques: A review on the manufacturing of tablets. RSC Adv. 11(14), 8323–8345 (2021). https://doi.org/10.1039/d0ra08030f

Sravan, J.S., Matsakas, L., and Sarkar, O.: Advances in biological wastewater treatment processes: Focus on low-carbon energy and resource recovery in biorefinery context. Bioengineering 11(3), 281 (2024). https://doi.org/10.3390/bioengineering11030281

Sridevi, B.: Application of Statistical process control in pharmaceutical manufacturing: a case study on tablet production. Int. J. Pharm. Sci. Res. 15(3), 1125–1132 (2024).

Swarbrick, B.: Quality by design in practice. In: Multivariate Analysis in the Pharmaceutical Industry. Academic Press, pp. 125–171 (2018).

Tesfay, Y. Y. (2021). Concurrent problem-solving models for industrial applications. In Developing Structured Procedural and Methodological Engineering Designs (pp. 213–237). Springer International Publishing.

Udugama, I.A., Lopez, P.C., Gargalo, C.L., Li, X., Bayer, C., and Gernaey, K.V.: Digital twin in biomanufacturing: Challenges and opportunities towards its implementation. Syst. Microbiol. Biomanuf. 1(3), 257–274 (2021). https://doi.org/10.1007/s43393-021-00024-0

Wang, X., Yang, L., Fan, M., Zou, Y., and Wang, W.: An unsupervised clustering method for selection of the fracturing stage design based on the Gaussian mixture model. Processes 10(5), 894 (2022). https://doi.org/10.3390/pr10050894

Wu, C., Xu, Y., Fang, J., and Li, Q.: Machine learning in biomaterials, biomechanics/mechanobiology, and biofabrication: State of the art and perspective. Arch. Comput. Methods Eng. (2024). https://doi.org/10.1007/s11831-024-10100-y

6 AIoT-Based Sustainable Engineering and Technology for Improvement in Manufacturing Efficiency

Lakshmi Prasad Panda, Kali Charan Rath, Debasis Mishra, Pradeep Padhamnath, and Jai Krishna Sanku

6.1 INTRODUCTION

The intersection of AI, IoT, and IIoT has been instrumental in improving the manufacturing process and efficiency, where growth is being driven by automation, digital transformation efforts, and lean practices. As a result, these improvements have translated to more efficient processes – making it possible for real-time monitoring and sustainable operations; in turn, boosting productivity while decreasing wastage. At the time of COVID-19, various industries were faced challenges related to integrating with legacy systems, supply chain disruptions, labor shortages, and maintaining consistent quality, yet it demonstrated resilience by adapting quickly and finding innovative solutions to overcome these obstacles. At the same time, they are faced with increasingly stringent regulations, cyber threats, and cost pressures alongside a requirement to invest in new technology areas to compete – emphasizing the importance of achieving a sustainable, resilient transformation that ensures long-term success and the evolution of their technology. A comprehensive whole systems framework that focuses on advancing in technology, workforce, and a 360-degree risk management approach is critical factor for sustainable efficiency competitive advantage against the world market domain [Lee et al. (2023), Mahajan et al. (2020), Menon et al. (2018)].

One of the main focus areas of this study is to address several existing inefficiencies within supply chain management and manufacturing processes, despite the continuing development in technology and organizational practice. One of the key problems with smart technology adoption is that it is for many manufacturing firms difficult to integrate in their conventional systems, causing inefficiency and higher costs, and operational disruptions [Javaid et al. (2021), Wu et al. (2016)] with supply chain vulnerabilities and a scarcity of skilled labor making it hard to deliver products with consistent quality. Regulatory compliance, cyber security and sustainability are

DOI: 10.1201/9781003503323-6

just a few of the many factors that manufacturers need to manage. The objective of this contribution is to identify and suggest ways of resolving the inefficiencies by investigating how smart operations as well as advanced supply chain management techniques [Fatima et al. (2022), Pasi et al. (2020)] can be seamlessly used to improve manufacturing efficiency, decrease costs, and enhance operational resilience.

6.1.1 Objectives

The following are the objectives fixed for this chapter:

a. To provide an understanding on how smart manufacturing technologies (e.g., IoT, AI, robotics and data analytics) could transform productivity and waste reduction in a manufacturing context.
b. Investigate how IoT, AI, robotics and data analytics can help smart manufacturing transform operations to achieve operational efficiency, waste reduction and productivity growth.
c. Analyze best practices involving Just-in-Time (JIT) inventory systems, agile methodologies, green sourcing, and application of block chain in the context of manufacturing and derive practical proposals for implementation.
d. Research prevailing problems in manufacturing efficiency, like technology integration problems, supply chain disturbances, worker shortages, and quality control issues.

6.2 SMART OPERATIONS IN MANUFACTURING

The application of modern technologies in manufacturing for its effective working platform measuring optimized usage of resources and reduced operation costs improve the productivity and enhances performance makes up a complex smart operations [Ang et al. (2017), Kandarkar & Ravi (2024)]. These may include but not be limited to cloud computing, big data analytics, and artificial intelligence (AI), and Internet of Things (IoT). Smart operations offer to make the entire production floor networked, automated, and intelligent [Muruganandam et al. (2023), Zhou et al. (2022)]. In this article, we delve into smart operations in manufacturing and examine different areas on which Industry 4 is based upon.

6.2.1 IoT in Manufacturing

It is the network of interconnected objects equipped with software, sensors, and other technologies to collect and exchange data over the Internet. In the manufacturing sector, the IoT fosters a more connected, automated, and intelligent production environment, resulting in significant enhancements in efficiency, productivity, and operational transparency. The key IoT devices used in the manufacturing processes are listed in Table 6.1.

6.2.2 Benefits of IoT in Manufacturing

IoT devices gather real-time data such as pressure, temperature, vibration, and others to continuously keep track of the machinery. Thus, in case of a major problem or any

TABLE 6.1
IoT Devices in Manufacturing Industry

IoT Devices Used in Smart Manufacturing		Usage
Sensors	Temperature Sensors	Monitor the temperature of equipment and environments.
	Vibration Sensors	Detect abnormal vibrations indicating potential equipment failures.
	Pressure Sensors	Measure the pressure in machinery and pipelines
	Humidity Sensors	Monitor the moisture levels in the air or materials
	Proximity Sensors	Detect the presence or absence of objects.
	Optical Sensors	Used for detecting light levels, presence of objects, and more.
	Gas Sensors	Monitor levels of gases like CO_2, O_2, or hazardous gas.
	Ultrasonic Sensors	Measure distances using ultrasonic waves.
	Accelerometers	Measure acceleration forces and vibrations.
Actuators	Electric Motors	Make mechanical motion out of electrical energy.
	Pneumatic Actuators	Use compressed air to produce motion
	Hydraulic Actuators	Use hydraulic fluid pressure to create motion
	Servo Motors	Provide precise control of angular or linear position.
	Solenoid Valves	Control the flow of liquids or gases.
Wearable's	Smart Helmets	Provide safety features and data collection for workers.
	Smart Gloves	Enhance manual tasks with data input and feedback.
	Health Monitoring Wearable's	Track heart rate, stress levels, and more.
	Location-Tracking Wearable's	Monitor the location of workers for safety and efficiency.
Asset Tracking Devices	RFID Tags and Readers	Track and manage inventory and assets.
	GPS Trackers	Monitor the location of vehicles and movable assets.
	Barcode Scanners	Scan barcodes for inventory and asset management.
Industrial Controllers	Programmable Logic Controllers (PLCs)	Control machinery and processes
	Distributed Control Systems (DCS)	Manage complex processes with multiple control loops.
	Remote Terminal Units (RTUs)	Collect data from sensors and send it to control systems.
Smart Meters	Energy Meters, Water Meters, Gas Meters	Monitor energy consumption, Track water usage,
Network and Communication Devices	Industrial Routers, Gateways, Edge Devices, Wi-Fi Modules, Bluetooth Modules	Provide robust network connectivity, Connect IoT devices to the cloud or other networks, Process data locally before sending it to the cloud, Enable wireless communication, Facilitate short-range wireless communication.

(Continued)

TABLE 6.1 ***(Continued)***
IoT Devices in Manufacturing Industry

IoT Devices Used in Smart Manufacturing		Usage
Robotics	Industrial Robots, Collaborative Robots (Cobots), Automated Guided Vehicles (AGVs), Autonomous Mobile Robots (AMRs)	Automate repetitive tasks in manufacturing, Work alongside humans to enhance productivity. Transport materials around facilities. Navigate and perform tasks without human intervention.
Environmental Monitoring Devices	Air Quality Monitors: Noise Level Monitors, Environmental Sensors	Measure pollutants and air quality. Track noise levels to ensure a safe working environment. Measure temperature, humidity, and other environmental factors.
Additive Manufacturing Devices	3D Printers and Additive Manufacturing Machines	Produce prototypes and parts using various materials. Build complex parts layer by layer

defect, immediate action can be taken which can avoid the failure of the equipment or interruption of the production process. As a result, totality and completeness of processes as well as timely and appropriate execution of operational tasks are ensured. Real-time monitoring and data gathering is one of the requirements of today's business enterprises which since it helps organizations enhance their operational efficiency, reduce idle time, and adhere to tight quality regulations. We look at the context of real-time monitoring and data collection with data about a factory and its operations. This example has different parameters as noted in table 6.2 with their concerned values.

Terminology:

- Timestamp : The date and time when the data was recorded.
- Machine ID : A unique identifier for each machine being monitored.
- Temperature (°C) : Temperature of the machine in degrees Celsius.
- Vibration (mm/s) : The vibration level of the machine is in millimetres per second.
- Pressure (kPa) : Pressure in kilopascals at the point of measurement.
- Humidity (%) : Relative humidity of the environment in which the machine is located.
- Noise Level (dB) : The noise level in decibels in the vicinity of the machine.
- Production Rate (units/ h) : The number of units produced per hour by the machine.
- Energy Consumption (kWh) : The amount of energy consumed by the machine in kilowatt-hours.
- Operator ID : Identifier of operator overseeing machine at the time of data collection.

TABLE 6.2
Real-time Monitoring and Data Collection System in a Manufacturing Unit

Time stamp	Machine ID	Temperature (°C)	Vibration (mm/s)	Pressure (kPa)	Humidity (%)	Noise Level (dB)	Production Rate (units/h)	Energy Consumption (kWh)	Operator ID
2024-06-23 08:00:00	M001	75.3	2.5	150	45	70	120	15.2	OP123
2024-06-23 08:05:00	M001	75.5	2.6	149	46	71	118	15.4	OP123
2024-06-23 08:10:00	M001	75.8	2.7	148	45	72	119	15.6	OP123
2024-06-23 08:15:00	M002	70.2	3.1	145	50	68	130	16.1	OP124
2024-06-23 08:20:00	M002	70.5	3.0	144	49	67	131	16.3	OP124
2024-06-23 08:25:00	M002	70.6	3.2	143	51	66	129	16.5	OP124
2024-06-23 08:30:00	M003	68.0	1.8	155	48	65	140	14.8	OP125
2024-06-23 08:35:00	M003	68.2	1.7	156	47	64	141	14.9	OP125
2024-06-23 08:40:00	M003	68.4	1.9	154	49	63	138	15.0	OP125
2024-06-23 08:45:00	M004	72.1	2.0	160	52	69	125	15.7	OP126
2024-06-23 08:50:00	M004	72.3	2.1	159	51	70	126	15.8	OP126
2024-06-23 08:55:00	M004	72.5	2.2	158	50	71	127	15.9	OP126

a. Table 6.2 serves several functions that add value to the manufacturing processes. With such information readily available to managers, more precise decisions on maintenance, scheduling production work, or placing resources are likely. In other words, it helps schedule maintenance before a breakdown by facilitating the monitoring of temperature, vibration, and pressure to foresee trouble ahead. It enhances operational performance by providing insights on output levels, energy consumption, operational delays and looking for ways on how to improve the operations. The table also makes it a point to keep everyone safe and in the legal bounds of the workplace by observing environmental parameters and noise, and following the set rules on workplace safety. Also, there is such a way as to know what operators are in charge of a machine at a given time which is helpful for staff skills development and assessment. Its aim is to take care of and optimize the manufacturing process by measuring and integrating this wide range of key parameters along the entire table.
b. Predictive Maintenance: Predictive maintenance aims to predict when equipment failure might occur so that maintenance can be performed just before it is needed. In a smart manufacturing scenario, predictive maintenance uses IoT sensors to monitor parameters such as temperature, vibration, and energy consumption in real-time. Machine learning models analyze these data to predict equipment failures before they occur, thereby enabling proactive maintenance scheduling. This approach minimizes downtime, reduces maintenance costs, and enhances overall operational efficiency, ensuring continuous production and a competitive advantage in the industry. Predictive maintenance uses IoT data to predict equipment failures prior to their occurrence.

Elevated temperatures and vibration data were collected on a real-time basis to establish a baseline and calculate measurements for predictive maintenance. Based on these average figures, we established some limits and started applying the ML for anomaly detection. Trendiness forecasting was accomplished in order to find out how the situation will develop in time and how probable it is that certain parts will fail. Perform the corrective action on-time and resource efficient predictive maintenance operations right before the accurate model predicted failure time. Incorporate IoT technology for better and faster data capturing and predictive analytics model performance enhancements based on changing information and evaluations. Predictive maintenance in smart manufacturing leverages advanced technologies and data analytics to optimize equipment reliability, minimize downtime, and enhance the overall operational efficiency. This scenario illustrates how predictive maintenance can be applied in a smart manufacturing environment.

6.2.2.1 Predictive Maintenance in Smart Manufacturing

Suppose that an operations manager at a smart manufacturing facility produces automotive components. The facility is equipped with IoT-enabled machinery and sensors to monitor various parameters, including the temperature, vibration, pressure, and operational metrics.

6.2.2.2 Case Study: Predictive Maintenance for a CNC Machine

Suppose that a manufacturing facility produces precision components using CNC (Computer Numerical Control) machines. These machines are critical to production line, and any downtime can lead to significant losses. Sensors were used to monitor the temperature and vibration levels of the CNC machine (Machine ID: CNC001).

- *Timestamps: Recorded every 5 minutes during operational hours.*
- *Temperature (°C): Measures the operating temperature of CNC001.*
- *Vibration (mm/s): Vibration levels were monitored during the machining process.*

6.2.2.3 Application of Predictive Maintenance

A comprehensive approach to maintaining machinery through data analysis and predictive modeling involves several key steps. Initially, historical data (as mentioned in Table 6.3) were analyzed to establish normal operating ranges for temperature and vibration, with the mean and standard deviation calculated to define abnormal condition thresholds. In case of possible reasons for alarm in anomaly detection, upper limits like temperature should be 75°C or vibration should be 3.0 mm/s simply are cut off and the pattern abnormality is learned through machine learning for failure prognosis. The forecast of temperature and vibration state for the future periods and some other procedures have been performed with the time series prediction approach. This is where component failure prediction model comes in handy of making forecasts of when it is safe to fix,

TABLE 6.3
CNC Machine Monitored Data

Date and Time	Machine ID	Temperature (°C)	Vibration (mm/s)
2024-06-23 08:00:00	M001	75.3	2.5
2024-06-23 08:05:00	M001	75.5	2.6
2024-06-23 08:10:00	M001	75.8	2.7
2024-06-23 08:15:00	M002	70.2	3.1
2024-06-23 08:20:00	M002	70.5	3.0
2024-06-23 08:25:00	M002	70.6	3.2
2024-06-23 08:30:00	M003	68.0	1.8
2024-06-23 08:35:00	M003	68.2	1.7
2024-06-23 08:40:00	M003	68.4	1.9
2024-06-23 08:45:00	M004	72.1	2.0
2024-06-23 08:50:00	M004	72.3	2.1
2024-06-23 08:55:00	M004	72.5	2.2

based on historical data and current statistics in order to avoid unnecessary cost incurred from Component downtimes. The permanently monitoring of the conditions is useful to alert the staff on abnormal situations so that early action can be taken. Predictive analytics is concerned with data so as to make it possible for the development models to be upgraded by new information and comments received in future. The last step involves autonomic data collection using IoT sensors, which enhances operational efficiency through an active monitoring system. This system is connected to a maintenance management system to enable proactive, data-driven maintenance, reducing downtime and optimizing resource use [Talamo et al. (2019)].

Implementing predictive maintenance based on temperature and vibration data ensures enhanced equipment reliability by proactively addressing potential issues to minimize unplanned downtime. This approach leads to cost savings through optimized maintenance schedules and reduces unnecessary costs. It also improves productivity by maintaining consistent production levels and by avoiding disruptions. Additionally, it enhances safety by preventing equipment failures that could lead to accidents or injuries and extends the lifespan of CNC machines through timely maintenance. Overall, predictive maintenance ensures high operational efficiency, reduces equipment failure risks, and supports continuous improvement in production capabilities.

6.2.2.4 Mathematical Equation for Predictive Maintenance

In the predictive maintenance of CNC machines, a linear regression model is typically used to predict the machine's health status or future parameter values based on current sensor readings.

The general form of the linear regression equation is:

$$y = \beta_0 + \beta_1 x_1 + \beta_2 x_2 + \beta_n + \epsilon$$

where

- *y is the target variable (e.g., temperature, vibration).*
- *β_0 is the intercept.*
- *$\beta_1, \beta_2, \ldots, \beta_n$ are the coefficients of the predictor variables.*
- *$x_1, x_2, \ldots, x_n$ are predictor variables (e.g., pressure and energy consumption).*
- *ϵ is the error term.*

From the Table 6.4 and analysis the following points were drawn as an output.

a. *Mean Squared Error (MSE): The MSE measure displays the average squared difference between the actual and expected temperatures. A decreased MSE suggested that the model's performance had improved.*
b. *R-squared (R^2): This metric shows how much of the variance in the independent variables – vibration, pressure, energy consumption, and production rate – can be attributed to the dependent variable, which is temperature. A better match is indicated by a higher R^2 score.*

TABLE 6.4
CNC Machine and Observed Data through Monitoring Process

Timestamp	Machine ID	Temperature (°C)	Vibration (mm/s)	Pressure (kPa)	Humidity (%)	Noise Level (dB)	Production Rate (units/hour)
2024-06-23 **08:00:00**	M001	75.3	2.5	150	45	70	120
2024-06-23 **08:05:00**	M001	75.5	2.6	149	46	71	118
2024-06-23 **08:10:00**	M001	75.8	2.7	148	45	72	119
2024-06-23 **08:15:00**	M002	70.2	3.1	145	50	68	130
2024-06-23 **08:20:00**	M002	70.5	3.0	144	49	67	131
2024-06-23 **08:25:00**	M002	70.6	3.2	143	51	66	129
2024-06-23 **08:30:00**	M003	68.0	1.8	155	48	65	140
2024-06-23 **08:35:00**	M003	68.2	1.7	156	47	64	141
2024-06-23 **08:40:00**	M003	68.4	1.9	154	49	63	138
2024-06-23 **08:45:00**	M004	72.1	2.0	160	52	69	125
2024-06-23 **08:50:00**	M004	72.3	2.1	159	51	70	126
2024-06-23 08:55:00	M004	72.5	2.2	158	50	71	127

c. *Plots: Line plot showing the actual versus predicted temperatures over time. Scatter plots showing the relationships between each predictor variable and the temperature.*
d. *MSE: Suppose the output MSE is 0.65. This would suggest, on an average, the squared difference between the predicted and the actual temperatures is 0.65°C. Suppose the output R^2 is 0.85. This would suggest that 85% of in the temperature variance can be explained by the given predictors.*

The line plot shows a comparison between the actual temperature trends and the predicted values. Any significant deviations between the red dots (actual test data) and blue crosses (predicted values) were easily noticeable. The scatter plots show how well each predictor variable correlates with temperature. Strong correlations appeared as clear linear patterns (Figure 6.1).

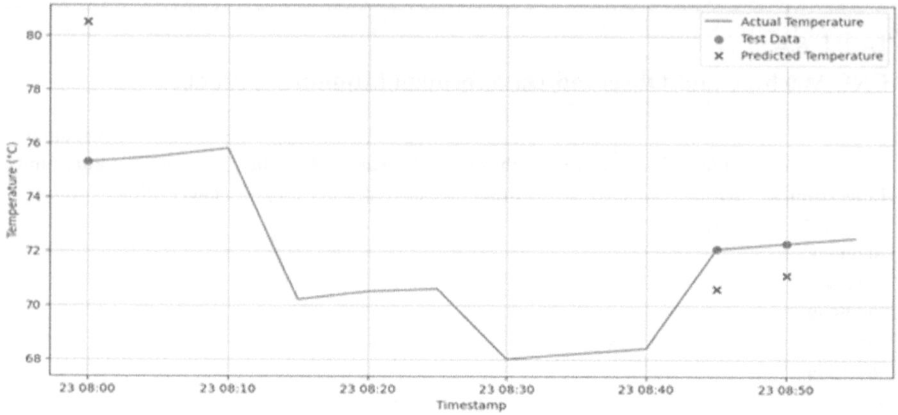

FIGURE 6.1 Actual versus predicted temperature.

6.3 SUPPLY CHAIN MANAGEMENT FOR IMPROVED MANUFACTURING EFFICIENCY

SCM plays an important role in enhancing manufacturing efficiency by ensuring smooth flow of materials, information, and finances. SCM practices have extensively explored their impact on efficiency in various industries. The key aspects of SCM practices that significantly enhance operational efficiency are given below.

6.3.1 Real-time Inventory Management

Inventory control is one of the sub-schemas of supply chain management that could be enhanced through IoT integration. IoT-driven real-time supply chain managerial tools for logistics, demand planning, and inventory transparency ensure the optimal stock rates and save money with IoT-driven inventory management systems. IoT real-time tracking of inventories is a game changer in inventory tracking where data from device sensors is compiled in real time and transmitted to hub systems. It does this to enhance the accuracy of data, manage inventory, and reduce inventory costs, increase operational excellence, and make decisions more sensitive to market changes. Some advantages of this approach include reduced costs and improved customer satisfaction due to timely updates of inventory status. The implementation of IoT sensors with analytics and ERP systems provided methodological approaches for industries' risk anticipation and more aligned operational processes [Ali et al. (2021), Rezaei et al. (2017)].

Variables and Parameters:

I_{act} – Actual Inventory Level
I_{opt} – Optimal Inventory Level
I_{low} – Low Inventory Threshold
$I_{replenish}$ – Replenishment Quantity

Inventory Alert : $\{\frac{\text{True if } I_{act} \leq I_{low}}{\text{False otherwise}}\}$
Actual Inventory Level (I_{act}) : 150 units
Low Inventory Threshold (I_{low}) : 200 units
Replenishment Quantity ($I_{replenish}$) : 100 units

Because $I_{act} = 150$ units is less than $I_{low} = 200$ units, an inventory alert is triggered. Here, IoT sensors provide real-time updates on the inventory levels. When the actual inventory (I_{act}) falls below the low inventory threshold (I_{low}), an alert is generated to replenish stock.

6.3.1.1 Supply Chain Optimization Using IoT Data

We built a comprehensive mathematical model that incorporated multiple variables. To demonstrate how IoT can optimize the supply chain.

List of Variables:

I_t – Current Inventory Level at time t
D_t – Demand Forecast at time t
I_{min} – Minimum Inventory Threshold
T_t – Expected Transit Time
T_{max} – Maximum Allowable Transit Time
T_{env} – Environmental Temperature
$[T_{min}, T_{max}]$ – Acceptable Temperature Range
k – Proportionality Constant for Production Rate
P_t – Production Rate at time t
C_{inv} – Inventory Holding Cost per unit per time period
C_{trans} – Transportation Cost
C_{prod} – Production Cost per unit

Optimization Objectives: Minimize Total Cost (CCC), Ensure Timely Delivery and Maintain Product Quality

Total cost calculation:

TC (C) includes inventory holding costs, transportation costs, and production costs:

$$C = C_{inv} \cdot I_t + C_{trans} + C_{prod} \cdot P_t$$

Constraints and Equations:

Replenishment Order : $R_t = D_t - I_t$
Inventory Threshold Condition : $I_t \geq I_{min}$
Transit Time Condition : $T_t \leq T_{max}$
Environmental Condition : $T_{min} \leq T_{env} \leq T_{max}$
Production Rate Adjustment : $P_t = k\,(D_t - I_t)$

By integrating IoT data into supply chain management, we developed a comprehensive real-time optimization model and considered the real time data as mentioned in Table 6.5. The calculations ensured that the inventory was maintained at optimal

TABLE 6.5
Hypothetical Data

Variables	Symbol	Hypothetical Value
Current Inventory Level	I_t	500 units
Demand Forecast	D_t	1000 units
Minimum Inventory Threshold	I_{min}	300 units
Expected Transit Time	T_t	5 days
Maximum Allowable Transit Time	T_{max}	7 days
Environmental Temperature	T_{env}	22°C
Acceptable Temperature Range	$[T_{min}, T_{max}]$	[15°C, 25°C]
Proportionality Constant	K	0.5
Inventory Holding Cost	C_{inv}	₹2 per unit per day
Transportation Cost	C_{trans}	₹1000 per shipment
Production Cost	C_{prod}	₹10 per unit
Maintenance Cost	C_{maint}	₹500 per day
Energy Consumption per unit	E_{prod}	0.5 kWh per unit
Quality Level	Q_{prod}	95%
Minimum Quality Level	Q_{min}	90%

levels, transit times were within acceptable limits, and environmental conditions were monitored to maintain product quality. The total cost of supply chain operations is minimized by adjusting production rates and managing inventory efficiently.

Optimization Calculation:

a) Replenishment Order : $R_t = D_t - I_t = 1000 - 500 = 500$ units
b) Inventory Threshold Check : $I_t = 500$ units, $I_{min} = 300$ units
Condition ($I_t \geq I_{min}$) : $\Rightarrow 500 \geq 300$ (True)
c) Transit Time Check : $T_t = 5$ days, $T_{max} = 7$ days
Condition : $T_t \leq T_{max} \Rightarrow 5 \leq 7$ (True)
d) Environmental Condition Check : Tenv = 22°C, Tmin = 15°C, Tmax = 25°C
Condition : Tmin ≤ Tenv ≤ Tmax ⇒ 15 ≤ 22 ≤ 25 (True)
e) Production Rate Adjustment : $P_t = k(D_t - I_t) = 0.5 \times (1000 - 500) = 250$
Total Energy Consumption : $E_{tot} = E_{prod} \cdot P_t = 0.5 \cdot 250 = 125$ kWh
Total Cost Calculation : $C_{inv} \cdot$ Ctrans = 1000 dollars per day
$C_{prod} \cdot P_t = 10 \cdot 250 = 2500$ dollars
$C_{maint} = 500$ dollars per

$$C = C_{inv} \cdot I_t + C_{trans} + C_{prod} \cdot P_t + C_{maint}$$

$$C = 1000 + 1000 + 2500 + 500 = 5000 \text{ dollars}$$

By integrating smart manufacturing into supply chain optimization, we achieved a highly efficient and responsive system. The total cost includes production, transportation, inventory, and maintenance costs, while also considering the energy consumption and product quality. IoT devices provide real-time data that help adjust production rates, ensure timely deliveries, and maintain optimal environmental conditions, thereby enhancing overall the supply chain performance.

6.3.2 Just-in-Time (JIT) Inventory Management

JIT minimizes inventory by ensuring that materials and products are received only when they are needed for production, thereby significantly reducing storage costs and the risks associated with obsolescence. This approach streamlines inventory management, ensuring that capital is not tied up in excess stock and that resources are allocated efficiently [Mukwakungu et al. (2019), Roy et al. (2023)].

Smart operations leverage real-time inventory tracking by using IoT sensors, which helps in the technologies that provide an accurate, up-to-the-minute inventory level data, helping to prevent both overstocking and stock outs. By maintaining optimal inventory levels, smart operations enhance the efficiency and responsiveness of the supply chain, ensuring that the right materials are available when needed, thus complementing the JIT approach. For example, Annual Demand (DDD) = 10,000 units, Ordering Cost (SSS) = ₹50 per order, holding cost (HHH) = ₹2 per unit per year. Table 6.6 shows the implementation of JIT principles with the EOQ model, that is, optimizing order quantities and minimizing inventory, result in lower total costs as mentioned in Table 6.6.

JIT manufacturing, IoT devices: Assume the following hypothetical data: Initial Inventory Holding Cost (IHC): ₹50,000 per year, Ordering Cost (OC): ₹10,000 per

TABLE 6.6
Different Inventory Models

JIT Principles	Inventory Models and Cost Equations	Results and Interpretations
Economic Order Quantity (EOQ)	$=\frac{\sqrt{2DS}}{H}$	$\text{EOQ}=\frac{\sqrt{2\times 10{,}000\times 50}}{2}=500$ units
Inventory Holding Cost (IHC)	$=\frac{Q}{2}\times H$	$=\frac{707}{2}\times 2=$ ₹707 per year
Ordering Cost (OC)	$=\frac{D}{Q}\times S$	$=\frac{10{,}000}{707}\times 50=$ ₹707.213 per year
Total Cost (TC)	TC = IHC + OC	= 707 + 707 = ₹1414 per year
A scenario without JIT principles, assuming a larger order quantity 1000 units	TC = IHC + OC	= 1000 + 500 = ₹1500 per year

Using the EOQ model, which is aligned with JIT principles, the total cost is ₹1414 per year compared to ₹1500 per year for ordering 1000 units at a time. This demonstrates that JIT principles, through optimizing order quantities and minimizing inventory, result in lower total costs. The mathematical models validate the effectiveness of JIT in reducing both holding and ordering costs, leading to enhanced manufacturing efficiency.

year, Cost of Downtime: ₹1000 per hour, Average Downtime without Predictive maintenance: 100 hours per year. Assume the IoT integration reduces holding and ordering costs by 20% owing to improved accuracy and efficiency.

- Reduced Holding Cost (H′) : $H' = 2 \times 0.8 =$ ₹1.6 per unit per year
- Reduced Ordering Cost (S′) : $S' = 50 \times 0.8 =$ ₹40 per order
- New EOQ Calculation : $\frac{\sqrt{2 \text{ x } 10{,}000 \text{ x } 40}}{1.6} = \sqrt{500{,}000} = 500$ units
- New Inventory Holding Cost (IHC) : $\frac{500}{2} x 1.6 =$ ₹400 per year
- New Ordering Cost (OC) : $\frac{10000}{500} x 40 =$ ₹800 per year
- New Total Cost (TC) : 400 + 800 = ₹1,200 per year

By integrating IoT devices with EOQ principles, the total cost was reduced from ₹1414 to ₹1200 per year. This reduction is achieved through more accurate real-time inventory tracking leading to reduced holding costs, improved demand forecasting resulting in lower ordering costs, enhanced efficiency in inventory management, minimizing of waste and ensuring timely replenishment.

Reorder Point (RP): A crucial indicator for inventory management, the RP indicates, when a fresh order has to be placed in order to refill stock before it runs out while taking lead time and demand volatility into account. This ensures that the inventory is reordered at the right time to prevent stock outs and maintain optimal stock levels.

RP = D (Average Daily demand rate) × LT (Lead time in days) + SS (Safety stock, i.e., Extra inventory to prevent stock outs due to variability in demand or lead time)

Data value taken as, D = 50 units/day, LT=7 days

$$\text{So, RP} = (50 \times 7) + 100 = 350 \text{ units} + 100 \text{ units} = 450 \text{ units and SS} = 100 \text{ units.}$$

When the actual inventory level I_{act} falls below the reorder point RP, it triggers a replenishment order to maintain the optimal stock levels. Here, the Reorder Point (RP) of 450 units means that when the inventory level drops to 450 units, a new order should be placed. This calculation ensures that there is sufficient stock to cover the average demand during the lead time, along with a buffer (safety stock) to account for any unexpected increases in demand or delays in supply.

6.3.3 Inventory Turnover Ratio (ITR)

ITR is one of the most important financial ratios measuring the efficiency of managing stocks that existed from the beginning of a certain period up to its end by determining the frequency rate of sales and purchases in the business. It is arrived at by using the following formula: ITR = COGS/Average Inventory. For illustration, if the Cost of Goods Sold is ₹200,000, beginning inventory is ₹50,000, and the ending inventory, is ₹70,000; the average inventory is ₹60,000. It translates to an ITR of 3.33, which means that the companies sell and restore its inventory 3.33 times in this period. This ratio allows enterprises to evaluate their performance in inventory management, plan and regulate costs, and compare with the industry benchmarks.

6.3.4 Vendor Managed Inventory (VMI)

This implies that vendors should be responsible for the management of inventory of products supplied to customers. This integration enhances the number of times inventory is circulated within a given period, ensures minimum stock out time, and decreases carrying cost all at the same time that ensures the supply chain is uninterrupted.

6.3.5 Lean Supply Chain Practices

SCM is all about reducing waste as well as enhancing the value delivery processes in its supply chain. Using such methods as kaizen and Kanban two major changes were brought about in the streamlining of processes, shortening of lead times.

6.3.6 Agile Supply Chain Strategies

It underlines the organized evolution of flexibilities of the Agile SCM to produce according to the customer requirements and the dynamic changes in the market. It includes tactics like FMS, RPDC and this approach offers several advantages, including lowering costs and enhancing customer satisfaction through timely inventory status updates.

6.3.7 Technology Integration (IoT, AI, Blockchain)

The use IoT for real-time tracking, AI for demand forecast and optimization and blockchain for improving traceability in SCM has brought about major change. These technologies enhance business decisions, provide better control of supply chain, and minimize operational risks.

6.3.8 Sustainable Supply Chain Practices

The main concerns of sustainability strategies in SCM deal with environmental concerns, energy consumption, and fairness. This approach provides several benefits, such as reducing costs and improving customer satisfaction by delivering timely updates on inventory status.

Research also identified about various challenges in implementing these practices, including technological barriers, organizational resistance to change, complexities in global supply chain networks, and the need for skilled talent. Overall, prior research underscores the critical role of SCM practices in driving efficiency improvements across the supply chain, from collection to the hands of ultimate consumers and customer service, ultimately contributing to enhanced competitiveness and profitability for businesses.

Integrating IoT into SCM can significantly improve various aspects of the supply chain, leading to optimized operations, reduced costs, and increased productivity. Here we provided an in-depth look at how SCM can be optimized for better manufacturing efficiency using the IoT, supported by mathematical equations and hypothetical data.

6.3.9 Demand Forecasting

An important parameter, that is, demand forecasting leverages mathematical models and data analysis to accurately predict future customer demand accurately [Zohra Benhamida et al. (2021), Frank & Henry (2024)]. Key demand forecasting methods include the moving average, exponential smoothing, and linear regression. For example, using the moving average method with actual demand data of 150, 180, 200, and 190 units over the last four months, the forecasted demand for the next period can be calculated as 190 units. The exponential smoothing method adjusts the forecast based on previous and historical demand, giving more weight to recent data. With a previous forecast of 200 units, historical demand of 180 units, and a smoothing constant of 0.4, the forecasted demand would be 188 units. The linear regression method uses historical demand data to predict future trends. For monthly demand data points (1,150), (2,180), (3,200), and (4,190), the forecasted demand for the fifth month is 175 units. These methods help maintain optimal stock levels, enhance resource planning, minimize costs, and improve customer satisfaction. By integrating demand forecasting with IoT-enabled real-time data collection, businesses can further enhance their accuracy and responsiveness, supporting agile strategies in dynamic market environments.

6.3.10 Quality Control and Assurance

IoT-enabled quality control processes can significantly enhance defect detection, process optimization, and traceability. By leveraging continuous monitoring and data analysis, manufacturers can maintain high-quality standards and ensure regulatory compliance. Let us discuss for the defect rate and quality score calculation process.

Variables and Parameters:

Qs	– Quality Score
Q_{min}	– Minimum Acceptable Quality Score
D	– Defect Rate
T_{corr}	– Time for Corrective Action
S_{def}	– Sensor Defect Detection Rate

Let's take an example to calculate the quality score.

Total Units Produced	: 1000
Total Defects Detected	: 50
Minimum Acceptable Quality Score	: 0.95
Sensor Defect Detection Rate	: 0.98

a. Quality Score Calculation $= \dfrac{\text{Total Number of Defects Detected}}{\text{Total Number of Units Produced}} = \dfrac{50}{1000} = 0.05$

b. Defect Rate $= \dfrac{\text{Total Number of Defects Detected}}{\text{Total Number of Units Produced}} = \dfrac{50}{1000} = 0.05$

$$\text{c. Corrective Action Time} = \frac{\text{Total Number of Corrective Actions}}{\text{Total Number of Defective Units}}$$

$$= \frac{10\,\text{hours}}{50} = 0.2\,\text{hours per defect}$$

IoT sensors detect defects at a high detection rate, enabling a quality score of 0.95. If the defect rate exceeds this threshold, immediate corrective actions are taken to maintain quality.

6.3.10.1 Process Optimization

In smart manufacturing and quality control and assurance through the application of the IoT device and big data, optimization of the process aims at enhancing the efficiency of production, the quality of end products as well as reducing wastage. Process efficiency is a measure of how well a production process is doing, by comparing how long the process should take (optimal time to produce a given amount of output under perfect conditions) against how long it took (actual time took to produce the output). Inefficiency Index gives energy loss fraction in the production process out of the total energy used. It can also be obtained if an efficiency ratio is to be made when measuring energy efficiency.

For Example: Optimal Production Time – 900 hours, Actual Production Time – 1000 hours, energy wasted – 100 kWh, Total Energy Consumed 1000 kWh

Process Efficiency (PE) = $\frac{\textit{Optimal Production Time}}{\textit{Actual Production Time}} = \frac{900}{1000} = 0.9$. This means the process is operating at 90% efficiency, with a 10% gap between optimal and actual production times.

I_{opt} – Inefficiency Index= $\frac{\textit{Energy Wasted}}{\textit{Total Energy Consumed}} = 1 - 0.9 = 0.1$. This means 10% of the total energy consumed is wasted, indicating areas where energy use can be optimized.

6.3.11 Energy Management

IoT improves controlling of energy in manufacturing plants since the energy consumption is monitored and controlled in real time. IoT devices are capable of tracking the energy consumption regularly and can detect the high usage or even savings that can be achieved; that is, IoT devices can assist the facilities in detecting inefficient areas [Bagdadee et al. (2020), Hossein Motlagh et al. (2020)]. This can enable real-time control of lighting, heating, cooling, and machinery while optimizing energy use based on actual operational needs. In doing so, it minimizes both wastage and costs, reduces the negative impacts of manufacturing activities, and supports overall sustainability objectives. The incorporation of the IoT in energy management can save various manufacturing companies a lot of cash, time, and energy while at the same time reducing environmental pollution.

The IoT greatly improves energy supervision in production units, organizing real-time energy intake, and fostering sustainability mechanisms. IoT devices assist in the monitoring of the energy usage in the different sections including lighting, heating, cooling techniques, and using machinery for which if taken an account, the total energy consumption that ranges to 3000 kWh may be determined. Other optimization techniques can then be applied to control these sections mechanically and in so doing will minimize energy consumption. For example, the efficient use of lights, heat, cool, and machinery translated to an ideal energy consumption of 2560 kWh with a saving of 440 kWh. This optimization does not only lead to the optimization of cost issues of the operation but also results to the optimization of the emission of carbon. From the assumption of a carbon emissions rate of 0.5 kg CO_2 per kWh this leads to an overall carbon emission saving of 220 kg CO_2. Therefore, IoT enhances efficiency in energy utilization and ensures that the manufacturing lines carry out their work in harmony with the objectives of sustainable development hence cutting on the impacts chucked to the environment.

6.3.12 Enhanced Worker Safety

Third, IoT creates smart apparel and items that can help increase worker safety in manufacturing facilities with smart and wearable technology. When using IoT sensors for monitoring the environment, any dangerous conditions including gas leakage or high noise levels can be seen and workers warned to take precautions. Other benefits of smart wearable devices are increase safety by monitoring health status and/or signs of tiredness, anxiety or stress to avoid accidents. Moreover, IoT collaborates with the automatic safety systems which pass the instructions for the shutdown of a certain equipment or the turning on of alarms in dangerous scenarios. By so doing it makes work environment safer since IoT technology can monitor, alert, and protect the workers in real time.

6.3.13 Smart Manufacturing Ecosystem

The IoT enables the creation of a smart manufacturing ecosystem in which all components – machines, people, and systems – are interconnected and communicate seamlessly. This leads to interoperability, where different machines and systems work together smoothly, enhancing overall efficiency and reducing downtime. Enhanced communication and data sharing through the IoT facilitate better collaboration across departments and with external partners, driving innovation in product design, process improvements, and business models. Key technological innovations contributing to smart operations include advanced sensors, machine learning algorithms, and real-time data analytics. Implementing smart operations in manufacturing involves strategies such as integrating IoT devices with existing systems, adopting a modular approach to upgrades, and investing in workforce training. The benefits of adopting smart operations are substantial and include increased efficiency, reduced downtime, and improved product quality, ultimately leading to more agile and competitive manufacturing processes.

6.4 CASE STUDIES AND PRACTICAL APPLICATIONS

Smart operations and supply chain management need to be supported by case studies and application examples because they are examples and practical applied knowledge that enables firms to enhance their supply networks and establish sound manufacturing processes. From the successful corporate examples and the use of practical strategies, the firms and organizations can identify the directions for growth, decrease expenses, increase the production volume, and increase performance efficiency. This knowledge assists them in their decision making and competitive supply chain management, and, therefore, make them more sustainable.

6.4.1 Case Study 1: Toyota Motor Corporation

Toyota Motor Corporation, which is among the leading world companies that build automobiles, has been recently developing smart operation and supply chain management. Toyota manufacturing system better known as Toyota production system (TPS) is famous for efficiency, quality, and improvement.

Already in the course of its development, it has become one of the largest manufacturers and traders in the world, striving for pre-eminence in the automotive industry which was launching in 1937. The company's philosophy revolves around two main pillars: JIT and Jidoka – automation with a human touch or automation with a sense of being human. Some of these principles have been key to Toyota's operations, and have adapted to improvements in technology.

Challenges Faced: There were a number of barriers for Toyota as it struggled to undertake the transformation of becoming a smart manufacturing plant and optimizing its supply chain. Global automaker competition demanded great improvement of efficiency and cost decreasing. Coordinating and integrating a large number of suppliers to ensure timely supply of parts is now a major management challenge in automobile industry. Also, it creates the problem of ensuring high quality, as increasing production volume means that the company has to save customer trust and increase production while maintaining Toyota's reputation for quality and reliability.

6.4.2 Smart Operations and SCM Implementation Overview, Rectified In The Title

a. *JIT System: Objective is to minimize inventory and reduce waste.*
b. *Implementation: Parts are produced and delivered in time for the production process, reducing the need for large inventories. This system relies heavily on precise scheduling and real-time data sharing with the suppliers.*
c. *Jidoka (Automation with a Human Touch): The objective is to enhance quality and efficiency by integrating automation with human oversight.*
d. *Implementation: Machines are equipped with sensors to detect anomalies and can automatically stop production to prevent defects. Human operators then inspect and address these issues to ensure high quality.*
e. *Kanban System: Objective is to improve workflow and inventory management.*
f. *Implementation: Visual scheduling system where production and supply are triggered by actual demand. This helps in maintaining optimal inventory levels and ensuring a smooth flow of materials.*
g. *Lean Manufacturing: Objective is to eliminate waste and enhance productivity.*

h. *Implementation: Continuous improvement (Kaizen) practices are ingrained in the company culture. All workers were motivated to identify inefficiencies and suggest improvements.*
i. *Supply Chain Integration: Objective is to enhance collaboration and coordination with suppliers.*
j. *Implementation: Toyota has developed strong relationships with its suppliers through long-term contracts and shared goals. Information sharing platforms have been established to provide real-time data on demand, inventory, and production schedules.*
k. *Smart Technologies: Objective is to leverage technology for better decision-making and efficiency.*

Toyota's adoption of IoT and AI technologies has revolutionized their production processes through predictive maintenance, real-time monitoring, and advanced analytics, enabling the identification of potential issues and optimization of operations [Hatani et al. (2017), Nandini et al. (2023)] These smart operations and SCM practices have significantly reduce lead times, minimize inventory costs, and enhance overall efficiency. The combination of automation and human oversight has maintained high-quality standards and boosted customer satisfaction. Lean manufacturing and JIT practices have effectively reduced waste and operational costs, while collaborative relationships with suppliers have improved coordination, ensured timely deliveries, and minimized disruptions. Furthermore, efficient resource utilization and waste reduction are aligned with Toyota's sustainability goals, reinforcing their commitment to environmental responsibility. The company's ability to adapt to changing market conditions and leverage advanced technologies has solidified its position as a leader in the automotive industry. Toyota's practices serve as a benchmark for other companies to enhance their operational and supply chain efficiency.

6.4.3 Case Study 2: Amazon

Smart operations and supply chain management are principles that Amazon, a customer-focused, customer-centric worldwide e-commerce corporation founded in 1994, adheres to. Amazon is a global marketplace that offers a wide range of products and services. It has transformed from an online bookstore to a global marketplace with a huge logistical network, cutting-edge technology, and new practices. Amazon has set the standard for smart supply chain management and operational excellence. The company's aim is to become the most customer-focused organization on Earth, and it works toward this objective by putting efficiency, innovation, and customer happiness front and center.

Current Challenges in Manufacturing Efficiency: Manufacturing efficiency therefore has several challenges such as technological integration, disrupted supply chains, lack of sufficient human resource, and quality problems. The majority of manufacturers face problems of outdated corporate infrastructures and data fragmentation, which hinders the implementation of new technologies and strengthens cybersecurity threats. That the global nature of supply chains shows that manufacturers assume certain risks such as natural calamities, geo-political risks, and transport hazards, which may affect the production timeline. This is made worse by labor shortage due to old age of current employees and scarcity of qualified personnel for new technical positions. Moreover, the global quality of the products and adherence to strict rules remains a continuous problem, aggravated by the gradual increase in the level of customer expectations. Solving these challenges demands major investments in IT, in employees, and in strong quality assurance mechanisms that will make a positive impact on manufacturing industry productivity and competiveness.

6.4.4 Smart Operations and SCM Implementation

6.4.4.1 Warehouse Automation

Smart operations and SCM implementations include warehouse automation, which is the use of robotics and automation technology to increase production rates and reduce labor costs. Amazon's Kiva robots, for example, have quickly shifted product shelves to human pickers, dramatically speeding up the picking process and significantly reducing labor costs. Through advanced forecasting and inventory management techniques that minimize inventory levels, stock-outs are prevented. Amazon's utilization of big data analysis and machine learning algorithms enables the anticipation of customer demands and the appropriate supply of its multiple warehouses. Kiva robots, for instance, transport product shelves to human pickers, significantly enhancing the efficiency of the picking process. Using machine learning algorithms and big data analytics, Amazon predicts customer demand ensure optimal inventory levels across its warehouses network. The company employs various strategies, including Prime Air (utilizing drones), Flex (engaging independent contractors as drivers), and Logistics (a network of distribution hubs). Additionally, the business uses robotic systems to transport product shelves to human workers, substantially accelerating the item selection process.

6.4.4.2 Supply Chain Visibility

In this context, the primary objective should be to enhance information flow across departments, divisions, organizations, and the entire supply chain, thereby increasing data accessibility. Real-time updates and information exchange with suppliers and logistics providers enable Amazon to maintain accurate records of inventory and shipments, thus optimizing delivery efficiency to the highest degree. Robotic systems, for instance, transport shelves of products to human pickers, thereby significantly expediting the picking process [Somapa et al. (2018)].

Amazon's successful implementation of smart operations and SCM highlights the importance of innovation, technological integration, and customer focus. The company's ability to foresee trends, adapt to changing market conditions, and continuously improve its operations has established Amazon as a leader in the global e-commerce and logistics sectors. Amazon's practices serve as a model for companies aiming to enhance efficiency, scalability, and sustainability in their supply chains and operational strategies.

To increase industrial efficiency, supply chain management and smart operations must be integrated. Organizations can get useful insights and expertise to optimize their operations, boost productivity, and cut expenses by analyzing case studies and practical implementations. Businesses may achieve unprecedented levels of efficiency, innovation, and sustainability by implementing smart technology and data-driven strategies. This will ultimately provide them a competitive advantage in the current fast-paced industrial scene.

6.5 CONCLUSION

This chapter demonstrated that the implementation of smart operations and supply chain management can significantly enhance the manufacturing efficiency. By harnessing advanced technologies such as IoT, IIoT, data analytics, and automation, manufacturers can achieve optimized production processes and improved operational performance. The integration of APS, ERP, and SCM systems offers comprehensive visibility and coordination, enabling better decision-making and resource utilization. Lean manufacturing principles and JIT inventory management further help minimize waste and reduce costs. Additionally, the use of predictive maintenance and real-time quality control ensures consistent production quality and minimizes the downtime. The proposed framework underscores the importance of a holistic approach to modernizing manufacturing processes, and shows how such strategies and technologies lead to substantial gains in productivity, cost savings, and sustainability.

REFERENCES

Ali, Syed Imran, Abdilahi Ali, Muhanad AlKilabi, and Michael Christie. "Optimal supply chain design with product family: A cloud-based framework with real-time data consideration." *Computers & Operations Research* 126 (2021): 105112.

Ang, Joo Hock, Cindy Goh, Alfredo Alan Flores Saldivar, and Yun Li. "Energy-efficient through-life smart design, manufacturing and operation of ships in an industry 4.0 environment." *Energies* 10, no. 5 (2017): 610.

Bagdadee, Amam Hossain, Li Zhang, and Md Saddam Hossain Remus. "A brief review of the IoT-based energy management system in the smart industry." *Artificial Intelligence and Evolutionary Computations in Engineering Systems* (2020): 443–459.

Fatima, Zainab, Muhammad Hassan Tanveer, Waseemullah Shehnila Zardari, Laviza Falak Naz, Hina Khadim, … and Midha Tahir. "Production plant and warehouse automation with IoT and industry 5.0." *Applied Sciences* 12, no. 4 (2022): 2053.

Frank, Edwin, and Elizabeth Henry. *AI-Driven Demand Forecasting for Optimized Inventory Management*. No. 13214. EasyChair, 2024.

Hatani, La. "Integrated supply chain management practices in the flow information Toyota car dealer in Kendari." *The International Journal of Engineering and Science* 6, no. 1 (2017): 36–43.

Hossein Motlagh, Naser, Mahsa Mohammadrezaei, Julian Hunt, and Behnam Zakeri. "Internet of Things (IoT) and the energy sector." *Energies* 13, no. 2 (2020): 494.

Javaid, Mohd, Abid Haleem, Ravi Pratap Singh, Shanay Rab, and Rajiv Suman. "Upgrading the manufacturing sector via applications of Industrial Internet of Things (IIoT)." *Sensors International* 2 (2021): 100129.

Kandarkar, Prajakta Chandrakant, and V. Ravi. "Investigating the impact of smart manufacturing and interconnected emerging technologies in building smarter supply chains." *Journal of Manufacturing Technology Management* 35, no. 5 (2024): 984–1009.

Lee, Khai Loon, Sze Yun Wong, H. M. Alzoubi, B. Al Kurdi, M. T. Alshurideh, and Mounir El Khatib. "Adopting smart supply chain and smart technologies to improve operational performance in manufacturing industry." *International Journal of Engineering Business Management* 15 (2023): 18479790231200614.

Menon, Sarath, Satya Shah, and Alec Coutroubis. "An overview of smart manufacturing for competitive and digital global supply chains." In *2018 IEEE International Conference on Technology Management, Operations and Decisions (ICTMOD)*, pp. 178–183. IEEE, (2018)

Mukwakungu, Sambil Charles, Matimba Davis Mabasa, Sihle Mankazana, Xolani Mzileni, and Sine Aurore Burakeye. "The impact of just in time (JIT) in inventory management–perspectives from two case studies in a South African environment." In *Proceedings of the International Conference on Industrial Engineering and Operations Management*, vol. 5, no. 7, pp. 1271–1278. 2019.

Muruganandam, S., Anas A. Salameh, Mohd Affendi Ahmad Pozin, S. V. Manikanthan, and T. Padmapriya (2023). "Sensors and machine learning and AI operation-constrained process control method for sensor-aided industrial internet of things and smart factories." *Measurement: Sensors*, *25*, 100668.

Nandini, T. S., M. MohanRam, and G. L. Shekar. "Modeling 4Ps of Toyota way principles in MSMEs supply Chain-an empirical approach." *Materials Today: Proceedings* 92 (2023): 284–290.

Pasi, Bhaveshkumar N., K. Subhash, Santosh B. Mahajan, and Rane. "Smart supply chain management: A perspective of industry 4.0." *Supply Chain Management* 29, no. 5 (2020): 3016–3030.

Rezaei, Mahdi, Mohsen Akbarpour Shirazi, and Behrooz Karimi. "IoT-based framework for performance measurement: A real-time supply chain decision alignment." *Industrial Management & Data Systems* 117, no. 4 (2017): 688–712.

Roy, Marc-Antoine, Georges Abdul-Nour, and Sébastien Gamache. "Implementation of an industry 4.0 strategy adapted to manufacturing SMEs: Simulation and case study." *Sustainability* 15, no. 21 (2023): 15423.

Somapa, Sirirat, Martine Cools, and Wout Dullaert. "Characterizing supply chain visibility–a literature review." *The International Journal of Logistics Management* 29, no. 1 (2018): 308–339.

Talamo, C., G. Paganin, and F. Rota. "Industry 4.0 for failure information management within proactive maintenance." In *IOP Conference Series: Earth and Environmental Science*, vol. 296, no. 1, p. 012055. IOP Publishing, 2019.

Wu, Lifang, Xiaohang Yue, Alan Jin, and David C. Yen. "Smart supply chain management: A review and implications for future research." *The International Journal of Logistics Management* *27*, no. 2 (2016): 395–417.

Zhou, L., Jiang, Z., Geng, N., Niu, Y., Cui, F., Liu, K., & Qi, N. "Production and operations management for intelligent manufacturing: A systematic literature review." *International Journal of Production Research* 60, no. 2 (2022): 808–846.

Zohra Benhamida, Fatima, Ouahiba Kaddouri, Tahar Ouhrouche, Mohammed Benaichouche, Diego Casado-Mansilla, and Diego López-de-Ipina. "Demand forecasting tool for inventory control smart systems." *Journal of Communications Software and Systems* 17, no. 2 (2021): 185–196.

7 AI- and IoT-Based CNC Optimization for Sustainable Production

Kali Charan Rath, Kamalakanta Muduli, Manas Ranjan Panda, Gopal Krushna Mohanta, and Sasank Sekhar Panda

7.1 INTRODUCTION

A revolution in manufacturing, computer numerical control (CNC) technology uses computer-aided design and production systems to precisely and quickly control machine equipment. With its accuracy, efficiency, and adaptability, CNC technology, which has been around since the 1940s, is now crucial in many different industries. Industry 4.0, the fourth industrial revolution, depends heavily on CNC. CNC incorporates Internet of Things (IoT) technology and automation data interchange with ease. CNC is still essential to Industry 4.0. It integrates IoT and automation data transmission with ease [Samanta et al. (2023)]. In order to optimize production lines and enable adaptive control, predictive maintenance, and real-time monitoring, modern CNC machines make use of advanced software and sensors. By using CNC, industries can produce complex parts and customized commodities with unprecedented levels of accuracy, flexibility, and efficiency. Innovation and competitiveness are increased in the automotive, healthcare, and aerospace industries by CNC's integration with Industry 4.0 technologies, which creates a collaborative environment [Khan et al. (2023)].

Industry 4.0, which transforms organizations through artificial intelligence (AI), IoT, robotics, and data analytics, enables real-time monitoring, predictive maintenance, increased efficiency, and customization [Patalas-Maliszewska et al. (2024)]. The ability of manufacturers to produce goods, reduce waste, and meet market demands through data-driven insights leads to a more flexible, sustainable, and interconnected manufacturing ecosystem [My (2021)].

Industry 4.0 is transforming manufacturing through the convergence of advanced technologies such as CNC machining and the IoT. IoT enables intelligent inventory management and adaptive manufacturing, while AI enhances CNC with real-time quality control and predictive maintenance [Laghari et al. (2023)]. This collaboration opens up new avenues for creative product design, shorter lead times, and more customer satisfaction by allowing manufacturers to achieve previously unheard-of levels of accuracy, adaptability, and efficiency.

A major contributor to sustainable manufacturing, CNC machining provides accuracy, productivity, automation, and adaptability. CNC machines optimize production

DOI: 10.1201/9781003503323-7

processes to decrease energy consumption and emissions while creating products with ultra-high accuracy, minimizing material waste and rework [Sundaresan et al. (2024)]. Additionally, CNC machines reduce labor expenses and increase workplace safety through automated processes. Furthermore, small-batch production is made possible by their flexibility, which lowers surplus inventory and promotes environmentally friendly manufacturing techniques.

AI and the IoT are redefining CNC machining for sustainable production in order to increase productivity, decrease waste, and use less energy [Kaur et al. (2023)]. CNC machining is being revolutionized by AI and IoT, which enables manufacturers to boost output, streamline processes, and establish an efficient and sustainable manufacturing environment [Mejía-Moncayo et al. (2023)]. To optimize efficiency, sustainability, and performance in CNC manufacturing, "Unlocking AI & IoT in CNC Machining for Sustainable Production" explores cutting-edge techniques and technology.

The chapter's goal is to create an AI- and IoT-based CNC machining optimization framework that will enable sustainable production by lowering energy usage, cutting down on material waste, and increasing overall efficiency. In order to address economic and environmental issues and support global sustainability goals, the authors want to improve CNC machining's sustainability, efficiency, and competitiveness with AI and IoT.

7.2 CURRENT CHALLENGES TO ADOPT CNC IN MANUFACTURING FLOOR

CNC machining poses a number of challenges these days. These include significant upfront costs, the need for highly qualified operators, and guaranteeing accuracy and precision. Additional challenges include managing various material properties, incorporating Industry 4.0 technologies, managing coolant and chip, and addressing chatter and vibration issues. In addition, there is a growing concern about environmental sustainability and a strong desire to reduce energy use, trash, and noise pollution. There are various tactics that might be used to lessen these difficulties in CNC machining. These include:

a. Upskilling and Reskilling: In-depth instruction and training programs are essential for giving operators the cutting-edge abilities needed to operate intricate CNC machines with ease and become skilled in their programming.
b. Investing in Advanced Technologies: The use of state-of-the-art CNC equipment, complex simulation programs, and sophisticated automation systems greatly improves manufacturing accuracy, precision, and productivity.
c. Process Optimization: Enhancing machining parameters for intricate forms and materials, as well as putting in place effective chip and coolant management systems, greatly increases manufacturing efficiency and decreases production downtime.
d. Industry 4.0 Integration: Combining CNC equipment with intelligent manufacturing technologies by utilizing IoT, AI, and data analytics.
e. Material Science Advancements: Advanced materials with predictable and consistent qualities are developed and used to reduce machining problems, increase manufacturing efficiency, and lower prices.

f. Vibration and Chatter Reduction: By combining sophisticated damping systems with well-designed tools and well-calibrated machining parameters, vibration and chatter are successfully reduced, greatly enhancing surface smoothness and overall machining accuracy.
g. Sustainable Practices: Manufacturing can be made more sustainable and ecologically conscious by adopting eco-friendly machining techniques, minimizing waste production, and cutting energy use through effective machine design and improved operating procedures.

Manufacturers can overcome the difficulties associated with CNC machining and achieve notable gains in productivity, efficiency, and sustainability by successfully putting these strategies into practice.

7.2.1 Problem Facing by Industries and it's Overcome Solution

Even while CNC technology has transformed production by increasing accuracy and efficiency, its general acceptance is hampered by the high initial cost, the requirement for trained operators, and the possibility of expensive machine downtime [Müller et al. (2020)]. Industries may, however, get around these obstacles by funding thorough employee training programs, putting preventative maintenance plans into place, and using cutting-edge diagnostic technologies to reduce downtime. While there are many benefits to CNC technology, its successful deployment frequently requires a phased approach, beginning with smaller-scale adoption to reduce the initial investment costs, train skilled labor gradually, and reduce the risk of prolonged machine downtime while achieving long-term cost savings and efficiency gains [Smith et al. (2018)]. Industries may optimize the advantages of CNC technology while minimizing any potential downsides by adopting a methodical and incremental approach.

7.2.2 AI- and IoT-Enabled CNC

The integration of AI and IoT into CNC machines has the potential to transform sustainable manufacturing by facilitating real-time monitoring of vital parameters such as vibration and temperature. Predictive maintenance, resource allocation, and downtime reduction are made possible by this data-driven strategy, which eventually results in a manufacturing process, that is, more eco-friendly and efficient [Lee et al. (2015)]. CNC machines with AI capabilities minimize material utilization, cut waste, and maximize energy consumption. Additionally, they make remote monitoring and control possible, which lowers the carbon footprint of manufacturing processes [Lu (2017)]. By integrating AI and IoT into CNC machines, productivity, sustainability, and environmental impact are all improved, as is profitability.

7.2.3 Benefits of Advanced Technology Adoption with CNC

It is possible to implement AI and IoT technologies with CNC machines in stages, beginning with the installation of IoT sensors to gather data on machine performance.

AI algorithms are then used to analyze and interpret the data. The creation of machine learning models can subsequently be applied to improve manufacturing schedules, predict maintenance needs, and improve product quality [Zhou et al. (2015)]. AI-powered CNC machines maximize production efficiency and reduce environmental impact by enabling data-driven decision-making, adaptive control, and real-time monitoring [Wang et al. (2016)].

Various benefits for AI and IoT integrated in CNC machines are:

a. In order to forecast any problems and minimize downtime, AI-powered predictive maintenance in CNC machines tracks machine health in real-time using IoT sensors.
b. AI-driven manufacturing process optimization is made possible by IoT sensors, which offer real-time data on CNC machine performance.
c. AI-integrated CNC machines use machine vision and IoT sensors to automatically identify flaws and irregularities throughout the manufacturing process, guaranteeing high-quality output and reducing the need for manual inspections.
d. Intelligent machining reduces material waste and improves accuracy by optimizing CNC machining parameters like feed rates and speed using AI algorithms.
e. CNC machines may be remotely monitored and controlled thanks to IoT connectivity, which enables real-time modifications and shorter production times.
f. Real-time tracking of tool wear and tear using IoT sensors and AI algorithms in CNC machines allows for preemptive replacement and reduces downtime.
g. AI-enabled CNC machines use adaptive machining to adjust to shifting material characteristics and machining circumstances, guaranteeing peak performance.
h. When AI and IoT are integrated into CNC machines, real-time assistance improves operator abilities, increasing workforce augmentation and overall productivity.

7.3 DEVELOP AN AI-DRIVEN OPTIMIZATION FRAMEWORK FOR CNC OPERATIONS

Manufacturing processes are being transformed by an AI-powered CNC optimization framework that analyzes production data and pinpoints opportunities for improvement using machine learning [Xu et al. (2017)]. Workflows, tool paths, and machine settings are all optimized in real time using AI-powered CNC optimization frameworks that make use of machine learning and predictive analytics [Qin et al. (2016)]. By enabling predictive maintenance and lowering downtime, the AI-powered CNC framework turns conventional manufacturing into intelligent, data-driven, and adaptable systems.

The development of an AI-driven optimization framework for CNC operations involves the use of several devices and components, as follows.

Hardware Components

- CNC machines (e.g., milling, turning, grinding)
- Sensors (e.g., temperature, vibration, acoustic emission)
- Controllers (e.g., PLC, CNC controller)
- Actuators (e.g., servo motors, stepper motors)
- Data acquisition systems (e.g., DAQ cards, industrial PCs)

Software Components

- AI and machine learning frameworks (e.g., Tensor Flow, PyTorch)
- Optimization algorithms (e.g., genetic algorithm, simulated annealing)
- CNC programming software (e.g., G-code, M-code)
- Data analytics and visualization tools (e.g., Tableau, Power BI)
- Integration platforms (e.g., OPC UA, MQTT) for connecting hardware and software components

Other Components

- Data storage solutions (e.g., databases, cloud storage)
- Networking infrastructure (e.g., Ethernet, Wi-Fi)
- Safety and security systems (e.g., firewalls, access control)
- Human-machine interface (HMI) devices (e.g., monitors, keyboards)
- These components work together to enable real-time data collection, AI-driven optimization, and intelligent control of CNC operations

7.3.1 Leverage IoT Sensors for Real-Time Data Acquisition and Analysis

In CNC machines, real-time data collection and analysis using IoT sensors enables better monitoring, accurate adjustments, and increased output and efficiency [Wan et al. (2017)]. In order to forecast maintenance requirements, maximize equipment efficiency, enhance product quality, and save production costs, CNC systems incorporating IoT sensors gather and evaluate data on temperature, vibration, and other factors [Zheng et al. (2017)]. The parts and tools utilized to leverage IoT sensors for real-time data collection and analysis in CNC machines include:

Sensors

- Temperature sensors (e.g., thermocouples, thermistors)
- Vibration sensors (e.g., accelerometers, vibration meters)
- Acoustic emission sensors (e.g., AE sensors, ultrasonic sensors)
- Force sensors (e.g., load cells, strain gauges)
- Displacement sensors (e.g., LVDT, proximity sensors)
- Vision sensors (e.g., cameras, machine vision systems)

Data Acquisition Devices

- Data acquisition cards (DAQ cards)
- Industrial PCs (IPCs)
- Programmable Logic Controllers (PLCs)
- IoT gateways (e.g., Raspberry Pi, Intel Edison)

Communication Protocols

- Wired protocols (e.g., Ethernet, Modbus TCP)
- Wireless protocols (e.g., Wi-Fi, Bluetooth, Zigbee)

Data Analytics and Storage

- Cloud-based data platforms (e.g., AWS IoT, Microsoft Azure IoT)
- Edge computing devices (e.g., smart gateways, fog computing nodes)
- On-premise data servers (e.g., relational databases, NoSQL databases)

CNC Machine Integration

- CNC controllers (e.g., Fanuc, Siemens)
- Machine monitoring software (e.g., CNCnetPDM, MachineMetrics)
- IoT-enabled CNC machines (e.g., Industry 4.0 compliant machines)

Together, these integrated elements like: predictive maintenance, real-time monitoring, automated quality control, and smart tooling enable real-time data collection, analysis, and insights from CNC machines, which boost manufacturing processes' overall productivity, efficiency, and decision-making.

7.3.2 Case Study: Leveraging IoT Sensors for Real-time Data Acquisition and Analysis in CNC

Real-time data gathering and analysis is made possible by IoT sensors in CNC machining, which boosts industrial operations' productivity and quality [Zhang et al. (2018)]. CNC machines were equipped with IoT sensors to track important variables like spindle speed, vibration, temperature, and power consumption. The data was sent to a central processing system for ongoing analysis.

Table 7.1 displays a sample dataset that was gathered from a single CNC machine over the course of five hours, recording variables such spindle speed, vibration, temperature, and power usage.

The CNC machine's IoT sensor data analysis based on Table 7.1, the analysis graphs as shown in Figure 7.1 shows steady and effective operation, with only slight variations in spindle speed and power consumption, which are explained by an increase in

TABLE 7.1
CNC Data Matrix

Time (HH)	Spindle Speed (RPM)	Vibration (mm/s)	Temperature (°C)	Power Consumption (kW)
8:00	3000	1.2	75	2.5
9:00	3200	1.1	78	2.6
10:00	3100	1.3	77	2.4
11:00	3000	1.4	76	2.5
12:00	3050	1.2	75	2.3

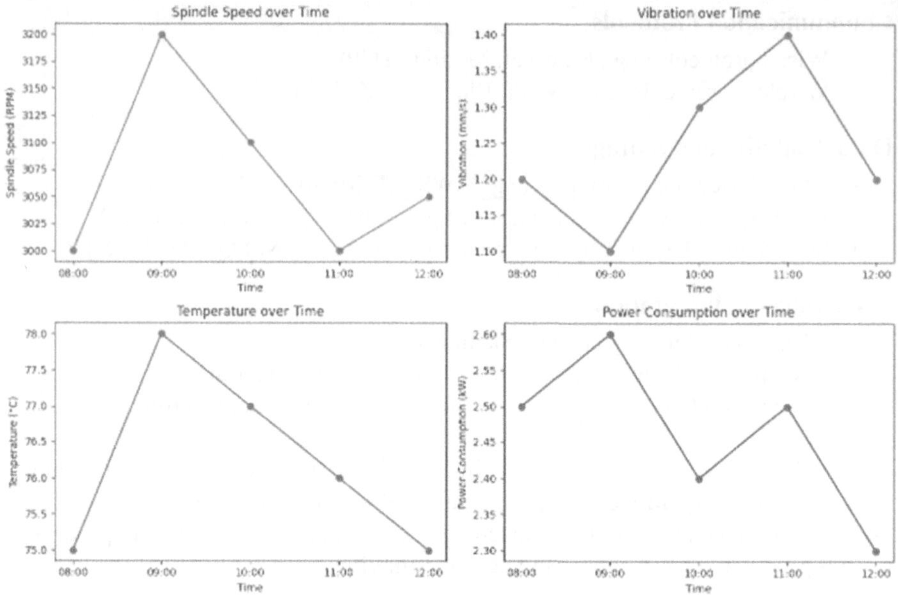

FIGURE 7.1 CNC machine performance.

workload at 9:00 and a quick maintenance break at 11:00. While temperature readings stayed within a small range of 75°C to 78°C, guaranteeing efficient thermal management and component longevity, vibration levels were continuously low, indicating smooth mechanical functioning, with a slight rise at 11:00 signaling routine maintenance inspections. Optimizing CNC machine performance, reducing downtime, and increasing energy efficiency are all made possible by real-time data collection and analysis, highlighting the advantages of a data-driven approach in CNC machining operations.

The CNC machine is functioning steadily and effectively, according to the graphs taken together. Spindle speed and power usage are expected to fluctuate slightly, which is consistent with typical operational variances. The equipment appears to be properly maintained and functioning under ideal circumstances, as seen by the steady temperature measurements and minimal vibration levels. The facility can proactively address problems, maximize machine performance, and increase productivity by utilizing IoT sensors. Data-driven decision-making also results in decreased downtime, better maintenance, and energy efficiency, all of which contribute to sustainable operations and ongoing improvement.

7.3.3 Improve Machining Precision and Reduce Energy Consumption

Improving machining accuracy and cutting energy use necessitates a multifaceted strategy. It is possible to reduce errors and waste by combining sophisticated algorithms with powerful control systems, like CNC machines. Additionally, unplanned downtime can be avoided using predictive maintenance that uses AI and IoT. Energy consumption and accuracy are balanced by optimizing machining parameters such

TABLE 7.2
Impact of Control Systems on Machining Precision and Energy Consumption

Control System	Precision Improvement (%)	Energy Consumption (kWh)	Energy Reduction (%)
Basic	0	500	0
Advanced	20	400	20
Predictive	25	380	24
Optimized	30	350	30
Energy-efficient	35	300	40

as feed rate and cutting speed. Further lowering energy usage through the use of energy-efficient motors and drives leads to a more accurate and environmentally friendly machining process [Jin et al. (2016)]. Manufacturers can attain more accurate machining procedures while drastically reducing energy consumption by combining these tactics, which will increase output and lessen their influence on the environment as listed in Table 7.2.

The figures illustrate (Figure 7.2) the impact of different control systems on machining precision and energy consumption. The precision increase percentage from a basic control system (0%) to an energy-efficient system (35%) is shown in the bar graph. This trend highlights the substantial gains in precision with more

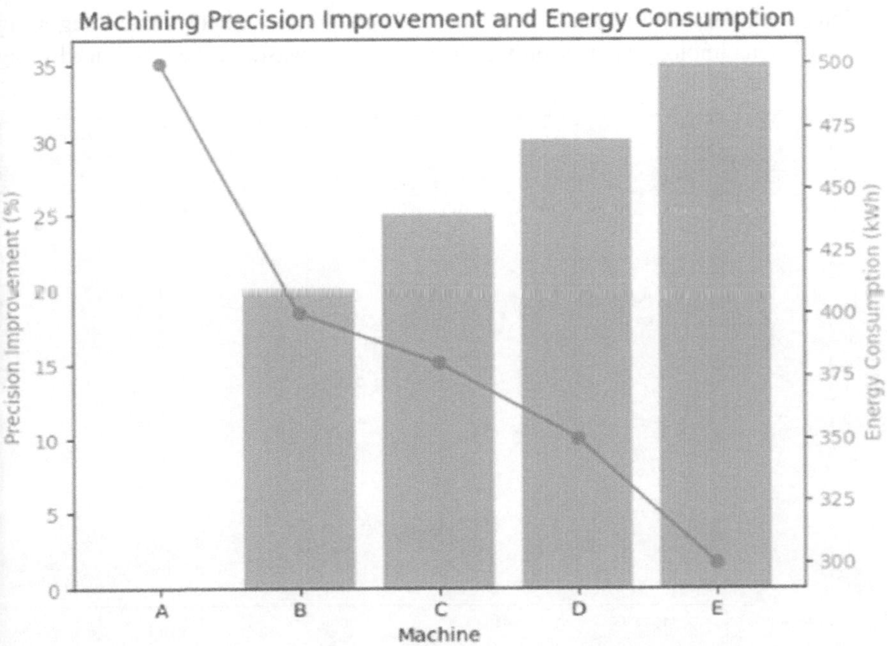

FIGURE 7.2 Machining precision improvement and energy consumption.

advanced technologies. The line plot depicts energy consumption, demonstrating a clear reduction from 500 kWh for the basic system to 300 kWh for the energy-efficient system. The dual-axis chart effectively conveys that as control systems become more sophisticated, both precision and energy efficiency significantly improve. The energy-efficient system, in particular, stands out by achieving the highest precision improvement and the lowest energy consumption, underscoring the value of integrating advanced, energy-efficient technologies in machining processes.

7.3.4 Minimize Material Waste, Aligning with Sustainable Manufacturing Goals

AI and IoT technologies can significantly enhance CNC machining processes by providing real-time monitoring, predictive maintenance, and optimized operation. These technologies contribute to reducing material waste in several ways:

- Predictive Maintenance: AI algorithms can predict machine failures before they happen, reducing downtime and defective products.
- Process Optimization: Machine learning models can optimize cutting paths and parameters to minimize waste.
- Real-time Monitoring: IoT sensors provide real-time data on machine performance and material usage, allowing for immediate adjustments.
- Quality Control: AI-powered vision systems can detect defects early, reducing the number of defective parts.

Interpretation result of analysis on the basis of Figure 7.3 shows that integrating AI and IoT technology significantly reduces material waste in the CNC machining

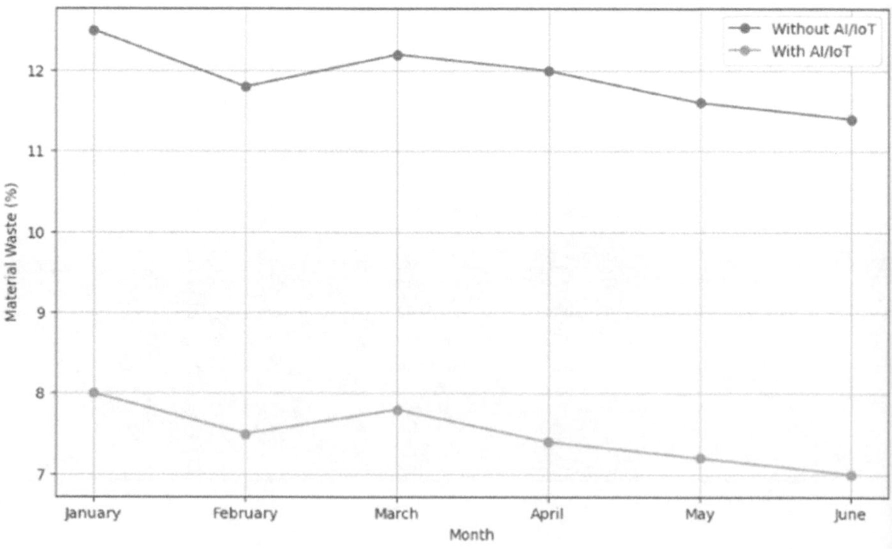

FIGURE 7.3 Material waste reduction with AI and IoT in CNC machining.

process. In particular, the average material waste dropped from about 12.08% to 7.48%. This steady improvement over the course of the six months demonstrates how dependable and efficient these technologies are in reducing waste. Reducing the use of materials makes the manufacturing process more environmentally friendly and helps the company save a significant amount of money.

Therefore, implementing AI and IoT technologies in CNC machining can greatly reduce material waste and encourage environmentally friendly production methods.

7.4 METHODS FOR INCORPORATING DATA INTO THE OPTIMIZATION FRAMEWORK

Data integration guarantees accurate manufacturing execution, improves process accuracy, and permits data-driven decision-making [Gharbia et al. (2018)]. Manufacturers may learn a great deal about the behavior of their machines by utilizing both historical and current data. They can thus spot patterns, forecast results, and streamline industrial procedures. Data integration in factory optimization frameworks promotes continuous monitoring, adaptive control, minimized material waste, optimal machining settings, and well-informed decision-making. Through process optimization and alignment with sustainability goals, these innovations improve operational efficiency, reduce expenses, and advance sustainable manufacturing practices [Chen et al. (2018)].

Algorithm for **Methods for Incorporating Data into the Optimization Framework**

Step 1: Data Collection and Preprocessing

- When IoT sensors are installed on CNC machines, important factors like material consumption, temperature, vibration, and tool wear may be monitored in real time. Predictive maintenance is made possible by this, which lowers downtime and boosts equipment efficiency overall.
- Manufacturers should make use of cloud-based storage systems in order to efficiently gather and preserve real-time sensor data. This method guarantees accessible and effective data management.
- Important processes in getting data ready for analysis include preprocessing and data cleansing. To guarantee correctness and consistency, this entails eliminating noise and superfluous data points, normalizing the data, and addressing missing values via interpolation or imputation. This improves the quality and reliability of the data, allowing for deeper insights and better decision-making.

Step 2: Feature Engineering

- The depth of cut, feed rate, cutting speed, and machine uptime all affect material waste in manufacturing. These variables have a direct effect on the rates of material removal, tool forces, and processing time, all of which can lead to more scrap.
- In order to increase predictive capacity, think about developing derived features such as past waste patterns, average tool wear rate, and uptime-to-downtime

ratio. The accuracy and dependability of the model can be improved by these elements, which can offer insightful information on waste generation trends, tool replacement requirements, and maintenance efficiency.

Step 3: Machine Learning Model Training

- Material waste can be accurately predicted using machine learning methods like random forest, gradient boosting, and neural networks. To produce precise forecasts, these models examine elements such as material kinds, production procedures, and waste data from the past. Businesses can drastically cut waste and increase resource efficiency by putting these strategies into practice and improving their performance.

Step 4: Optimization Algorithms

- An objective function is created to minimize material waste, balancing waste reduction with limitations such as production schedules and machine availability. The best operational parameters are then determined by using optimization approaches like genetic algorithms and particle swarm optimization.

Step 5: Real-time Implementation

- Adaptive control allows real-time machining parameter adjustments by integrating machine learning model with CNC machine control system. This system continuously learns from fresh data, updating models and parameters in a feedback loop for continuous performance improvement. It also uses sensor data and model predictions to reduce waste.

Step 6: Visualization and Monitoring

- In addition to providing warnings and notifications for anomalies, dashboards that display real-time data, model predictions, and optimization outcomes enable operators to keep an eye on the process, make well-informed decisions, and quickly handle departures from ideal parameters.

As indicated in Table 7.3, the data was gathered from four CNC machines. The output graph represented by Figure 7.4 shows a favorable link between temperature and vibration (0.6), vibration and tool wear (0.8), and spindle speed and feed

TABLE 7.3
CNC Machine Operating Matrix

Machine ID	Temperature (°C)	Vibration (mm/s)	Spindle Speed (RPM)	Feed Rate (mm/min)	Tool Wear (mm)
1	55	1.2	2000	500	0.05
2	60	1.5	2100	520	0.07
3	58	1.3	1950	480	0.04
4	62	1.6	2200	530	0.06

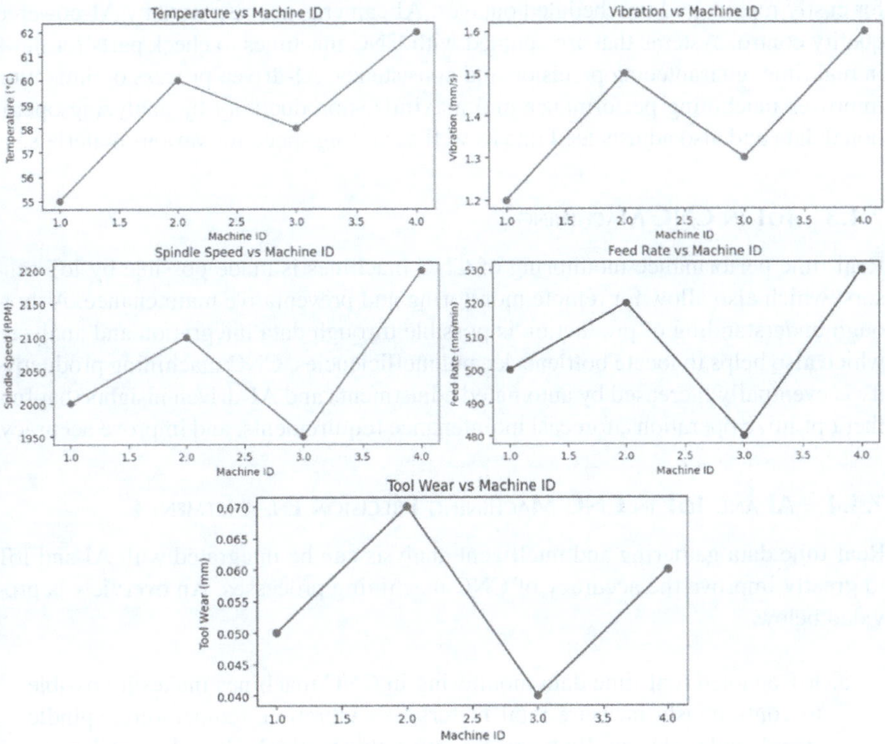

FIGURE 7.4 Various parameters versus Machine ID.

rate (0.7) is also revealed by the data analysis, along with trends and correlations between machine characteristics and machine ID. While higher feed rates are linked to quicker spindle speeds, higher heat and vibrations are linked to increased tool wear. By keeping an eye on these variables, machine performance can be improved and maintenance requirements can be anticipated. All things considered, the data indicates that monitoring temperature, vibration, and tool wear is crucial to preserving ideal machine conditions, and that associations among these variables can guide data-driven choices.

7.4.1 Operational Efficiency Improvements

CNC machining efficiency is increased by AI and IoT technologies through increased precision, less downtime, and improved production process optimization. Through the use of predictive analytics and real-time data, these technologies facilitate data-driven decision-making, which raises output and lowers waste.

7.4.2 AI in CNC Machining

AI systems facilitate predictive maintenance by evaluating data from CNC machines and projecting maintenance requirements based on past patterns. This reduces the need for costly repairs and unscheduled outages. AI cameras are also used by AI-powered quality control systems that are coupled with CNC machines to check parts for flaws in real time, guaranteeing precision and consistency. AI-driven process optimization improves machining performance and maximizes productivity by analyzing operational data and also adjusts feed rate as well as cutting speed for various materials.

7.4.3 IoT in CNC Machining

Real-time performance monitoring of CNC machines is made possible by IoT sensors, which also allow for remote monitoring and preventative maintenance. A thorough understanding of production is possible through data integration and analysis, which also helps to locate bottlenecks and inefficiencies. CNC machining productivity is eventually increased by automated adjustments and AI-driven insights that further optimize operations, forecast maintenance requirements, and improve accuracy.

7.4.4 AI and IoT in CNC Machining Precision Enhancement

Real-time data gathering and intelligent analysis can be integrated with AI and IoT to greatly improve the accuracy of CNC machining processes. An overview is provided below.

a. IoT-enabled real-time data monitoring in CNC machines makes it possible to continuously measure vital factors like vibration, temperature, spindle speed, and tool wear. By being proactive, this method allows for quick reactions to abnormalities, including lowering cutting speeds or modifying cooling systems when a temperature sensor identifies excessive heat production. In addition to reducing downtime, real-time monitoring greatly increases overall productivity by avoiding overheating and preserving accuracy.
b. In order to anticipate maintenance requirements, AI-driven predictive maintenance examines data from IoT sensors, averting unforeseen problems and guaranteeing accurate machining procedures. Timely tool replacement is made possible by this proactive strategy, which also boosts productivity and decreases downtime.
c. Using IoT sensor data and AI algorithms, adaptive control systems dynamically modify CNC machine parameters in real time. For example, on a CNC lathe, an AI system can dynamically adjust feed rates and cutting speeds to compensate for changes in material hardness or tool wear.
d. AI-powered visual inspection finds flaws or deviations from specifications, allowing for quick quality control. AI vision systems guarantee high-quality output by scanning CNC machine parts for dimensional accuracy and sending out notifications for user intervention or automatic corrections.
e. To optimize factors like feed rate and depth of cut, AI systems examine past machining data. Through the analysis of data from multiple operations with different materials and equipment, the AI system determines the ideal feed rate for every combination. This method guarantees the perfect balance between precision and machining speed, leading to far better efficiency and performance.

The combination of AI and IoT makes CNC machining far more accurate and predictive. Optimal performance is ensured by real-time monitoring of crucial parameters, which enables prompt modifications. Additionally, adaptive control systems that are powered by AI algorithms and data from IoT sensors allow the machine to dynamically adjust for things like tool wear and material variances, resulting in continuously high accuracy and fewer errors.

7.4.5 Energy Consumption Reduction Metrics

AI-driven CNC analysis and IoT technology optimize energy consumption in CNC machining, promoting sustainability and innovation. Real-time tracking and data-driven insights reduce energy waste and encourage innovative and efficient production methods. This cooperation changes the industry and ensures a more creative and sustainable future [Santos et al. (2017)]. AI algorithms and IoT sensors help CNC machines become more energy efficient, reducing costs and waste without compromising productivity. Energy savings are significant because opportunities for improvement can be promptly identified through real-time monitoring and analysis [Liu et al. (2018)]. After implementing AI-powered predictive maintenance, a CNC machine manufacturer saw a 15% reduction in energy use and a 20% decrease in downtime, which increased output and profitability [Fernández-Caramés and Fraga-Lamas (2018)].

Operators can adjust settings and reduce energy waste by using IoT sensors to identify idle times of CNC machines. As seen in CNC milling, where energy use is regulated, this data-driven strategy enables operators to make knowledgeable choices that maximize energy efficiency and reduce waste [Salama & Al-Turjman (2025)]. AI-driven analysis optimizes cutting parameters in CNC milling, reducing energy consumption by 12% without compromising product quality. This innovative strategy aids enterprises in finding a balance between production and sustainability by promoting eco-friendly practices and fostering innovation.

7.4.6 Waste Reduction and Sustainability

The use of AI and IoT in CNC analysis has transformed production by reducing waste and boosting sustainability. Precision and efficiency, anomaly detection, maintenance prediction, and parameter optimization – all made possible by real-time monitoring and optimization – all contribute to significant material and energy savings.

The integration of AI and IoT in CNC analysis enables real-time monitoring and optimization, supporting eco-friendly production practices. When IoT sensors detect inefficiencies, AI-powered alerts and improvement recommendations are generated [Kasiviswanathan et al. (2024)]. Waste production is reduced and the environmental effect of manufacturing operations is lessened thanks to this proactive approach. AI systems may also analyze historical data to identify trends and possible areas for improvement, further bolstering sustainability initiatives.

Example:

Sustainable CNC Machining in Aerospace Manufacturing

The use of AI- and IoT-driven CNC analysis by an Indian aerospace manufacturing company resulted in a significant decrease of 22% in material waste, 15% in energy consumption, and 30% in machine downtime. An illustration of how AI and the IoT might transform the encouragement of ecologically conscious conduct [García et al. (2020)].

The industrial industry is undergoing a change thanks to the incorporation of AI into CNC machining. Manufacturers may drastically reduce waste and energy usage in their machining operations by utilizing AI-driven insights. A key component of this process is predictive analytics, which skillfully modifies important cutting parameters like feed rate and spindle speed to get the best results. Additionally, sophisticated AI algorithms delve deeper, revealing hidden opportunities for enhancement and enabling producers to optimize their workflows, which eventually results in increased output, lower expenses, and a more sustainable future.

By optimizing cutting parameters through AI-driven CNC analysis, material waste and energy usage are greatly decreased. AI algorithms forecast the ideal feed rates and spindle speeds for every machining process by evaluating input from IoT sensors. Energy efficiency is increased and waste production is reduced with this data-driven strategy. This method led to an impressive 18% reduction in energy consumption and a 25% reduction in material waste for a precision components firm.

Example:

Predictive Maintenance in CNC Machining

An AI-powered CNC analysis was used by an aircraft component maker to find irregularities in the machining process. AI algorithms examined data to forecast possible problems, while IoT sensors tracked machine performance. The technology prevented waste and guaranteed sustainable practices by initiating a prompt remedial action when an abnormality was detected. This proactive strategy decreased energy consumption by 12% and machine downtime by 40%, showcasing the ability of AI-powered CNC analysis to promote environmentally friendly production techniques.

Example:

Sustainable CNC Machining in the Automotive Industry

A well-known Indian automaker serves as an example of how AI- and IoT-based CNC analysis can be used to support sustainable operations. They were able to reduce material waste by 18% and energy usage by 12% by utilizing AI-driven predictive maintenance and IoT sensor monitoring. This success story demonstrates how CNC analysis powered by AI and IoT can promote sustainability in CNC machining, allowing firms to reduce their environmental footprint while preserving efficiency and productivity.

7.4.7 Material Waste Minimization Strategies

CNC analysis is undergoing a revolution in material waste reduction due to the confluence of AI and the IoT. AI and IoT sensors are combined to optimize machining processes, which significantly reduces waste and increases productivity. Because AI-driven predictive maintenance analyzes machine performance data, finds anomalies, and allows for prompt interventions, it significantly reduces material waste. Furthermore, AI algorithms can optimize machining processes in real-time thanks to IoT-enabled real-time monitoring of crucial factors including material feed rates, machine axis alignment, and cutting tool temperature and vibration. In the end, this changes the industrial landscape by resulting in optimal cutting tool performance, less material waste, increased machine efficiency, and intelligent material management.

Intelligent material management is also made possible by the integration of AI and IoT. AI algorithms can optimize material utilization by evaluating production schedules, material inventories, and machining requirements. This reduces the need for unnecessary material procurement and the waste that comes with it. IoT sensors can also monitor the flow of materials, identifying possible obstructions and allowing for preventative actions to reduce material waste.

Real-time monitoring has been transformed by the incorporation of IoT sensors into CNC machines, which has also greatly increased machining efficiency and reduced material waste. These sophisticated sensors carefully measure variables like vibration, temperature, and sound emissions in order to track important characteristics, including cutting tool wear. This makes it possible to identify wear and tear early, which speeds up maintenance and minimizes downtime. In addition, materials are fed into the machine at optimal speeds thanks to IoT sensors, which improve material feed rates and reduce waste. In order to avoid overheating, which can harm tools and lead to the production of scrap, machining temperatures are also regularly monitored. Manufacturers may cut expenses, increase production, and streamline processes by utilizing IoT sensors.

A revolutionary method for optimizing machining parameters has emerged as a result of the collaboration of IoT sensors and AI algorithms. AI-driven systems can instantly spot optimization opportunities and analyze intricate patterns by utilizing the massive volumes of data produced by IoT sensors. In order to reduce material waste during cutting operations, this makes it possible to dynamically modify machining parameters like feed rates, depths, and cutting tool speeds. AI systems can also anticipate and stop tool failure, which lowers downtime and related expenses. AI-driven solutions may maximize productivity, reduce waste, and improve product quality by optimizing machining operations through ongoing learning and improvement. This closed-loop strategy, in which real-time modifications are informed by AI-driven data, has the potential to completely transform the manufacturing industry and allow businesses to embrace more efficient and sustainable production methods.

Material utilization and procurement are revolutionized by the combination of inventory management systems and IoT-enabled CNC machines. Manufacturers can precisely measure material use, optimize procurement, and reduce excess waste by utilizing real-time data from IoT devices. Because of this smooth connection, just-in-time material ordering is possible, which lowers the expense of keeping goods on hand and saves the environment. Manufacturers are able to take a more comprehensive and sustainable approach to production as a result of the streamlined operations, increased efficiency, and substantial cost savings.

7.4.7.1 Case Study: Automotive Component Manufacturing

Through the use of AI and IoT into CNC analysis, manufacturers may enhance sustainability and drastically cut down on material waste. IoT sensors provide real-time data that AI algorithms may use to optimize machining operations, reducing mistakes and increasing material use. This proactive strategy lessens the negative effects of manufacturing activities on the environment while simultaneously cutting expenses.

7.4.8 Alignment With Sustainable Manufacturing Goals

The manufacturing industries are revolutionized with the integration of AI and IoT in CNC machining that helps to optimize real-time processes, lowers waste, energy consumption and improves product quality. By anticipating possible malfunctions, AI-powered CNC analysis and predictive maintenance save downtime and energy waste, and IoT sensors offer real-time energy consumption data, enabling optimum energy use and a smaller carbon footprint. Additionally, AI-based CNC analysis improves resource efficiency and minimizes material waste by optimizing cutting tools and machining parameters. According to studies, AI-powered CNC optimization can cut material waste by 25% and energy consumption by 30%, indicating its potential for sustainable and profitable manufacturing. Siemens and other companies have seen notable drops in both material waste and energy expenses.

7.4.9 Predictive Maintenance and Adaptive Control

In the field of CNC machining, the combination of AI and the IoT has completely changed the game. Through the smooth integration of AI algorithms with real-time data streams from IoT sensors integrated into CNC machines, manufacturers may achieve previously unheard-of levels of automation, efficiency, and accuracy. The optimization of machining parameters, preventative equipment maintenance, and a notable increase in overall operational efficiency is made possible by this potent combination.

IoT sensor data is used by AI-powered predictive maintenance to continuously check the condition and functionality of CNC machines. AI systems can anticipate possible equipment breakdowns before they happen by examining patterns and spotting abnormalities in real time. Manufacturers may reduce expensive downtime and guarantee smooth production cycles by utilizing AI-driven predictive maintenance. Timely maintenance is made possible by this proactive approach, and adaptive control systems optimize vital machine parameters like feed rate and cutting speed in real time by using real-time IoT sensor feedback. This dynamic optimization increases machining precision, prolongs the life of expensive cutting tools, and guarantees optimal performance under a variety of circumstances.

7.4.9.1 Aerospace Manufacturing: A Case Study

The use of AI and IoT technology has transformed production processes in the rigorous field of aerospace manufacturing, where accuracy and dependability are critical. Through the use of AI-powered predictive maintenance, manufacturers can keep an eye on the condition of CNC machines in advance, predicting possible malfunctions and planning repair for when output is low. This proactive strategy ensures steady and uninterrupted production flows while reducing expensive downtime. Additionally, producers can now dynamically modify machining parameters in real time by implementing IoT-enabled adaptive control systems. These systems look for departures from ideal circumstances by continuously analyzing sensor data, including temperature, vibration levels, and tool wear. Through rapid adjustments to cutting speeds, feed rates, and other crucial parameters, these systems guarantee outstanding accuracy and continuously satisfy the exacting standards for quality required by the aerospace sector.

There are several advantages to this technological collaboration. In addition to minimizing unscheduled downtime, predictive maintenance optimizes maintenance schedules, lowers labor costs, and increases the overall performance of equipment. Manufacturers may prolong the life of their priceless equipment and further cut capital expenditures by proactively resolving possible problems. Additionally, combining AI and IoT encourages a data-driven manufacturing strategy that offers insightful information on production procedures. By analyzing this data, bottlenecks may be found, procedures can be improved, and total efficiency can be consistently raised. Finally, the seamless integration of AI and IoT technology enables aircraft producers to attain previously unheard-of levels of accuracy, dependability, and productivity, spurring innovation and competition in this vital industry.

Manufacturing has significantly improved as a result of the application of AI and IoT in CNC machining. Predictive maintenance and adaptive control are now considerably more successful thanks to these cutting-edge technologies, which have improved industry operations and reduced costs. Machines can now anticipate when they need maintenance thanks to AI and IoT, which lowers the likelihood of unplanned malfunctions. In order to preserve high-quality output, they can also automatically modify their settings. As a result, factories can decrease costs, increase productivity, and create superior goods. Because of this, manufacturers who use these technologies are better able to compete in the competitive market of today while still maintaining high standards of production.

7.4.10 Energy-efficient Machining Strategies

The integration of AI and IoT within CNC machining has transformed factory floors by enabling more efficient and energy-conscious operations. IoT sensors constantly monitor crucial performance indicators of CNC machines, creating a continuous stream of data. This data is carefully examined by AI algorithms to find trends, abnormalities, and areas for improvement. Manufacturers may dynamically modify machining parameters, including feed rates, depths, and cutting speeds, thanks to this real-time data analysis, which reduces energy usage while maintaining strict quality requirements. AI also optimizes toolpaths by cutting down on idle time and energy-intensive movements. AI predicts possible machine faults by using machine learning for predictive maintenance. This enables proactive maintenance scheduling and drastically lowers expensive downtime. This AI and IoT synergy promotes a more effective and sustainable production process, reducing waste and optimizing energy savings throughout the manufacturing environment.

Example:

Aerospace Industry

The industrial sector, especially the aerospace industry, is changing as a result of the incorporation of AI and the IoT into CNC machining. Manufacturers may minimize their environmental impact and reduce energy usage by optimizing machining operations through the use of real-time data from IoT sensors and AI-driven predictive analytics. An AI-powered analysis of IoT sensor data from CNC machines, for example, allowed a manufacturer of aircraft engine parts to reduce energy consumption by 15% without sacrificing product quality. Significant energy savings and a decrease in waste generation are made possible by this synergy, which makes energy-efficient machining procedures possible. The future of sustainable production will be greatly influenced by AI- and IoT-based CNC analysis as the manufacturing sector develops further. This will enable manufacturers to simplify operations, anticipate maintenance needs, and distribute resources more effectively, ultimately fostering a more ecologically conscious sector and opening the door to a more sustainable future.

7.5 CONCLUSION AND FUTURE SCOPE

This chapter combines CNC machining with AI and IoT to present a revolutionary method to sustainable production. The framework ushers in a new era of intelligent and environmentally conscious manufacturing by dramatically improving production efficiency and sustainability parameters through the use of IoT sensors to gather real-time data and AI for optimization.

Manufacturing has changed as a result of the incorporation of cutting-edge technology including adaptive control, predictive maintenance, and energy-efficient machining methods. A more sustainable and effective manufacturing future is now possible for enterprises by utilizing CNC machining, AI, and IoT to optimize processes, cut waste, and improve overall performance. Hence, the manufacturing industry is at an exciting moment due to this creative approach, which highlights the importance of using cutting-edge technologies to accomplish sustainable production.

The future of this work is bright and full of opportunities for investigation. The efficiency and sustainability of CNC machining can be improved by incorporating cutting-edge technology like blockchain and augmented reality. Furthermore, these manufacturing industries could undergo a revolution if this AI-driven framework is applied to robotics and 3D printing. Predictive maintenance and remote optimization would be transformed by establishing a cloud-based platform for real-time monitoring and control, which would result in notable increases in productivity and decreased downtime. Furthermore, investigating renewable energy sources to power

CNC machines can lower their carbon footprint, so contributing to global sustainability objectives and laying the groundwork for a more environmentally friendly future. A more innovative and sustainable industrial landscape will result from following these paths.

REFERENCES

Chen, Xingzheng, Congbo Li, Ying Tang, and Qinge Xiao. "An Internet of Things Based Energy Efficiency Monitoring and Management System for Machining Workshop." Journal of Cleaner Production 199 (2018): 957–968.

Fernández-Caramés, Tiago M., and Paula Fraga-Lamas. "A Review on the Application of AI to the IoT." IEEE Access 6 (2018): 24379–24393.

García, Eva, et al. "IoT-Based Monitoring and Control of CNC Machines for Sustainable Production." Sensors 20, no. 4 (2020): 1207.

Gharbia, Khaled, and Fawzi Benammour. "Application of AI in Predictive Maintenance and Quality Control in CNC Machining." Procedia CIRP 72 (2018): 212–217.

Jin, Xiaoning, Jun Ni, and Samuel B. Murray. "A Digital Twin Approach for Cutting Force Control in CNC Machining." Manufacturing Letters 8 (2016): 5–8.

Kasiviswanathan, S., Gnanasekaran, S., Thangamuthu, M., & Rakkiyannan, J. "Machine-learning-and Internet-of-Things-driven techniques for monitoring tool wear in machining process: A comprehensive review." Journal of Sensor and Actuator Networks 13, no. 5 (2024): 53.

Kaur, Rupinder, Raman Kumar, and Himanshu Aggarwal. "Role of Machine Learning in Sustainable Manufacturing Practices: An Outline." In *International Conference on Intelligent Manufacturing and Energy Sustainability*, pp. 541–551. Singapore: Springer Nature Singapore, 2023.

Khan, Sohail Imran, Chamandeep Kaur, Mohammed Saleh Al Ansari, Iskandar Muda, Ricardo Fernando Cosio Borda, and B. Kiran Bala. "Implementation of Cloud Based IoT Technology in Manufacturing Industry for Smart Control of Manufacturing Process." International Journal on Interactive Design and Manufacturing (IJIDeM) 19, no. 2 (2023): 773–785.

Laghari, Rashid Ali, and Samir Mekid. "Comprehensive Approach Toward IIoT Based Condition Monitoring of Machining Processes." Measurement 217 (2023): 113004.

Lee, Jay, Behrad Bagheri, and Hung-An Kao. "A Cyber-Physical Systems Architecture for Industry 4.0-Based Manufacturing Systems." Manufacturing Letters 3, no. 1 (2015): 18–23.

Liu, Chen, Ronghua Hu, and Xiaobo Zhang. "Artificial Intelligence and Machine Learning for Smart Manufacturing." Journal of Manufacturing Systems 48 (2018): 248–260.

Lu, Yang. "Industry 4.0: A Survey on Technologies, Applications and Open Research Issues." Journal of Industrial Information Integration 6 (2017): 1–10.

Mejía-Moncayo, Camilo, Jean-Pierre Kenné, and Lucas A. Hof. "On the Development of a Smart Architecture for a Sustainable Manufacturing-Remanufacturing System: A Literature Review Approach." Computers & Industrial Engineering 180 (2023): 109282.

Müller, Benjamin, and Johannes Vom Brocke. "How AI and IoT Are Transforming CNC Machining for Sustainable Production." Journal of Manufacturing Systems 51 (2020): 204–215.

My, Chu Anh. "The Role of Big Data Analytics and AI in Smart Manufacturing: An Overview." *Research in Intelligent and Computing in Engineering: Select Proceedings of RICE 2020* (2021): 911–921. https://doi.org/10.1007/978-981-15-7527-3_87

Patalas-Maliszewska, Justyna, Hanna Łosyk, Ewa Dostatni, and Sławomir Kłos. "IoT-Based Monitoring the Level of Sustainable Production: A Case of Energy Consumption in Turning Process." In *International Scientific-Technical Conference MANUFACTURING*, pp. 213–222. Cham: Springer Nature Switzerland, 2024.

Qin, Jiang, Ying Liu, and Roger Grosvenor. "A Categorical Framework of Manufacturing for Industry 4.0 and Beyond." Procedia CIRP 52 (2016): 173–178.

Salama, Ramiz, and Al-Turjman, Fadi. "A novel design for 3D drawing machine using "computer numerical control" for high precision and accuracy." In *Quaternion-Based Sparse Image Processing* (pp. 311–324). Elsevier, 2025. https://www.sciencedirect.com/science/article/abs/pii/B9780443292446000235

Samanta, Hiranmoy, Pradip Kumar Talapatra, Kamal Golui, and Pritam Chakraborty. "IoT-based Intelligent Manufacturing System: A Review." In *Intelligent Manufacturing Management Systems: Operational Applications of Evolutionary Digital Technologies in Mechanical and Industrial Engineering*, pp. 59–84. Scrivener Publishing LLC, 2023

Santos, Marcio Y., Diogo A. R. da Costa, and Rui M. Lima. "A Framework for Big Data Analytics in Manufacturing Environments." Procedia CIRP 63 (2017): 117–122.

Smith, John D., and Alice R. Johnson. "Integration of AI and IoT in CNC Systems for Enhanced Efficiency." International Journal of Advanced Manufacturing Technology 97, no. 1–4 (2018): 1123–1137.

Sundaresan, Jithin, Michael Siek, and Samuel Mahatmaputra Tedjojuwono. "Sustainable Manufacturing Through Empowering Machine Tools With IoT Innovations: a Literature Review." In *IOP Conference Series: Earth and Environmental Science*, 1324, no. 1, p. 012088. IOP Publishing, 2024.

Wan, Jiafu, Houbing Song, Qingrong Zhang, and Changan Sun. "Industrial Internet of Things: Principles and Design." International Journal of Distributed Sensor Networks 13, no. 3 (2017): 1550147717696397.

Wang, Shiyong, Jiafu Wan, Di Li, and Chunhua Zhang. "Implementing Smart Factory of Industrie 4.0: An Outlook." International Journal of Distributed Sensor Networks 12, no. 1 (2016): 3159805.

Xu, Lihui, and Andrew Newman. "Smart Manufacturing Systems Based on Cyber-Physical Systems for IoT Environment." Procedia CIRP 63 (2017): 235–240.

Zhang, Yong, Youyun Zhang, and Yu Zhou. "A Framework for Smart Production-Logistics Systems Based on CPS and Industrial IoT." IEEE Access 6 (2018): 12252–12264.

Zheng, Peng, Hai Wang, and Zhongjie Xu. "Smart Manufacturing and Intelligent Manufacturing: A Comparative Review." Engineering 3, no. 5 (2017): 613–620.

Zhou, Kai, Taigang Liu, and Lifeng Zhou. "Industry 4.0: Towards Future Industrial Opportunities and Challenges." In *2015 12th International Conference on Fuzzy Systems and Knowledge Discovery (FSKD)*, pp. 2147–2152. IEEE, 2015.

8 Recent Advancements in AIoT for Predicting GFRP-Properties

Alak Kumar Patra and Arun Kumar Singh

8.1 INTRODUCTION

Glass fiber-reinforced polymer (GFRP) composites are widely used (Morampudi et al. 2021) because of their high strength, low weight, and ability to withstand corrosion in a variety of sectors for improving their reliability and performance in different applications, including aerospace, automotive, and civil engineering (Sousa et al. 2014). Recent advancements in general-purpose AI for the Artificial Intelligence of Things (AIoT) have revolutionized how we predict and optimize these features compared to what was possible with unprecedented accuracy, soon enough processing speed (Baduge et al. 2022). This introduction gives a great summary of recent action in AIoT when utilized to forecast the mechanical break and thermal attributes of GFRP composites.

So, in the modern-day society GFRP composites are of great importance. GFRP composites have unique combination of properties, that is, high strength-to-weight ratio, corrosion resistance, and design flexibility which are needed for modern engineering applications. Nonetheless, due to the intricate relationships between material composition and production method (and exposure conditions as aforementioned), their mechanical fracture and thermal behavior cannot be reliably predicted (Broughton et al. 2011).

Artificial Intelligence of Things (AIoT) is the integration of AI technologies with Internet of Things (IoT) infrastructures and devices (Zhang and Tao 2020). Broader AIoT can get real-time data collection, processing and decision-making, with the power of IoT sensors and AI algorithms. This enables faster, more reliable, and smarter performances in a variety of fields.

Recently, AIoT systems dedicated for predicting mechanical fracture and thermal properties of fiber-reinforced polymer composites (Qiu and Yang 2022) have made great progress. Prediction models for GFRP composites are increasingly being formulated using various approaches including prediction analytics, machine learning (ML), and deep leaning as well as other relevant AI methods. Using extensive datasets on manufacturing properties, material data, and environmental conditions, these models can predict how heat will behave and how fracture will occur mechanically. Data-driven approaches are combined with physics-based simulations in AIoT systems to interface with FEA tools. This integration allows more comprehensive predictions by capturing the intricate relationships among material microstructures, external loads, and temperature gradients.

DOI: 10.1201/9781003503323-8

IoT sensors provide real-time data on mechanical stresses, temperature variations, and environmental aspects (Rocha et al. 2021) of GFRP structures. Using this data, AI systems are able to predict structural health and anticipate problems before they arise, making proactive upkeep and safety management a more straightforward process.

Digital twins (Gao et al. 2023) enabled such monitoring of a series of GFRP structures in real time – virtual replicas at the modeling level of actual GFRP structures is feasible by AIoT. Real-time sensor data and simulation results, ready to synchronize from continuous updates ensure accurate thermal material models and simulation of mechanical fracture under all operating conditions.

Materials informatics guided by AI (Gomez et al. 2022) employ big data analytics and ML approaches to expedite the identification and optimization of new GFRP composites. They use databases of material properties and composition to predict the most appropriate mechanical and thermal behavior for advanced materials in a wide range of application areas.

AIoT Technologies Convergence for Mechanical Properties, Thermal Stability and Durability of GFRP Composites – Virtualization Toward Actualization (Baduge et al. 2022). Through accurate predictions and insights into material behavior, AIoT helps engineers and researchers design safer structures which would be less prone to failures due to reduced operating conditions reducing product development time and costs.

Current developments in AIoT have revolutionized the prediction of mechanical fracture and thermal properties of GFRP composites (Azad et al. 2024). These new semiconductors will offer scientists a fundamentally unexpected toolbox in 2024 to guide and improve their performance across a range of different applications. This will provide a huge platform for creating innovative sustainable solutions in several domains and this marks progress as we proceed toward Material Science Engineering.

Recent Advances in AIoT (Bibri and Jagatheesaperumal 2023, Chang et al. 2021; Huang et al. 2023; Rehman et al. 2024, Shen et al. 2023, Shi et al. 2021) have played a vital role in improving the predictive accuracy and efficiency for these properties. Table 8.1 mentions a few of the recent notable advancements that are discussed later as well.

8.2 MACHINE LEARNING MODELS

Similarly, ML almost fit all types of mechanical fracture parameters (tensile strength, flexural strength, and fracture toughness) of GFRP composites by AI approach for predictive analysis. The algorithms that have been utilized are Random Forest, Support Vector Machines, and Neural Networks. These models use manufacturing features and material properties from big datasets to make accurate predictions.

ML models are the heart and soul of AIoT systems, which use these models to process and analyze data from IoT devices, make decisions, and automate a broad range of tasks.

The applications of ML models have been made to investigate large-scale data related to material properties, production parameters, and environmental variables

TABLE 8.1
Remarkable Improvements of AIoT

Sl. No.	Advancements in AIoT	Key Features	Remarks
1.	Machine Learning Models	Flexibility in Response to Varying Data, Making Decisions in Real Time, Ongoing Education and Adjustment and Interoperability	Together, these characteristics allow AIoT machine learning models to improve the intelligence, effectiveness, and dependability of IoT systems across a variety of sectors and applications.
2.	Finite Element Analysis (FEA) Integration	Modeling of intricate structures, forecasting their behaviors, real-time monitoring, data-driven decision-making	The total efficacy and efficiency of structural health monitoring systems in guaranteeing the dependability and safety of vital infrastructure are improved by this integration.
3.	Sensor Integration and IoT Connectivity	Real-time information on mechanical loads, temperatures, and other environmental parameters by IoT sensors embedded in GFRP constructions	Data is analyzed with AI algorithms so that predictions on how the health and performance of a structure will change in time can be made, making preventive maintenances allowing catastrophic breakdowns to be avoided.
4.	Digital Twin Technology	Virtual copies of actual GFRP structures, optimization of maintenance schedules and material selection	This digital twin can forecast mechanical fracture and thermal qualities under various operating situations with real-time sensor data and simulation.
5.	Materials Informatics	Big data, machine learning, AI-driven materials informatics, discovery, development of new materials for particular usage.	It is possible by predicting more accurate mechanical and thermal properties of novel GFRP composites from AI-driven large databases of properties and composition.
6.	Multiscale Modeling	Modeling at different length scales, incorporating data from microstructural aspects to macroscopic behavior by AIoT	More accurate forecasts of mechanical fracture and thermal characteristics in GFRPs are possible by this AI-assisted modeling.

for the prediction of the mechanical fracture and thermal properties of GFRP composites. Several AIoT integration standard ML models that are popular on the market currently including:

8.2.1 Regression Models

Tensile strength, flexural strength, fracture toughness, and thermal conductivity of GFRP composites are some more of the output properties that can be predicted widely through regression models. The relationships between input data (like material composition and processing parameters) and output attributes are developed through the techniques like polynomial, linear, Support Vector Regression (SVR) regression, etc.

8.2.2 Classification Models

We will develop classification models based on input features, to predict discrete outcomes (material types or failure modes [e.g., delamination, fiber breakage]). Algorithms such as logistic regression, decision tree, random forest, and neural network can be used to classify data into predetermined categories (problem diagnosis, quality control, or material identification) in GFRP manufacturing processes.

8.2.3 Deep Learning Models

Deep learning is a subset of ML, which involves training multi-layered neural networks to extract complex patterns and deep features from data. Sequence prediction, time-series analysis, and image recognition in AIoT applications on GFRP composites mainly employ Convolutional Neural Networks (CNNs), Recurrent Neural Networks (RNNs), and Long Short-Term Memory (LSTM) networks. Deep learning algorithms are very powerful tools for analyzing sensor data and images from GFRP structures as they are good at capturing high-dimensional, complex correlations in the data which is suitable for our research.

8.2.4 Ensemble Methods

Ensemble methods are designed to combine multiple ML models to increase prediction accuracy and robustness. For instance, bagging (Random Forest) and boosting (AdaBoost, Gradient Boosting) are methods that combine predictions from several base learners to perform better than the individual learners. The results shown that ensemble methods are a good way to cope with the noise and overcome overfitting problems of AIoT systems used in mechanical fracture and thermal properties prediction of GFRP composites.

8.2.5 Anomaly Detection Models

Anomaly detection models that detect abnormal behavior in sensor data can be used to detect deficiencies and/or anomalies in GFRP structures. Anomaly detection models (Isolation Forest, One-Class SVM, etc.) are used in supervised, semi-supervised, and unsupervised settings to detect outliers in sensor readings allowing for prognostic maintenance which improves structural reliability.

8.2.6 Transfer Learning

Transfer learning is a strategy to adapt a pre-trained ML model to address a specific task or domain with only limited data by utilizing big datasets. Transfer learning helps to speed up model training, and improve generalization performance, even in cases when labeled data for GFRP composite is scarce by utilizing knowledge from other related domains. In summary, machine loads are the heart export model of AIoT systems that anticipate the mechanical fracture and thermal properties of GFRP composites, which makes it possible to directly interrogate, model and control material-level phenomena across a range of applications. By leveraging numerous algorithms and methodologies designed to extract meaningful information from data collected by IoT sensors, these models enhance the safety, reliability, and efficiency of GFRP structures in a multitude of sectors.

8.3 INTEGRATION OF FINITE ELEMENT ANALYSIS (FEA)

The AIoT system could be integrated with finite element analysis (FEA) software for predicting mechanical properties (Sun et al. 2024). These approaches enable more accurate predictions that account for the complex coupling of material microstructures with external loading conditions by unifying physics-based simulations with data-driven ML algorithms.

By combining FEA with AIoT, many benefits accrue, especially in design optimization, predictive maintenance, and structural health monitoring. Here are a few ways how FEA can be integrated into AIoT.

8.3.1 Data Acquisition

For factors such as loads, boundary conditions, material properties, AIoT systems use sensors embedded in equipment or building to capture real-time data on that impact the operation of system.

8.3.2 Data Preprocessing

AI algorithms are used for preprocessing the raw sensor data collected before feeding them into the FEA model. Feature extraction, data normalization, and data cleaning – this is where the data gets prepared to be analyzed.

8.3.3 Modeling and Simulation

The FEA models are generated according to the system's design geometry and material properties. There are a lot of different methods to collect real-world data.

AI approaches can further refine these models automatically by altering the model parameters to have a best fit to real-world data. This operation reduces the computational cost and increases the simulation precision.

8.3.4 Predictive Maintenance

In predictive maintenance, the failure or need of work is predicted in advance using majorly historical data combined with AI algorithm analysis of FEA simulation results. Prepare the maintenance work, and knowledge whether or not the system has a failure by AIoT will benefit to save inspection fee.

8.3.5 Design Optimization

Design Optimization FEA along with AI can be used to optimize product designs by analyzing different configurations iteratively and picking the ones that satisfy certain performance criteria. This is called generative design, and it can produce structures or parts that are lighter and stronger than those achievable through traditional manufacturing.

8.3.6 Real-time Monitoring

The AIoT systems constantly monitor the health of the building by evaluating sensor data in real-time. A departure from normal is identified in behavior and an alarm goes off, so failure can be averted before it reaches the catastrophic level.

8.3.7 Edge Computing

To minimize latency and speed up response times, AIoT systems may be able to conduct data processing and analysis at the edge (near the source of data) rather than relying on cloud-centric solutions alone. This is very good approach to literal time application like and structural dynamic monitoring.

8.3.8 Incorporation of Digital Twins

Digital twins are the concept of virtual image part to model real-world systems or assets. Featuring FEA and AIoT, real-time data integration improvements can help to improve digital twins by allowing for more accurate estimations and even as a predictive measure. Digital twins are mentioned as computerized iterations of a physical systems, asset or operations and utilize historical and real-time data to mimic its functioning in the actual world. By including FEA, digital twins can accurately represent the mechanical and structural characteristics of physical systems, offering detailed numerical simulations of how structures behave in different conditions. FEA integration with digital twins facilitates complete analytics enabling data fusion from numerous sources (sensors, historical data, and simulation). Sophisticated AI algorithms can subsequently analyze this combined data and generate actionable insights toward achieving end-to-end operational reliability and efficiency. SHM – Additional value is added with AIoT technology, leveraging FEA and digital twins to increase the accuracy and predictive power of a Structural Health Monitoring System including advanced proactive maintenance implementation throughout its lifecycle for optimized asset performance lifecycle decision-making. To keep our critical infrastructure and industrial assets safe, reliable, and sustainable requires an integrated solution. In Table 8.2, the possibilities of integrating FEA with in AIoT are summarized.

In summary, the integration of FEA with AIoT help in providing improved decision-making power, enhanced operational efficiency, and reliable system performance and security.

TABLE 8.2
Advantages of Integrating FEA with AIoT

Sl No.	The Items of FEA and AIoT	The Procedure of Integration of FEA and AIoT Features
1.	Data Acquisition	FEA items: Information on loads, boundary conditions, material properties and other parameters. AIoT feature and integration: **Sensors** embedded into equipment or buildings can continuously collect real-time data about what influences the system-behavior.
2.	Data Preprocessing	FEA item(s): Data fit for FEA. AIoT feature and integration: **AI algorithms** pre-process the raw data collected by sensors before being supplied into the FEA model. In this stage, feature extraction, normalization, and data cleaning are performed to ensure that the data is prepared for analysis.
3.	Modeling and Simulation	FEA item(s): System's geometry and material parameters. AIoT feature and integration: **AI techniques** have the potential to improve these models and yield the greatest fit for real-world data by autonomously changing parameters. This process increases simulation accuracy while reducing computational expenses.
4.	Predictive maintenance	FEA item(s): Output of FEA simulations. AIoT feature and integration: **AI algorithms** can use historical data and output of FEA simulations to predict potential anomalies in the system's behavior to anticipate maintenance needs reducing repairing costs.
5.	Design Optimization	FEA item(s): Optimized or Generative product design. AIoT feature and integration: **Repeated evaluation** can be performed by FEA and AI used together to choose one from many configurations that satisfy predetermined performance standards resulting in the creation of stronger, lighter, and more effective components or structures.
6.	Real-time Monitoring	FEA item(s): Deviation from usual pattern of behavior. AIoT feature and integration: **AIoT systems** continuously monitor the structural health by analyzing sensor data in real-time enabling prompt intervention to prevent catastrophic failures.
7.	Edge Computing	FEA item(s): FEA results of dynamic systems. AIoT feature and integration: **Edge Computing** by the AIoT systems or processing and analyzing data closer to the data source without exclusive dependence on cloud-based solutions improves responsiveness and decreases latency. Applications needing in-the-moment decision-making, such as structural monitoring in dynamic situations, benefit greatly from this methodology.
8.	Integration with Digital Twins	FEA item(s): FEA-based prediction on digital twins, AIoT feature and integration: **Digital twins** can be improved with real-time data through the integration of FEA and AIoT, enabling more exact simulations and predictive capabilities.

8.4 IoT CONNECTIVITY AND SENSOR INTEGRATION

Real-time data on mechanical stresses, temperatures, and other environmental parameters can be obtained from IoT sensors installed in GFRP constructions (Cheruvu et al. 2020). With the use of AI algorithms, this data is analyzed to forecast changes in structure health and performance over time, allowing for preventive maintenance and averting catastrophic breakdowns.

This data is examined with AI algorithms to model the evolution of structural health and performance over time, predicting when they will fail before they do so and can cause catastrophic failure. Given the aforementioned limitations, the purpose here is that there are many benefits to analyzing GFRP from a sensor-equipped intelligent perspective on mechanical fracture and thermal behavior using IoT connectivity. Next, we provide an overview of how this combination can enhance GFRP material inspection.

8.4.1 Real-time Monitoring

They can accommodate sensors for real-time monitoring of mechanical loads, strains, and temperature into GFRP materials. In this way, you have continuous monitoring of the activity of the material under different conditions and you can see where it has started to show a weak point or any points of failure.

8.4.2 Data Collection

These sensors can measure a number of mechanical and thermal parameters such as tensile strength, modulus of elasticity, fracture toughness, and thermal conductivity. This data can be transmitted through IoT connectivity to a centralized database for further analysis or it can be stored in the edge.

8.4.3 Predictive Maintenance

IoT connectivity, through the examination of sensor data using ML algorithms allows predicting when GFRP component will need to replace or maintenance. Reducing maintenance costs and downtime associated with unanticipated failures.

8.4.4 Optimized Design

For data analysis and real-time monitoring, it is meaningful for the design of GFRP components. The structural design and material composition are optimized with empirical data to enhance mechanical performance and thermal stability, instead of relying solely on theoretical models.

8.4.5 Remote Monitoring and Control

Utilization of IoT connectivity can assist with remote control and monitoring GFRP structures, which is particularly advantageous for infrastructural, aerospace applications. Engineers can evaluate the condition and behavior of GFRP components remotely, supporting their interventions.

8.4.6 Safety and Reliability

This helps identify possible safety hazards before things fail catastrophically, by real-time tracking of mechanical fracture and thermal attributes. This increases reliability and security of GFRP systems and structures in general.

8.4.7 Environmental Monitoring

Over time environmental factors such as humidity, UV exposure, and chemical exposure may affect the mechanical and thermal properties of GFRP. Data can be recorded through IoT-enabled sensors as well. Knowledge of these environmental influences provides better long-term performance predictions and material durability assessments.

Using sensors and IoT connectivity in the mechanical fracture and thermal property analysis of GFRP materials has many benefits and facilitated remote operation, design improvements, predictive maintenance, real-time monitoring, and more innate safety and reliability.

8.5 DIGITAL TWIN TECHNOLOGY

AIoT will facilitate the creation of detailed digital twins – that is, a virtual representation of real GFRP constructions. On the one hand, these digital twins (Botín-Sanabria et al. 2022) allow for maintenance schedules and material selection to be optimized by staying always current on sensor data and simulations. It can then use this to predict the mechanical or thermal properties of a powertrain under different operating conditions.

The combination of digital twins technologies and AIoT could be groundbreaking for GFRP materials examined. How to do it is as follows:

8.5.1 Virtual Modeling

With the use of digital twin technology, real-world GFRP structures or components can be virtually modeled. Without the need for actual testing, engineers may study mechanical fracture and thermal qualities and simulate a variety of scenarios thanks to these digital twins that replicate the behavior of their real-world counterparts.

8.5.2 Real-time Monitoring and Feedback

8.5.2.1 Virtual Modeling

Digital twin technology allowed the virtual modeling of real-world GFRP structures or components. Through these digital twins – actions like mechanical fracture and thermal qualities can be studied and a broad array of scenarios can be simulated without conducting any actual test by mimicking the conduct of their physical counterparts.

8.5.3 Real-time Monitoring and Feedback

The ability to monitor variables like stress, strain, temperature, and environmental conditions in real time by using IoT sensors can be added to both the physical GFRP structures and their digital twins. Through this data analysis, AI systems can provide insight into the behavior of the material and predict potential failures or areas that should be watched closely beforehand.

8.5.4 Predictive Analytics

Using data obtained by digital twins and physical sensors, AI can predict the behavior and performance of GFRP materials in future scenarios. Predictive models detect these patterns and correlations from the data and can identify if there could be an impending mechanical failure or points to a shift in thermal characteristics. This allows for maintenance and risk reduction strategies to be delivered up front. Figure 8.1 presents an example where a skilled engineer is pictured in a modern control room and scrutinizing big data, which are projected onto screens using moving pictures and complex graphics. Even when it just sits there, bathed in the cool blue light of monitors from the future, making a nice place for an insightful decision to be made. The ICT/IoT Dz is the location at which customers are delivered with the end product from human discernment to cutting-edge tech, presenting a workspace in which real- and digital-worlds come together for a traffic of progress.

8.5.5 Optimization and Design Upgrade

Engineers can experiment with different material compositions, structural designs, and manufacturing processes using digital twins. The performance of these virtual

FIGURE 8.1 A skilled engineer is pictured in a modern control room scrutinizing big data projected onto screens.

prototypes can be used to analyze and optimized the GFRP components for better mechanical strength, thermal stability, and overall performance utilizing the AI algorithms. In use cases including offshore wind turbines or aircraft components when physical access to GFRP structures is limited, AIoT paired with digital twins enable remote diagnostics and troubleshooting. Engineers save time and resources by being able to remotely monitor the condition and state of GRFP assets without having to go on site themselves, readily detecting any issues upfront so these can be tackled accordingly.

8.5.6 Lifecycle Management

Digital twins record the full lifecycle of GFRP material from its production through to end-of-life. AI algorithms maximize the durability and efficiency of GFRP components by optimizing maintenance plans based on predicting the remaining useful life of these components as well as identifying trends and patterns in data.

8.5.7 Machine Learning for Improvement

By constantly resupplying the AI algorithms with data, you can train ML models to learn better over time – both those residing in digital twins and physical sensors. This iterative method may offer a way to increase the accuracy of predictions and decision-making concerning the study of GFRP materials. The application of digital twin and AIoT technologies allows to develop a comprehensive framework for GFRP material analysis as demonstrated in this chapter. In this framework, the real-time monitoring and predictive analytics should be involved for design optimization, remote diagnostics, and continuous improvement of GFRP structures/services in their lifecycle.

8.6 MATERIALS INFORMATICS

AI-driven materials informatics (Ramprasad et al. 2017) utilizes big data and ML strategies. This allows the design for new application-specific materials, by predicting mechanical and thermal properties of novel GFRP composites with high precision from extensive databases of material properties and composition. The science of materials informatics is growing fast – using data-driven methods to understand, predict, and create materials properties and functions. The utilization of AI models over GFRP is demonstrated with the application context in the field of AIoT where IoT structures converge with AI techniques for real-time monitoring and analysis of GFRP structures. And this is what it might look like in practice.

8.6.1 Data Collection

IoT sensors that are attached or incorporated into GFRP constructions provide continuous measurements of temperature, humidity, strain, vibration, and other characteristics as well. Then the data is forwarded to a central system for analysis. *Preprocessing of data:* The initial data collected from Io T sensors is routinely corrupt, so the first step to effective analytics process will reduce this noise, removing duplications or outliers, and should be normalized for later data mining.

8.6.2 Processing of Data

In addition to this, it confirms the conformance of data as consistency of result. The above stage is for ensuring that the input will meet the contraction if and only if key features regarding the health and function of GFRP structures are extracted from these pre-processed datasets. These could be patterns of strain, distributions of temperature, frequency spectra of vibrations, etc. Model Development AI methods, including ML and deep learning as well as symbolic AI, are employed to develop models that can predict. These models learn how to link input variables to specific outcomes, such as structural integrity, fatigue life or probability of failure, by using past data.

8.6.3 Predictive Analysis

They are then trained to forecast the behavior of GFRP structures in different scenarios or setups. For example, they may be able to predict impending failures by appreciating an environmental condition and level of stress.

8.6.4 Enable Real-Time Monitoring and Decision Support

An AIoT system monitors the status and conditions of GFRP structures in real time. It triggers alerts if an anomalous behavior is detected and corrective actions (such as reinforcement or maintenance) can be applied to prevent catastrophic failures.

8.6.5 Iterative Improvement

With more and more data collected and studied in due course, the AI models could be continuously updated to provide better prediction values to maintain GFRP structures more effectively. By leveraging AIoT and materials informatics, stakeholders can essentially balance the safety, reliability, and longevity of GFRP structures with an effective maintenance strategy that uses real-time data analytics to optimize operations in a predictive manner.

8.7 MULTISCALE MODELING

At the same time, the IoT allows for application-level models to consider material properties across length scales from macroscopic behavior to microstructure aspirations (Chang 2013; Jhon 2009). This will establish the basis for the development of predictive models that can predict mechanical fracture and thermal characteristics in fiber-reinforced polymer composites with higher accuracy by using data from several scales. Multiscale modeling, as used in AIoT and new GFRP analysis, combines computer models that operate at different length and time scales to provide a holistic understanding of the material, from the molecular level to the macrostructural level. Size scale of multiscale modeling and simulations are depicted in Table 8.3.

TABLE 8.3
Different Length Scales of Multiscale Modelling and Simulations

	Description	Different Length Scales		
Systems/Simulations	Distances/scales	Microscale	Mesoscale	Macroscale
		Nano (10^{-9}) to micro (10^{-6})	Micro (10^{-6}) to milli (10^{-3})	Milli (10^{-3}) to kilo (10^{3})
Molecular dynamics simulations	Interatomic distances	0.1–1 nm		
Quantum mechanical simulations	Atomic and subatomic scales	0.01–0.1 nm		
Material defects	Grain boundaries		0.1–10 μm	
Biological cells and organelles	Cell sizes		10–100 μm	
Continuum mechanics	Structural components			1 mm–1 m
Ecological systems	Population dynamics			1 m–100 km

A more comprehensive understanding of how microlevel phenomena contribute to macroscopic behaviors can be achieved by bridging scales. Models at different scales often require different computational techniques and resources due to varying levels of complexity and detail. Scale is important for modeling because microscale models must capture atomic interactions, while macroscale models focus on bulk properties.

Coupling Scales: Multiscale modeling frequently uses techniques like upscaling, which integrates microscale data to inform mesoscale models, and downscaling, which uses macroscale data to inform mesoscale models.

Validation: To guarantee correctness across scales, each scale's model needs to be verified against higher-fidelity models or experimental evidence. A brief outline of multiscale modeling is provided below:

8.7.1 Molecular Dynamics (MD)

At the smallest scale (i.e., at nm scales), MD simulations can provide insight into individual molecule behavior within the GFRP matrix and interaction with embedded glass fibers. This contributes toward understanding concepts like the mechanics of stress transfer, interfacial bonding, and polymer chain mobility.

8.7.2 Finite Element Analysis (FEA)

FEA is generally used for the analysis of the behavior of composite materials like GFRPs at the Mesoscale level in which all fibers, matrix, and interfaces are segregated. This entails modeling techniques such as resin curing, fiber orientation, and mechanical stress to predict properties including stiffness, strength, and failure modes. At the macroscopic level, problem-specific continuum mechanics models are used to replicate the overall response of fiber-reinforced polymer systems in different loading scenarios. This includes applying methods like boundary element or finite element analysis to examine failure processes, deformation, and structural integrity.

8.7.3 Data-Driven Models

To capture intricate correlations between input parameters (such as material composition, processing conditions) and output responses (such as mechanical properties, damage initiation), data-driven methodologies, such as ML, can be utilized in addition to physics-based models. These models can be used in conjunction with physics-based simulations, particularly when computationally expensive or difficult analytical solutions are involved.

8.7.4 Predictive Maintenance and Optimization

Stakeholders can create predictive maintenance plans for GFRP structures by combining multiscale models with AIoT systems. This entails employing IoT sensors to continually monitor structural health, interpreting the data using multiscale models, and making defensible decisions to maximize performance, avoid failures, and prolong service life.

Figure 8.2 shows a professional sits in front of bright, glowing screens that show a complex network of data nodes and analytics. The strategically placed monitors foster a collaborative environment where important information is shared and decisions are made in real time. The atmosphere is one of concentration and expertise as each person plays a vital role in ensuring the smooth operation of a sophisticated digital infrastructure.

FIGURE 8.2 Real-time monitoring of complex infrastructure with AIoT.

8.8 IMPLICATIONS OF THE STUDY

The chapter facilitates both the academicians as well as the professionals. The implications of studying the chapter for academicians and professionals are provided in separate sections below.

8.8.1 Implications for Academicians

Academicians now have unique option of further research on the mechanical fracture and thermal properties of GFRP composites with the recent developments in AIoT. Researchers may improve material design, spur innovation, and get a deeper knowledge of composite materials across a range of applications by utilizing AI-driven insights from IoT data.

8.8.1.1 Improved Forecasting Models

By combining AI methods (such as ML and deep learning) with IoT data from sensors implanted in GFRP composites, AIoT makes it possible to create sophisticated predictive models.

Through these models, researchers are able to accurately predict the thermal (heat capacity, heat of phase transition), thermal (thermal-distributed strengths and expansion coefficients), and mechanical (bending strength, tensile strength fracture, etc.) properties of GFRP composites according to different environment applications. *Data-driven insights:* These tools collect real-time IoT data from thousands of sensors about the behavior and performance of GFRP composites in wide-ranging operational conditions. AI systems analyze this data for patterns, links, and outliers conventional methods may overlook – providing academics insights into the complex relationships between material composition, structural design, environment conditions, and mechanical and thermal properties.

8.8.1.2 Enhanced Material Design

Application in GFRP composites to optimize the composition and design of the materials that are used in GFRP composites, AIoT can be a very helpful tool. Enormous datasets and simulation results can be used to identify the perfect combinations of glass fiber types, polymer matrices, and reinforcement orientation that will drastically enhance mechanical strength as well as thermal stability. This form of rapid iteration leads to swift advancements in the academic research and development process.

8.8.1.3 Simulation and Virtual Testing Solutions

The simulation and virtual testing of AIoT can be applied to GFRP composites under multi-loading and super aging environments. Academician can mimic fracture mechanics and thermal behavior without an overburden of experimental observations, which saves costs as well as time of research work. These simulations generate valuable insights of failure mechanisms, durability, or performance characteristics which is critical for applications in practical fields.

8.8.1.4 Multidisciplinary Cooperation

AIoT promotes interdisciplinary cooperation between data analysts, computer scientists, mechanical engineers, and materials scientists. In order to promote innovation and idea cross-pollination, academicians can work together to integrate skills in AI, IoT, materials science, and mechanical engineering.

8.8.1.5 Possibilities for Education

Enhancing student learning chances is achieved through the integration of AIoT technology into academic programs. Students can prepare for future professions in advanced materials research and engineering by gaining practical exposure with cutting-edge AI technologies and IoT devices.

8.8.2 Implications for Practitioners

The field of GFRP composites practitioners who anticipate mechanical fracture and thermal properties would be greatly impacted by recent developments in AIoT. These are a few significant ramifications for practitioners:

8.8.2.1 Enhanced Predictive Capabilities

With the use of AIoT, practitioners may create prediction models for GFRP composite thermal and mechanical fracture that are incredibly accurate. By combining AI algorithms with data coming in from IoT sensors of an asset, practitioners are enabled to predict tensile strength, fatigue life, thermal conductivity, and expansion coefficients under various loads and external environments. This practically enhances decision-making procedures in the material selection, design optimization, and performance assessment of GFRP composites.

8.8.2.2 Real-time Status Monitoring and Maintenance

IoT sensors embedded in GFRP constructions make it possible to monitor the mechanical and thermal performance in real time. AI systems review streaming data for signs of structural deficiencies, temperature stress, and mechanical wear. Practitioners can leverage predictive insights to schedule maintenance and repair/replace activities at the right time enabling effective asset management and service life extension of GFRP composites.

8.8.2.3 Efficiency and Cost Cutting

AIoT-enabled predictive maintenance drives down costs from sudden breakage and reactive maintenance. They could perform condition-based maintenance procedures instead of regular inspections, focusing on components that are showing signs of deterioration. This strategy not only leads in lifecycle cost reduction of GFRP composites structures, but also enhances operational effectiveness and resource allocation throughout the lifecycle phase.

8.8.2.4 Improved Performance/Increased Design Solutions

Simulation and analysis of virtual scenarios related to GFRP composites behavior is performed using AIoT approach to perform iterative design optimization. For improving mechanical strength, thermal stability and all-round performance, practitioners can experiment with a wide variety of design parameters, material compositions, and reinforcement configurations. This iterative approach drives innovation cycles faster than ever before, and empowers practitioners to implement solutions that are based on the unique needs of their world or for a certain use-case.

8.8.2.5 Risk Reduction and Adherence

AIoT enhances risk management strategies as it provides detailed data analytics and predictive insights on GFRP composite failure modes. Practitioners can use real-time data and prediction models to address safety problems, regulatory compliance needs, and sustainability issues in an evidence-based way. Being proactive allows us to decrease the likelihood of structural failures, increases operational safety, and ensures compliance with regulations and industry standards.

8.8.2.6 One Piece of Secure Working: Collaboration and Skills Development

They develop expertise in the AI technologies and integrate it with IoT, acquiring specialization within data analytics, ML as well as sensor technologies for practitioners. Interdisciplinary collaboration across the crosscutting and foundational domains of materials science and engineering as well as data analytics drive innovation within companies and between industries. Furthermore, this approach speeds up the ongoing development of GFRP composites and enhances problem-solving abilities.

All these capabilities will be available to the practitioners directly in finding mechanical and thermal properties of GFRP composites, with an unimaginable accuracy and efficiency due to AIoT advancements. This can help in driving innovation, reducing risks, improving performance, and thus cost savings over the whole lifecycle of GFRP composite structures designed, maintained, and applied by practitioners using AI driven insights from IoT data.

By using AIoT technologies together with multiscale modeling, a holistic approach to GFRP analysis can be taken, providing insights into material behavior across length and time scales and leveraging real-time data for enhancing predictive capacity and maintenance strategies in practice.

8.9 CONCLUSION

These AIoT advances make accurate predictions for both mechanical fracture and thermal properties, if adopted will greatly benefit the design, production, and performance of GFRP composites thereby leading to more secure infrastructures in numerous fields. This way AIoT enables sustainable manufacturing strides acting naturally. Integrating AIoT in forecasting mechanical fracture and thermal characteristics of GFRP materials enhances structural health monitoring and predictive maintenance across various sectors. By using AI algorithms for analysis and collecting real-time data from IoT devices this method provides many important features. Since AIoT systems could monitor the structural integrity of GFRP components continuously, any potential mechanical defects or abnormalities can be detected at an early stage before they cause major failure. By analyzing historical data with real-time sensor inputs, the AI models are able to predict likely mechanical fracture occurrences on milling and thermal behavior of GFRP constructions under different environmental conditions. This feature allows for proactive maintenance scheduling and operational performance optimization. The developed data-driven approaches using ML algorithms are able to provide better predictions of mechanical fracture and thermal characteristics compared to traditional analytical models given the availability of sufficient training datasets that encompass different environmental conditions and material parameters. Predictive maintenance methods that are supported by AIoT will help to minimize operational costs related to the management of GFRP assets through preventing unforeseen failures and downtime. Adapting AIoT systems to specific GFRP applications and environments allows them learn over longer periods of time, increasing prediction accuracy. Customization and Flexibility: However, a plethora of challenges has to be addressed in order for AIoT-based GFRP quality forecasting to reach full potential including data quality and availability, interpretability, and explainability of AI model training for better accuracy as well as security and privacy concerns and integration with existing infrastructure. To emphasize the necessity for this, it is essential to ensure a continuous and reliable supply of relevant and accurate data generated by IoT sensors and other connected devices. The AI algorithms in predictive analytics used to build confidence and trust within the stakeholders, mainly the safety-critical applications need to be transparent and interpretable. Data protection against cyber threats and illegal access of sensitive data collected by IoT sensors and AI systems is essential to maintain the integrity and confidentiality of GFRP asset information. AIoT solutions need to work hand-in-hand with existing operational workflows, along with maintenance management systems to be adopted and scaled. Ultimately, the use of AIoT to predict the mechanical fracture and thermal behavior of GFRP materials has substantial potential for increasing asset reliability, optimizing maintenance scheduling and ensuring the long-term sustainability of infrastructure across multiple sectors. However, it does require addressing organizational, technological, and regulatory challenges if we are to truly unlock its transformative power.

REFERENCES

Azad, Muhammad Muzammil, Sungjun Kim, Yu Bin Cheon, and Heung Soo Kim. "Intelligent structural health monitoring of composite structures using machine learning, deep learning, and transfer learning: A review." *Advanced Composite Materials* 33, no. 2 (2024): 162–188.

Baduge, Shanaka Kristombu, Sadeep Thilakarathna, Jude Shalitha Perera, Mehrdad Arashpour, Pejman Sharafi, Bertrand Teodosio, Ankit Shringi, and Priyan Mendis. "Artificial intelligence and smart vision for building and construction 4.0: Machine and deep learning methods and applications." *Automation in Construction* 141 (2022): 104440.

Bibri, Simon Elias, and Senthil Kumar Jagatheesaperumal. "Harnessing the potential of the metaverse and artificial intelligence for the internet of city things: Cost-effective XReality and synergistic AIoT technologies." *Smart Cities* 6, no. 5 (2023): 2397–2429.

Botín-Sanabria, Diego M., Adriana-Simona Mihaita, Rodrigo E. Peimbert-García, Mauricio A. Ramírez-Moreno, Ricardo A. Ramírez-Mendoza, and Jorge de J. Lozoya-Santos. "Digital twin technology challenges and applications: A comprehensive review." *Remote Sensing* 14, no. 6 (2022): 1335.

Broughton, William, M. R. L. Gower, M. J. Lodeiro, G. D. Pilkington, and R. M. Shaw. "An experimental assessment of open-hole tension–tension fatigue behaviour of a GFRP laminate." *Composites Part A: Applied Science and Manufacturing* 42, no. 10 (2011): 1310–1320.

Chang, Fu-Kuo, ed. *Structural health monitoring 2013: A roadmap to intelligent structures: Proceedings of the ninth international workshop on structural health monitoring, September 10–12*. DEStech Publications, Inc, 2013.

Chang, Zhuoqing, Shubo Liu, Xingxing Xiong, Zhaohui Cai, and Guoqing Tu. "A survey of recent advances in edge-computing-powered artificial intelligence of things." *IEEE Internet of Things Journal* 8, no. 18 (2021): 13849–13875.

Cheruvu, Sunil, Anil Kumar, Ned Smith, and David M. Wheeler. *Demystifying internet of things security: Successful IoT device/edge and platform security deployment*. Springer Nature, 2020.

Gao, Yan, Haijiang Li, Guanyu Xiong, and Honghong Song. "AIoT-informed digital twin communication for bridge maintenance." *Automation in Construction* 150 (2023): 104835.

Gomez, Cristian, A. Guardia, J. L. Mantari, A. M. Coronado, and J. N. Reddy. "A contemporary approach to the MSE paradigm powered by Artificial Intelligence from a review focused on Polymer Matrix Composites." *Mechanics of Advanced Materials and Structures* 29, no. 21 (2022): 3076–3096.

Huang, Manjuan, Minglu Zhu, Xiaowei Feng, Zixuan Zhang, Tianyi Tang, Xinge Guo, Tao Chen, Huicong Liu, Lining Sun, and Chengkuo Lee. "Intelligent cubic-designed piezoelectric node (iCUPE) with simultaneous sensing and energy harvesting ability toward self-sustained artificial intelligence of things (AIoT)." *ACS Nano* 17, no. 7 (2023): 6435–6451.

Jhon, Mark. "Model studies of instabilities in heterogeneous materials systems." PhD diss., UC Berkeley, 2009.

Morampudi, Priyadarsini, Kiran Kumar Namala, Yeshwanth Kumar Gajjela, Majjiga Barath, and Ganaparthy Prudhvi. "Review on glass fiber reinforced polymer composites." *Materials Today: Proceedings* 43 (2021): 314–319.

Qiu, Cheng, and Jinglei Yang. "Machine learning applications in composites: Manufacturing, design, and characterization." In *Machine learning in materials informatics: Methods and applications*, pp. 65–85. American Chemical Society, 2022.

Ramprasad, Rampi, Rohit Batra, Ghanshyam Pilania, Arun Mannodi-Kanakkithodi, and Chiho Kim. "Machine learning in materials informatics: Recent applications and prospects." *Npj Computational Materials* 3, no. 1 (2017): 54.

Rehman, Faisal, Muhammad Anwar, Anees Ul Mujtaba, Hanan Sharif, and Naveed Riaz. "Big data analytics for AIoT network." In *Artificial Intelligence of Things (AIoT)*, pp. 140–163. CRC Press, 2024.

Rocha, Helena, Christopher Semprimoschnig, and João P. Nunes. "Sensors for process and structural health monitoring of aerospace composites: A review." *Engineering Structures* 237 (2021): 112231.

Shen, Meng, Aijing Gu, Jiawen Kang, Xiangyun Tang, Xiaodong Lin, Liehuang Zhu, and Dusit Niyato. "Blockchains for artificial intelligence of things: A comprehensive survey." *IEEE Internet of Things Journal* 10, no. 16 (2023): 14483–14506.

Shi, Qiongfeng, Zixuan Zhang, Yanqin Yang, Xuechuan Shan, Budiman Salam, and Chengkuo Lee. "Artificial intelligence of things (AIoT) enabled floor monitoring system for smart home applications." *ACS Nano* 15, no. 11 (2021): 18312–18326.

Sousa, João M., João R. Correia, Susana Cabral-Fonseca, and António C. Diogo. "Effects of thermal cycles on the mechanical response of pultruded GFRP profiles used in civil engineering applications." *Composite Structures* 116 (2014): 720–731.

Sun, Limin, Haibin Sun, Wei Zhang, and Yixian Li. "Hybrid monitoring methodology: A model-data integrated digital twin framework for structural health monitoring and full-field virtual sensing." *Advanced Engineering Informatics* 60 (2024): 102386.

Zhang, Jing, and Dacheng Tao. "Empowering things with intelligence: A survey of the progress, challenges, and opportunities in artificial intelligence of things." *IEEE Internet of Things Journal* 8, no. 10 (2020): 7789–7817.

9 Elevating Self-Service Business Intelligence through AI and IoT in Sub-Saharan Africa

Insights from Sustainable Manufacturing Companies

Timilehin Olasoji Olubiyi, Omolade Sunday Adeyemi, Oluwatoyin Damilola Adeyemi, and Kehinde O. Oketola

9.1 INTRODUCTION

Digital technology has changed the Business Intelligence (BI) environment in a big way, opening up amazing opportunities for making decisions based on data and getting strategic insights. Right now, businesses are getting a lot of data from a lot of different places, and they need advanced analytics to find the most important information. Big Data, Artificial Intelligence (AI), and the Internet of Things (IoT) are all coming together to make this change happen. This lets businesses utilise data to improve their operations and go ahead of their competitors. Self-service business intelligence empowers non-technical people to autonomously comprehend data and generate insights without relying on IT departments, hence transforming the utilisation of data in enterprises. Consumer satisfaction is complicatedly impacted by the process of making data more standardised or accessible, even though it provides greater flexibility and cost savings (Izzuddin & Abidin, 2025; Naidu et al., 2025). The effectiveness with which teams convert data into customer-focused decisions is unbalanced, as evidenced by studies showing that user-associated issues, such as inconsistent skill levels and irregular tool use, frequently hinder the adoption of self-service business intelligence (Arokodare & Olubiyi, 2023; Bakır & Sak, 2025). Lennerholt et al., 2021). This poses important questions: Does the availability of self-service business intelligence and the moderating role of non-technical users inevitably impact the organization's performance? Although self-service drivers, such as economy, convenience, and system dependability, are identified in the literature as sources of consumer value, the important role that user independence plays in mitigating these effects is not taken into account (Dang & Nguyen, 2023; Su, 2021; Wang, 2024).

 DOI: 10.1201/9781003503323-9

Organisations have uneven user adoption and misinterpretation of information that might unintentionally degrade service quality, even while self-service business intelligence may enable faster customer response (Su, 2021). According to Borodako et al. (2023) and Ibrahim et al. (2025), technological functionality alone is insufficient to produce desired customer results, and system design components like response time and aesthetics are crucial for maintaining user attention (Ellitan et al., 2022; Wang, 2024). These discrepancies reveal a fundamental disconnect: businesses invest in self-service business intelligence infrastructure, yet they lack the frameworks needed to align the drive for customer satisfaction with the analytical autonomy of non-technical users. (Andrade & Tumelero, 2022; Soeharso, 2024). The purpose of this study is to close this gap by examining the effect of self-service business intelligence and customer satisfaction the moderating role of non-technical users in the organisation. In order to maximise customer experiences and employee empowerment, this study examined how the independence of non-technical users interacts with self-service business intelligence systems. A noticeable gap in scholarly discourse that mostly looks at self-service business intelligence's operational advantages rather than its interactions with customers is filled by this dual focus (Izzuddin & Abidin, 2025; Lennerholt et al., 2021). The results of this study could inform future implementation strategies by ensuring that improved customer satisfaction, rather than a disjointed or misdirected analytics endeavour, is a reflection of a more standardised, accessible data with greater flexibility and cost savings.

9.2 LITERATURE REVIEW

Business Intelligence systems offer historical, present, and predictive insights into corporate operations, mostly using data aggregated in a data warehouse or data mart, and occasionally leveraging operational data. Software components provide reporting, interactive pivot-table analysis, visualisation, and statistical data mining (Mallikarjuna, et. al., 2024). Applications address sales, production, finance, and several other business data sources for the goal of company performance management. Information is frequently collected on rival firms within the same industry, a process known as benchmarking, as they compete in the same domain or offer items created or presented in a comparable marketplace. Organisations are starting to acknowledge that data and content should not be seen as separate components of information management, but rather should be governed via a cohesive organisational strategy. Enterprise information management amalgamates Business Intelligence with Enterprise Content Management (Arokodare et al., 2023; Mallikarjuna et al., 2024).

Self-service business intelligence is a paradigm shift in how businesses utilise and consume data to make decisions (Maghsoudi & Nezafati, 2023). In contrast to traditional BI, which mostly depends on IT or data specialists to provide reports and analytics, self-service business intelligence enables business users of all technical backgrounds to create, edit, and evaluate reports on their own utilising simple, low-code or no-code solutions (Lennerholt et al., 2021).

Decision-making becomes quicker and more flexible as a result of this data democratisation, which also removes constraints brought on by a lack of IT resources (Ramasamy et al., 2024).

The accessibility and flexibility of self-service business intelligence are its main strengths, enabling data professionals to analyse trends, create dashboards that are tailored to their specific needs, create ad hoc searches, and interact with data instantaneously (Zungu et al., 2025). Data consumers receive fast, accurate, and consistent information via robust data infrastructures with appropriate governance and security, which underpin effective self-service business intelligence environments (Andrade & Tumelero, 2022). In order to improve user experience and foster a data-driven culture of cooperation, new ways of thinking place a higher priority on integrating cooperative features like shared spaces and repositories (Buhasho et al., 2021). IT governance is not eliminated by self-service business intelligence, even though it makes available standardised, accessible data with greater flexibility and cost savings (Borodako et al., 2023). However, it necessitates a two-pronged approach in which business users are trained and motivated to use data in their daily work while IT teams handle data governance and quality (Brunner et al., 2025).

Žigienė et al. (2022) say that the rise of AI, the IoT, and Data Analytics in Business Intelligence (BI) is a big step forward for business strategy and decision-making. This merging has given rise to intelligent business analytics, a field that blends conventional business intelligence with the advanced features of big data and AI. AI, IoT, and Data Analytics have changed corporate intelligence in many ways, influencing many parts of company operations, including decision-making and strategic planning. Žigienė et al. (2022) say that intelligent business analytics is a new area of business intelligence that combines AI and big data technology. This connectivity has helped businesses make better and smarter decisions by letting them use the huge amounts of data that are accessible online. Modern data processing technologies, AI algorithms, and powerful data visualisation tools are all part of intelligent business analytics. Together, they provide companies a lot of information and help them plan for the future. Ilieva et al. (2021) look at how AI may improve BI systems, especially when it comes to helping people make decisions. They say that AI will be necessary for the next generation of analytics, which will improve corporate intelligence via features like predictive modelling, machine learning, and automated business process optimisation. Combining AI with IoT in BI systems provides several advantages such as higher productivity, more accurate data analysis, and the capacity to find patterns and trends that other techniques of data analysis could miss.

The entire quality of the system profoundly influences an organisation's capacity to fulfil customer requirements. Dependability, usability, and the correctness of the analytical conclusion are all components of system quality (Izzuddin & Abidin, 2025). Self-service business intelligence tools that are easy to use and provide quick responses will be utilised more frequently by non-technical users, according to a study (Ellitan et al., 2022). However, a slow or too complex system irritates users and erodes their confidence in the data, which eventually reduces customer satisfaction. A good self-service business intelligence system helps users make better decisions by acting as a bridge between unprocessed data and useful information (Soeharso, 2024).

Training is yet another essential element, having self-service business intelligence tools alone is insufficient; users also need to receive adequate training that goes beyond technical issues (Amiruddin et al., 2024). The primary objectives of effective training programs are to increase data literacy and teach users how to use insights in actual business settings (Andrade & Tumelero, 2022; Brunner et al., 2025). When users receive the right training, they will be able to clarify data, ask more insightful questions, and steer clear of errors that could damage relationships with customers (Wang, 2024). Conversely, insufficient or inconsistent training can result in unequal self-service business intelligence tool use, with some employees making the most of the technology while others are unable to, which would lower the quality of customer service (Izzuddin & Abidin, 2025).

Data accessibility refers to how easily users may access and utilise the information they require within the self-service business intelligence system. Users who are unable to get timely and relevant data will not benefit from the greatest training or tools (Lennerholt et al., 2021). Users may be unable to perceive the broad picture due to issues like fragmentation of single-data systems or restricted access regulations, which leads them to make decisions that do not completely take consumer needs into account (Maghsoudi & Nezafati, 2023). To reduce customer satisfaction, a marketing team without immediate access to stock information could, for instance, offer customers out-of-stock items (Ellitan et al., 2022). On the other hand, organisations with open, integrated data access empower users to make better-informed, customer-focused decisions (Ramasamy et al., 2024).

The level of independence non-technical users has while utilising self-service business intelligence technologies determines this. When people can study and gain insights without continuously relying on IT support, decision-making speeds up and creativity is stimulated (Kumah et al., 2025; Shojaei & Burgess, 2022). However, freedom has two sides. Insufficient coordination or monitoring may cause users to misinterpret data or overlook important details, which could result in less-than-ideal behaviours that have a detrimental effect on customer satisfaction (Maghsoudi & Nezafati, 2023; Naidu et al., 2025). Research indicates that the best results occur when users have the right balance of independence enough flexibility to actively explore data, while yet having access to validation and support to ensure correctness (Amiruddin et al., 2024; Lennerholt et al., 2021).

Customer satisfaction, which is often measured in terms of repeat purchase intent and service quality ratings, is typically considered the cornerstone of business success. When consumers' requirements are consistently met, whether through prompt or personalised service, they are more likely to remain loyal and come back (Wang, 2024). Research by Aguiar-Costa et al. (2022) and Brunner et al. (2025) indicate that satisfaction is not just a result but also a continuous process that fosters trust and encourages recurrent engagement. This connection gets stronger in self-service business intelligence (Su, 2021). Organisations must make sure the insights produced by SSBI technologies closely match the needs and values of their customers because decisions are data-driven rather than human-to-human. Even the best data analytics can fail to influence positive customer experiences in the absence of such alignment (Zungu et al., 2025).

9.3 DYNAMICS AND CUSTOMER SATISFACTION IN BUSINESSES

Two fundamental theories are well-founded in the knowledge of SSBI dynamics and customer satisfaction in businesses.

9.3.1 Expectancy-Disconfirmation Theory

The Expectancy-Disconfirmation Theory is often used to explain client satisfaction levels. The idea posits that customers formulate expectations before to using a service or product, and their satisfaction with performance is contingent upon the fulfilment or surpassing of these expectations (Kumah et al., 2025). Favourable behavioural intentions, such as repeat patronage and pleasure, are probable when performance aligns with or surpasses expectations (Ramasamy et al., 2024). This suggests that to positively influence customer satisfaction outcomes, the quality and usability of business intelligence solutions must align with consumer expectations for self-service business intelligence.

Technology Acceptance Model (TAM): The TAM is the second relevant theory that elucidates how people adopt and use new technology. According to the TAM, perceived usefulness and perceived ease of use are the primary determinants of technology adoption. The perceived ease of use of SSBI technologies is a determining factor in their acceptability and autonomous use by non-technical organisational users. End users will utilise self-service business intelligence tools independently if they find them helpful for making decisions and simple to use (Borodako et al., 2023; Izzuddin & Abidin, 2025). This will increase customer satisfaction by facilitating quicker and more insightful responses to clients. In these theoretical contexts, the study concentrates on a few essential concepts. The dependent variable, customer satisfaction, is operationalised through perceptions of service quality and intentions to make additional purchases, which show how well the company can satisfy the needs of its clients (Brunner et al., 2025). The self-service business intelligence system quality, which includes dependability, interface design, and data accuracy; user training adequacy, which deals with how well users are equipped to use self-service business intelligence tools; and data accessibility, which refers to how simple it is for users to access timely and pertinent data, are the independent variables (Izzuddin & Abidin, 2025; Lennerholt et al., 2021). Non-technical user independence is the moderating variable, which is determined by how comfortable and self-sufficient users feel when conducting data queries and coming up with insights without the assistance of IT (Kumah et al., 2025; Shojaei & Burgess, 2022). This is a crucial concept since it determines how much improved customer satisfaction is a result of the self-service business intelligence system's quality and other elements.

9.3.2 Theoretical Framework

In order to explain how self-service business intelligence affects customer satisfaction the moderating role of independent non-technology user, the study's theoretical framework combines the Expectancy-Disconfirmation Theory with the TAM. The Expectancy-Disconfirmation Theory states that when expected and perceived service performance align, customer satisfaction is the outcome. In this instance, the quality level of insights obtained by self-service business intelligence instruments acts as a mediator. This is complemented by the TAM, which emphasises the significance of perceived utility and usability, which influence the attitudes and capacities of non-technical users to utilise self-service business intelligence systems independently

(Andrade & Tumelero, 2022; Lennerholt et al., 2021). The self-service business intelligence system's quality, the effectiveness of user training, and the availability of data are the main factors influencing how well the self-service business intelligence tools are perceived in this context (Izzuddin & Abidin, 2025). The impact of these forces on customer satisfaction can be moderated by non-technical user independence. Customer satisfaction is strengthened by system quality and training when users feel empowered and secure enough to use self-service business intelligence tools on their own (Amiruddin et al., 2024). However, because users depend considerably on IT assistance and may encounter delays or misconceptions that have a detrimental influence on customer results, limited user independence may negate these benefits.

9.3.3 Empirical Review

Amiruddin et al. (2024) examined how organisational environment and non-technical elements affect the quality of systems, information, and services, as well as how user satisfaction is affected in petrochemical manufacturing companies. Questionnaires were used to gather quantitative data from ERP users at five different managerial levels, ranging from general managers to supervisors. The study, which used Structural Equation Modelling Partial Least Squares (SEM-PLS) analysis, found that organisational climate influences the important success criteria for ERP adoption. User happiness was found to be directly impacted by system and information quality, demonstrating the value of a supportive organisational context in guaranteeing ERP implementation success. According to Buhasho et al. (2021) for most firms, business intelligence remains a priority, and significant investment is being attracted to this field. Nevertheless, current research provides fragmented and ambiguous insights into the extent to which this technology is efficiently translated into useful business learning. The study examined the relationship between business intelligence capability and firm performance and how organisational capabilities affect it. Supported by theories of organisational learning and information systems capability, the study employed a mixed-methods approach, using thematic analysis for qualitative data and structural equation modelling (SEM-PLS) for quantitative data. Organisational competence positively supports this link, according to the results, providing managers, policy-makers, and researchers in the field with new information.

Borodako et al. (2023) examine the relationship between innovation orientation (IO), which is comprised of six dimensions market, organisational culture, technology, human resources, and structural process and organisational performance (OP). Knowledge management (KM) serves as a mediator and technological readiness (TR) as a moderator. The model was validated by experimenting with data from business service firms using route analysis and multiple regression. The findings supported six out of eight hypotheses, demonstrating that KM is positively impacted by strategy, technology, organisational culture, and market factors, which partially mitigate their impact on performance. Additionally, TR makes the connection between KM and company performance easier to understand, emphasising the need for effective KM and better results. According to Izzuddin and Abidin (2025), customers can independently purchase services using self-service technology (SST) via equipment such as automated stations, kiosks, and self-checkout devices. Businesses employ SST to boost competitive advantage, productivity, and service quality. However, it's important to comprehend how customers view the quality of SST services and how that affects loyalty. Izzuddin and Abidin (2025) used a quantitative methodology based on the SSTQUAL model and cognition-affective-behaviour theory. Via online surveys, 285 McDonald's SST users between the ages of 18 and 55 were polled. Functionality, enjoyment, assurance, design, and convenience are the five aspects that makeup SST service quality, according to the report. Good customer satisfaction increases customer loyalty.

Wang (2024) examined how consumer satisfaction in the post-COVID-19 economy is affected by digitalisation in global business. The SERVQUAL model of service quality was used in the research to operationalise satisfaction using a mixed quantitative approach that included surveys and trials. Digital improvements increased data security, order processing speed, packing, product range, alerts, and reliability, according to 3,796 customer reviews in four AliExpress stores. When chatbots were introduced, customer satisfaction dropped because they provided less individualised care and empathy. The study advances our knowledge that not all forms of digitalisation are equally popular with consumers, highlighting the significance of gauging satisfaction before fully implementing new digital solutions for global trade. According to Su (2021), a lot of research has been done on service quality, particularly since self-services have changed how consumers interact with businesses to get positive outcomes including ease, security, and positive behaviour. Su (2021) examines the driving forces behind consumers' intentions to use self-service, with a particular emphasis on the role that customer value plays as a mediator. 35 Taiwanese self-service consumers' data were analysed using structural equation modelling. The results show that consumer value is greatly impacted, albeit to differing degrees, by convenience, cost savings, security, and ease of use. It was also shown that the relationship between these and behavioural intentions was mediated by customer value. These findings imply that self-service should be a priority for Taiwan's service industry in order to encourage repeat business and consumer loyalty.

9.4 METHODOLOGY

This study employs a quantitative survey research design, drawing upon the methodologies established in past studies by Darda et al. (2025), Adeyemi and Olubiyi (2023, 2024), Olubiyi (2024b), and Adeoye et al. (2023). The research examines enterprises that prioritise self-service activities and have implemented self-service business intelligence tools across the retail, banking, manufacturing, and healthcare sectors in Lagos State, Nigeria's economic hub and a significant consumer market in Africa. The research used a stratified random sampling method to choose a representative sample from sustainable manufacturing firms in Lagos State, Nigeria. A sample size of 378 was determined using the Research Advisor Table, corresponding to a 95% confidence level and a 5% margin of error. To address the issue of non-response, appropriate measures were executed, leading to the inclusion of an additional 113 respondents, representing 30% of the original sample. The alteration resulted in a final sample size of 491 representing diverse sectors, including retail, banking, manufacturing, and healthcare. A standardised closed-ended questionnaire, created using Google Forms, was delivered to respondents online with a Likert scale answer format.

The questionnaire tool underwent reliability testing. The outcomes for each variable and portion of the study instrument exceeded 0.70, thereby affirming the questionnaire's excellent reliability. The data was assessed using user training adequacy, data accessibility, and system quality as independent factors, with customer satisfaction serving as the dependent variable. The moderating variable, non-technical user, is assessed by self-reported degrees of autonomy in generating insights without IT assistance and formulating enquiries based on data. Descriptive analysis was conducted using descriptive methods, whereas inferential statistics were used to investigate effects and correlations. The study was conducted using version 26 of the Statistical Package for Social Sciences (SPSS).

9.5 RESULT AND DISCUSSION

The researcher and trained research assistants sent out 491 questionnaires, and 480 of them were filled up and sent back for analysis. This means that 98% of sustainable manufacturing enterprises in Nigeria responded. The response rate is the percentage of people that filled out and sent back the survey questionnaire. Some of the surveys were not returned, and some of the answers were missing. However, there were enough questionnaires received to represent the population, and they were then evaluated. Table 9.1 shows the details of the answers.

9.5.1 Demographic Data Interpretation (Table 9.2)

The demographic information gives an in-depth understanding of the research participants, capturing their varied experiences and backgrounds in relation to self-service IT products in sustainable manufacturing companies in Lagos State Nigeria. The distribution across industries is balanced, with participants selected from manufacturing, healthcare, retail, finance, and other areas, guaranteeing that the results reflect a diverse organisational environment. Operations specialists, marketing managers, data analysts, and customer service agents were among the many employment types that would provide a thorough grasp of how various activities interact with self-service business intelligence technologies.

Table 9.2 shows the demographic information of the sample for the self-service business intelligence use study is broad and evenly distributed, with the majority of participants falling into the 26–35 age range (41.3%) and the 36–45 age range (30%),

TABLE 9.1
Response Rate

	Frequency	Percentage (%)
Usable validated questionnaire	480	98
Invalidated questionnaire	11	2
Total received	491	100

Source: Researchers' computation (2025).

TABLE 9.2
Demographic Information of Respondents

Variables	Categories	Frequency	Percentage
Age	18–25	33	11.0
	26–35	124	41.3
	36–45	90	30.0
	46–55	53	17.7
	Total	**300**	**100.0**
Gender	Male	153	51.0
	Female	147	49.0
	Total	**300**	**100.0**
Educational Qualification	Secondary	15	5.0
	Bachelor's Degree	57	19.0
	Master's Degree	160	53.3
	Others	68	22.7
	Total	**300**	**100.0**
Job Role	Data Analyst	124	41.3
	Marketing Manager	90	30.0
	Operation Specialist	53	17.7
	Customer Support	33	11.0
	Total	**300**	**100.0**
Experience with SSBI Tools	Less than 6 months	52	17.3
	6–12 months	117	39.0
	1–3 years	93	31.0
	Over 3 years	38	12.7
	Total	**300**	**100.0**
Industry	Retail	71	23.7
	Finance	118	39.3
	Manufacturing	56	18.7
	Healthcare	24	8.0
	Others	31	10.3
	Total	**300**	**100.0**

Source: Research Field Survey (2025).

which represents a young to mid-career workforce. With men making up 51% and women 49%, the gender distribution is almost equal, guaranteeing that the results are inclusive of all genders. Nearly one-fifth of the respondents had a bachelor's degree, and more than half (53.3%) have a master's degree, indicating a highly competent sample that will be able to use SSBI tools efficiently. As would be expected of regular users of SSBI systems, data analysts make up the largest category for job functions (41.3%), followed by marketing managers (30%). Experience with the SSBI tool varies; 39% of respondents have used it for 6–12 months, while 43.7% have used it for more than a year. This suggests that there is a mix of novice and seasoned users. Retail (23.7%) and finance (39.3%) are the two industries with the greatest

TABLE 9.3
Regression Analysis

Model	R	R Square	Adjusted R Square	Std. Error Estimate
1	0.966[a]	0.934	0.933	0.28130

[a] Predictors: (Constant), Non-Technical User Independence, Data Accessibility, System Quality, and User Training Adequacy

[b] Customer Satisfaction

Source: Researcher's Field Survey (2025).

industry participation; both sectors are known for using a lot of data, but manufacturing, healthcare, and other sectors also contribute insightful viewpoints. Hence, this demographic profile offers a solid starting point for investigating how various user characteristics modify the association between customer satisfaction and SSBI use.

Table 9.3 presents the regression analysis findings, demonstrating a significant association between customer satisfaction and the predictor variables of self-service business intelligence: system quality, data accessibility, non-technical user autonomy, and the sufficiency of user training. The model's correlation coefficient (R) of 0.966 indicates a robust association between these factors and the outcome variable. The model's predictors account for about 93.4% of the variance in customer satisfaction, as shown by an R Square value of 0.934. The model's robustness and generalisability are further corroborated by its adjusted R Square of 0.933, considering the number of predictors and sample size. Moreover, the model's predictions are precise and dependable, as shown by the very low predicted standard error of 0.281. Consequently, our findings indicate that the selected factors together provide a robust framework for elucidating variations in customer satisfaction about the utilisation of self-service business information.

Table 9.4 shows the regression analysis's ANOVA that the model as a whole is a very significant predictor of customer satisfaction. The overall impact of the

TABLE 9.4
Regression Showing the Significance of Each Predictor to Customer Satisfaction

Model		Sum of Squares	Df	Mean Square	F	Sig.
1	Regression	330.053	4	82.513	1042.739	0.000
	Residual	23.344	295	0.079		
	Total	353.397	299			

[a] Predictors: (Constant), Non-Technical User Independence, Data Accessibility, System Quality, and User Training Adequacy

[b] Customer Satisfaction

Source: Researcher's Field Survey (2025).

TABLE 9.5
Contribution of Each Predictor to Customer Satisfaction

Model		Unstandardised Coefficient		Standardised Coefficient		
		B	Std. Error	Beta	T	Sig.
1	(Constant)	2.277	0.016		140.180	0.00
	System Quality	0.216	0.035	0.198	6.188	0.00
	User Training Adequacy	0.845	0.044	0.777	19.359	0.00
	Data Accessibility	0.681	00.028	0.627	24.208	0.00
	Non-Technical User Independence	–0.666	0.054	–0.612	–12.250	0.00

Source: Researcher's Field Survey (2025).

predictors' non-technical user independence, data accessibility, system quality, and user training adequacy on customer satisfaction is statistically significant, according to the F-value of 1042.739 and the associated p-value of 0.000. This indicates that there is very no probability that these outcomes could have happened by accident, proving that the model accurately explains the shift in consumer satisfaction. The regression sum of squares' high value (330.053) in relation to the residual (23.344) further supports the idea that the predictors are mostly responsible for the outcome's variability. This analysis demonstrates how the chosen indicators have a significant impact on understanding and raising customer satisfaction levels within the scope of self-service business intelligence implementation.

Table 9.5 regression coefficients show how significantly each predictor contributes to customer satisfaction on its own. A standardised coefficient (beta) of 0.198 indicates a substantial and positive relationship between system quality and customer satisfaction, suggesting that improvements in are linked to higher levels of customer satisfaction. The highest positive effect is produced by user training adequacy, which has a beta of 0.777, indicating that successful training programs significantly increase user empowerment and satisfaction. The significance of having simple access to pertinent data is demonstrated by the positive relationship between data accessibility and customer satisfaction (beta = 0.627). Unexpectedly, the moderating role of non-technical user independence exhibits a substantial negative correlation (beta = –0.612), indicating that, in this situation, higher degrees of user independence are linked to worse customer satisfaction. This would imply that greater independence will result in incorrect interpretation and abuse of knowledge with detrimental effects if neither proper assistance nor direction is provided. In the self-service business intelligence environment, all of the predictors have substantial roles in forecasting customer satisfaction, as evidenced by the fact that they are all $p < 0.001$.

9.5.2 Discussion of Results

The study explored the impact of self-service business intelligence on customer satisfaction, highlighting the moderating role of non-technical users in sustainable manufacturing companies in Lagos State, Nigeria.. The results of this study give valuable insights into the factors that influence customer satisfaction in self-service business intelligence settings. Customer satisfaction is positively impacted by system quality, user training effectiveness, and data availability, all of which are supported by the research. Accordingly, businesses are better able to fulfil clients when self-service business intelligence technologies are dependable, users are well-trained, and data is accessible. According to these findings, user support and technology infrastructure are important factors that contribute to user satisfaction. The benefits of data availability and training emphasise the organisational and human elements that support technological spending. This finding is consistent with the findings of research scholars Su (2021), Izzuddin and Abidin (2025), Amiruddin et al. (2024), and Lennerholt et al. (2021). Intriguingly, though, the study also finds a negative correlation between the moderating effect of non-technical user independence and consumer satisfaction. According to this, while user independence in the use of self-service business intelligence technologies can liberate people, too much freedom combined with inadequate training or skills might result in misconceptions or mistakes that eventually lower customer satisfaction. This result runs against previous studies of Shojaei and Burgess (2022), Borodako et al. (2023), and and Kumah et al. (2025) that tend to link increased user independence to higher levels of technological performance and adoption. The inequity results from the intricacy of self-service business intelligence systems and the possibility that, in the absence of assistance, technical illiterates may make decisions based on inaccurate or partial information (Borodako et al., 2023). These findings crucially highlight the necessity of implementing self-service business intelligence holistically. Empowering users is important, but in order to avoid the dangers of unmoderated independence, appropriate training, support networks, and regulations are also required. The fact that this nuance highlights that user independence is not always desirable and needs to be handled carefully to maximise customer satisfaction adds to the literature's richness. Hence, the null hypothesis was rejected.

9.5.3 Implications of The Study

These results have substantial implications for the organisations implementing self-service business intelligence. Researchers stress that, while important, investing in top-notch technology and thorough training alone is insufficient. Organisations must also continue to supervise and support non-technical users in order to help them use data ethically. In the absence of this equilibrium, greater independence may unintentionally lower consumer satisfaction. Consequently, companies ought to endeavour to establish a setting in which users feel both empowered and encouraged to utilise self-service business intelligence products.

9.5.4 Contribution to Knowledge

This research enhances the existing literature by elucidating the relationship between self-service business intelligence and customer satisfaction, specifically focusing on the moderating influence of non-technical users within sustainable manufacturing firms in Lagos State, Nigeria. The research investigated the complex link between the moderating influence of non-technical user autonomy in self-service business intelligence environments and customer happiness, which has been undervalued. It enhances the existing understanding of the interplay between human factors and technology in influencing customer satisfaction, offering a more profound insight into the uptake and utilisation of self-service business intelligence tools. This knowledge is beneficial to scholars and practitioners seeking to optimise data-driven decision-making processes.

9.6 CONCLUSION AND RECOMMENDATIONS

This study examines effect of self-service business intelligence and customer satisfaction the moderating role of non-technical users in sustainable manufacturing companies in Lagos State, Nigeria. It emphasises that the main factors influencing customer satisfaction in businesses are the quality of self-service business intelligence systems, efficient user training, and information accessibility. These elements enable users to make wiser choices that enhance client experiences. Additionally, according to the study, giving non-technical users too much independence without adequate assistance lowers customer satisfaction, maybe as a result of miscommunications or information abuse. This emphasises how crucial it is for businesses to strike a delicate balance between promoting user independence and offering sufficient direction. In the end, human factors – which allow users to utilise data responsibly are more important for effective self-service business intelligence adoption than technological ones. Businesses can increase their chances of gaining a competitive edge and satisfying customers by serving both.

9.6.1 Recommendations

Based on the findings of this study, the following recommendations were made:

i. In order for users to effectively develop precise insights that will increase customer satisfaction, organisations should invest in making self-service business intelligence tools more responsive, useful, and dependable.
ii. Provide thorough and continuous training programs that address data literacy and usage in addition to technical abilities so that users can utilise and comprehend self-service business intelligence results efficiently.
iii. Give departments easy, quick, and safe access to high-quality, well-integrated information so that decisions can be made that satisfy customers.
iv. Enable independence for non-technical users, and establish clear policies, support mechanisms, and validation protocols. This will lower mistakes and raise the standard of insights that lead to satisfied customers.

9.6.2 Limitations and Future Study Direction

Although the findings were satisfying for the concept, the research recognised numerous limits, which are deemed valuable predecessors for further investigation. The data was mostly collected from a restricted sample within the technology-oriented small company sector. The chapter's exclusive emphasis on sustainable manufacturing enterprises in Nigeria limits the application of its findings. The findings and implications of this chapter pertain specifically to Nigeria and primarily address the small company sector, thereby limiting the generalisability of the conclusions. The outcomes of this study, however restricted, should motivate researchers to do more extensive research on self-service business intelligence and consumer satisfaction. The chapter's cross-sectional design constrains the author's capacity to establish causality. Subsequent research should include constructs that were not used in this study.

REFERENCES

Adeoye, O. O., Olubiyi, T. O., Ajiteru, W. O., & Adaranijo, L. O. (2023). Unveiling Capital Structure Determinants, Business Characteristics and High-Growth Performance of Small and Medium Sized Enterprises: Evidence from Nigeria. *Journal of Interdisciplinary Socio-Economic and Community Study*, *3*(2), 86–98.

Adeyemi, O. S., & Olubiyi, T. O. (2023). The Impact of Brand Awareness on Customer Loyalty in Selected Food and Beverage Businesses in Lagos State Nigeria. *Jurnal Multidisiplin Madani*, *3*(3), 541–551. https://doi.org/10.55927/mudima.v3i3.3095

Adeyemi, O. S., & Olubiyi, T. O. (2024). An Analysis of Investment Trends, Innovation and Its Impact on Cabotage and Local Content Regime: Nigerian Experience from Businesses in Oil and Gas and Maritime Industries. *ABUAD Journal of Social and Management Sciences*, *5*(2), 285–301.

Aguiar-Costa, L. M., Cunha, C. A. X. C., Silva, W. K. M., & Abreu, N. R. (2022). Customer Satisfaction in Service Delivery with Artificial Intelligence: A Meta-Analytic Study. *Revista de Administracao Mackenzie*, *23*(6), 1–29. https://doi.org/10.1590/1678-6971/eRAMD220003.en

Amiruddin, D., Raharja, S. J., & Purnomo, M. (2024). The Influence of Organisational Climate in ERP Implementation on Information System Service Quality and User Satisfaction: A Study on an Indonesian Petrochemical Company. *Review of Integrative Business and Economics Research*, *14*(1), 276–295. https://lembarsaham.com/daftar-emiten/industrial/B11/barang-kimia?page=2

Andrade, I. M., De, & Tumelero, C. (2022). Increasing Customer Service Efficiency Through Artificial Intelligence Chatbot. *Revista de Gestao*, *29*(3), 238–251. https://doi.org/10.1108/REGE-07-2021-0120

Arokodare, M. A., Falana, R. B., & Olubiyi, T. O. (2023). Strategic Agility, Knowledge Management and Organisational Performance in a Post-Pandemic Era: A Mediation Analysis. *Journal of Family Business and Management Studies*, *27*, 83–108.

Arokodare, M. A., & Olubiyi, T. O. (2023). Crises as Harbingers of Opportunities: An Empirical Insight into the Moderating Effect of Knowledge Management on the Absorptive Capacity-Strategic Agility Relationship. *Malaysian Management Journal (MMJ)*, *27*, 83–108.

Bakır, M., & Sak, F. S. (2025). Charting the Landscape of Customer Experience Research in the Business Field: A Bibliometric Exploration. *ESIC Market*, *56*(1), 1–26. https://doi.org/10.7200/esicm.56.379

Borodako, K., Berbeka, J., Rudnicki, M., & Łapczyński, M. (2023). The Impact of Innovation Orientation and Knowledge Management on Business Services Performance Moderated by Technological Readiness. *European Journal of Innovation Management*, *26*(7), 674–695. https://doi.org/10.1108/EJIM-09-2022-0523

Brunner, M., Tripathi, S., Gundolf, K., Bachmann, N., Thienemann, A. K., Tüzün, A., Pöchtrager, S., & Jodlbauer, H. (2025). Enhancing Customer Satisfaction through

Digitalization: Past, Present and Possible Future Approaches. *Procedia Computer Science*, *253*, 2929–2940. https://doi.org/10.1016/j.procs.2025.02.017

Buhasho, E., Wausi, A., & Njihia, J. (2021). Moderating Effect of Organisational Capability on the Relationship between Business Intelligence Capability and Performance Among Public Listed Firms in Kenya. *European Scientific Journal (ESJ)*, *17*(1), 335–352. https://doi.org/10.19044/esj.2021.v17n1p335

Dang, T. D., & Nguyen, M. T. (2023). Systematic Review and Research Agenda for the Tourism and Hospitality Sector: Co-Creation of Customer Value in the Digital Age. *Future Business Journal*, *9*(1), 1–4. https://doi.org/10.1186/s43093-023-00274-5

Darda, P., Olubiyi, T. O., Ayodeij, O. T., & Obisanya, A. R. (2025). Leveraging Disruptive Technology on Profitability of Small Businesses in Industry 5.0. In T. Olubiyi, S. Suppiah, & C. Chidoko (Eds.), *The Future of Small Business in Industry 5.0* (pp. 115–132). IGI Global Scientific Publishing. https://doi.org/10.4018/979-8-3693-7362-0.ch005

Ellitan, L., Sugiyanto, S. A., & Risdwiyanto, A. (2022). Increasing Repurchase Intention through Experiential Marketing, Service Quality, and Customer Satisfaction. *Jurnal Maksipreneur: Manajemen, Koperasi, Dan Entrepreneurship*, *12*(1), 164–182. https://doi.org/10.30588/jmp.v12i1.1022

Ibrahim, N. M., Abiduzzaman, S. M., Abdul Raziff, A. R. B., & Shah, A. (2025). A Collaborative Filtering Approach Using Machine Learning and Business Intelligence: A Critical Review. *International Journal on Perceptive and Cognitive Computing (IJPCC)*, *11*(1), 41–49. https://journals.iium.edu.my/kict/index.php/IJPCC/article/view/501

Ilieva, R., Ivanova, M., Peycheva, T., & Nikolov, Y. (2021). Modelling in support of decision making in business intelligence. In *Integration Challenges for Analytics, Business Intelligence, and Data Mining* (pp. 115–144). IGI Global. https://dx.doi.org/10.4018/978-1-7998-5781-5.ch006

Izzuddin, M. D., & Abidin, Z. (2025). Service Quality; Self-Service Technology; Customer Loyalty. *Cognition Affective Behavior. Journal of Advances in Information Systems and Technology*, *7*(1), 1–14.

Kumah, R., Han, S., Okunola, A., & Mondal, S. (2025). Designing an Intuitive Real-Time Financial Fraud Detection System for Non-Technical Users. *Research Gate*, 1–11. https://www.researchgate.net/publication/388036717

Lennerholt, C., Van Laere, J., & Söderström, E. (2021). User-Related Challenges of Self-Service Business Intelligence. *Information Systems Management*, *38*(4), 309–323. https://doi.org/10.1080/10580530.2020.1814458

Maghsoudi, M., & Nezafati, N. (2023). Navigating the Acceptance of Implementing Business Intelligence in Organisations: A System Dynamics Approach. *Telematics and Informatics Reports*, *11*, 1–12. https://doi.org/10.1016/j.teler.2023.100070.

Mallikarjuna, P., Nitin, L. R., & Jayesh, R. (2024). Big Data Analytics, Artificial Intelligence, Machine Learning, Internet of Things, and Blockchain for Enhanced Business Intelligence. *Partners Universal Multidisciplinary Research Journal*, *1*(2), 110–133. https://doi.org/10.5281/zenodo.12827323

Naidu, P., Jais, J., & Khudari, M. (2025). Probing into Digitalisation's Influence on Internal Customer Satisfaction: An Angle from Internal Customers of Shared Service Organisations. *Edelweiss Applied Science and Technology*, *9*(5), 282–299. https://doi.org/10.55214/25768484.v9i5.6827

Olubiyi, T. O. (2024b). Improving Customer Experience Through Differentiation Capability: The Rise of Competition among Africa's Consumer Goods Companies. *PaperASIA*, *40*(5b), 230–240. https://doi.org/10.59953/paperasia.v40i5b.233

Ramasamy, G., Ramasamy, G. D., & Ramasamy, P. (2024). Conceptual Review of Consumer Satisfaction Theories with Expectation-Confirmation and Disconfirmation Paradigm for Business Sustainable Growth and Decision Making. *F1000Research*, *13*, 1–16. https://doi.org/10.12688/f1000research.158612.1

Shojaei, R. S., & Burgess, G. (2022). Non-technical Inhibitors: Exploring the Adoption of Digital Innovation in the UK Construction Industry. *Technological Forecasting and Social Change*, *185*, 1–13. https://doi.org/10.1016/j.techfore.2022.122036

Soeharso, S. Y. (2024). Customer Satisfaction as a Mediator between Service Quality to Repurchase Intention in Online Shopping. *Cogent Business and Management*, *11*(1), 1–11. https://doi.org/10.1080/23311975.2024.2336304

Su, H. T. (2021). Impact of Self-Service Factors on Customer Value: How Customer Behavioral Intentions are Formed? *International Journal of Management, Economics and Social Sciences*, *10*(1), 32–48. https://doi.org/10.32327/ijmess/10.1.2021.3

Wang, G. (2024). The Issue of Interconnections and Mutual Influence of Digital Transformation in International Trade: How Digitization Affects Customer Satisfaction? *Heliyon*, *10*(18), 1–14. https://doi.org/10.1016/j.heliyon.2024.e36854

Žigienė, G., Rybakovas, E., Vaitkienė, R., & Gaidelys, V. (2022). Setting the grounds for the transition from business analytics to artificial intelligence in solving supply chain risk. *Sustainability*, *14*(19), 1827. https://doi.org/10.3390/su141911827

Zungu, N. P. G., Amegbe, H., Hanu, C., & Asamoah, E. S. (2025). AI-driven Self-Service for Enhanced Customer Experience Outcomes in the Banking Sector. *Cogent Business and Management*, *12*(1), 1–21. https://doi.org/10.1080/23311975.2025.2450295

10 AIoT-Driven Robotics System in Smart Manufacturing Sectors

Pradeep Padhamnath

10.1 INTRODUCTION

A major shift in the industrial sector is represented by smart manufacturing, which uses innovative digital technologies to improve conventional production processes. To develop highly automated and data-driven production environments, smart manufacturing fundamentally combines cyber-physical systems, the Internet of Things (IoT), artificial intelligence (AI), machine learning (ML), and big data analytics [Haldorai (2023)]. The objectives of Industry 5.0 are realized as factories operate more effectively, adaptably, and responsively thanks to this digital transformation. Intelligent digital technology and physical production have come together to allow for smooth communication between humans, machines, and systems [Radanliev et al. (2021)].

Increasing customer demands, shorter product life cycles, rising operating expenses, and the need for sustainability are some of the contemporary issues that have prompted the development of smart manufacturing. Agile production, predictive maintenance, and real-time monitoring are areas where traditional manufacturing techniques frequently fall short, notwithstanding their historical effectiveness [Lee et al. (2020)]. By allowing real-time data collection from equipment and processes, smart manufacturing gets beyond these restrictions. The collected data is then evaluated to optimize operations, cut waste, and enhance product quality. This ecosystem is further supported by cloud and edge computing, which guarantees quick access to voluminous data and facilitates decision-making.

To stay competitive and achieve sustainable industrial growth, governments and businesses around the world are investing in smart manufacturing. Industrial acceptance of smart manufacturing processes are being pushed by initiatives such as India's "Make in India" campaign, the United States' "Advanced Manufacturing Partnership," and "Industry 5.0" with a focus on digital transformation. The manufacturing workforce is changing because of these initiatives, necessitating the development of new skill sets centered on automation, systems integration, and data science [Ullah et al. (2025)]. Smart manufacturing is a strategic change that rethinks how goods are created, manufactured, and delivered in the contemporary economy, and has more intrinsic value than being just a technical breakthrough.

The transition from Industry 4.0 to Industry 5.0 signifies a substantial change in industrial paradigms, going beyond efficiency and automation to a more sustainable

 DOI: 10.1201/9781003503323-10

and human-centered strategy. Industry 4.0, which is characterized by the integration of big data, AI, the IoT, and cyber-physical systems, revolutionized manufacturing [Yao et al. (2019)]. As a result, "smart factories" were developed, which used substantial automation and interconnection to optimize production processes, boost flexibility, and reduce costs. These technologies opened the door for intelligent manufacturing systems that could quickly adjust to changing conditions and customer needs by enabling real-time communication, data sharing, and autonomous decision-making among machines and systems.

The ability to use data from raw materials to the delivery of the finished product is the cornerstone of smart manufacturing. Cloud platforms, linked devices, and embedded sensors enable ongoing data collecting and analysis, enabling real-time quality control, adaptive scheduling, and predictive maintenance [Abdulshaheed et al. (2019)]. This change improves product traceability and customization while cutting down on waste and downtime. Additionally, firms improve their capabilities to respond to consumer preferences and market developments by making rapid yet well-informed decisions thanks to AI-driven data analysis [Kim et al. (2022)].

Smart manufacturing offers a strategic answer as industries throughout the world are under increasing pressure to improve sustainability and preserve their competitiveness in the global market. Through the integration of innovative technologies with lean concepts [Varriale et al. (2025)], it promotes resource optimization, energy efficiency, and a lessened environmental effect. To facilitate the adoption of smart manufacturing, public and the private sectors are aggressively encouraging digital transformation through finance and policies. In addition to being technological, this revolution is organizational in nature, necessitating process reengineering, staff upskilling, and promoting a constant innovation culture.

But as technology develops further, Industry 5.0 is starting to take shape. To develop more individualized, robust, and sustainable production systems, Industry 5.0 places a strong emphasis on human collaboration with cutting-edge technologies including AI, robotics, and biotechnology. In Industry 5.0, the emphasis is on human-centric innovation, creativity, and well-being rather than just efficiency and automation. To tackle complicated challenges, improve customer experience, and make sustainable progress, integrating human intelligence with technical breakthroughs is of prime importance. Using AI and other technologies, Industry 5.0 allows humans to concentrate on high-value jobs that call for empathy, creativity, and problem-solving abilities, while robots take care of routine and repetitive tasks. It is anticipated that this constructive collaboration will spur innovation, enhance work-life balance, and open up fresh avenues for social and economic advancement.

By combining the sophisticated analytical powers of AI with the vast data collection capabilities of IoT devices, Artificial Intelligence of Things (AIoT) is transforming contemporary industrial scene [Rahmayanti (2024)]. Industries can advance from simple data monitoring to intelligent, self-governing decision-making and real-time optimization because of these constructive interactions. AIoT enables predictive maintenance by evaluating enormous datasets from linked sensors and machinery, spotting equipment faults before they happen and drastically cutting downtime. Intelligent automation, dynamic manufacturing process modifications depending on current conditions, and improved quality control through AI-powered inspections all contribute to the optimization of operational efficiency [Chander et al. (2022)]. In addition, AIoT enhances supply chain visibility, resource management, and overall system resilience, bringing in the age of self-optimizing "smart factories" and generating notable cost reductions and productivity increases in a range of industrial sectors.

To meet the urgent industrial needs for automation, accuracy, and real-time adaptability, this chapter aims to investigate the revolutionary integration of AIoT with robotic systems for smart manufacturing [Jayadatta (2023)]. Designing and implementing intelligent, networked robotic architectures that use AI algorithms, edge computing, cloud analytics, and IoT-enabled sensors to facilitate autonomous operations and predictive maintenance is included in the scope [Kavitha et al. (2025)]. Digital twin simulations for process improvement, ML methods for anomaly detection and decision-making, and the creation of adaptive manufacturing processes via low-latency communication protocols are critical areas of attention. By looking at these elements, the chapter shows how AIoT-driven robotics improve operational responsiveness, flexibility, and efficiency, establishing it as a key Industry 5.0 enabler.

10.2 LITERATURE REVIEW AND METHODOLOGY

Early 20th-century automation and mechanical engineering pioneers established the theoretical foundation for the historical evolution of industrial robots. In 1954, George Devol created Unimate, the first industrial robot. This robotic arm represents a major turning point in the use of robotics in industrial processes since it was created to conduct repetitive operations in production. Industrial robots gained popularity in the 1960s and 1970s, especially in the automobile sector. Manufacturing environments have become safer and more efficient because of the early use of robotic systems by companies like General Motors for operations like welding and assembling. As the potential of industrial robots to revolutionize production lines became increasingly apparent, their use increased dramatically by the end of the 1970s. Industrial robots continued to develop into the 1980s and beyond, propelled by developments in automation, AI, and computer technology. Automated systems have been used for more complicated activities like quality control and precision assembly as they have become more advanced. Industrial robotics is now a fundamental component of contemporary manufacturing, with millions of robots in use globally that are constantly changing the face of industry and production.

An era of previously unheard-of efficiency, intelligence, and flexibility is being ushered in by the rapid transformation of the manufacturing landscape brought about by the convergence of AI and the IoT [Lele (2018); Haleem et al. (2019)]. The extensive use of predictive maintenance, in which IoT sensors continuously track the functionality and health of equipment and provide data to AI algorithms, is a prominent trend. These algorithms analyze patterns and predict potential failures before they occur, enabling manufacturers to schedule maintenance proactively, minimize costly downtime, and extend asset lifespans. This shift from reactive to initiative-taking maintenance strategies is significantly boosting operational efficiency and reducing unforeseen production disruptions.

The widespread use of digital twins and real-time operational optimization is another significant trend. Real-time data from IoT sensors is used to continuously update digital twins, which are virtual representations of real assets, procedures, or entire factories. AI then analyzes this data to detect bottlenecks, optimize production workflows, simulate different scenarios, and even evaluate new product designs or process modifications without affecting ongoing business operations [Feldt, et al. (2019); Subramaniyan et al. (2021)].This combination gives producers rich insights and the capacity to quickly make data-driven decisions, which improves quality control, lowers waste, and boosts productivity all around. Additionally, the growing incorporation of AI-driven analytics straight into IoT devices—which is frequently made possible by edge computing—allows for quicker reactions and real-time decision-making, lowering latency and need on centralized cloud processing.

Supply chain management is being drastically changed by AI and IoT, which is also encouraging increased sustainability in production. With the help of IoT devices, inventory and shipment tracking can be done in real time, giving supply chain visibility never before possible. AI then makes use of this data to estimate demand, improve supply chain resilience to interruptions, and facilitate logistical optimization. This connection is important for sustainability initiatives in addition to efficiency. Manufacturers may lower their carbon footprint and comply with increasingly stringent sustainability laws by using AI-powered systems that are fed data from the IoT. These systems can analyze environmental effects, uncover inefficiencies that result in waste, and monitor and optimize energy consumption. This all-encompassing strategy, powered by intelligent connection, is opening the door for manufacturing processes that are more intelligent, flexible, and ecologically conscious.

Advanced robotic solutions are quickly relying on the confluence of AI and the IoT, sometimes referred to as AIoT. Because of this synergy, robots can embrace true intelligence and autonomy and transcend pre-programmed jobs [Khang et al. (2025)]. Robots powered by AI can gather enormous volumes of data from their surroundings in real time, interpret it at the edge, and make wise judgments without continual human assistance by directly integrating AI algorithms into IoT-connected devices.

Because they can manage complicated problems that traditional robotics or standalone AI systems cannot, AIoT-driven robotic solutions are becoming increasingly necessary. By evaluating sensor data and responding to shifting conditions, AIoT robots in manufacturing can optimize production lines, provide predictive maintenance, and conduct exact quality control. AIoT is used by autonomous mobile robots (AMRs) and autonomous mobile industrial robots (AMIRs) in warehousing and logistics industries to improve route planning, manage inventory effectively, and collaborate safely with humans [Rath et al. (2025)]. Furthermore, the emergence of smart factories is converting conventional manufacturing settings into intricately linked ecosystems. IoT devices that are integrated into tools, equipment, and even worker wearables provide smooth data flow and communication, improving production process visibility. In addition to increasing operational effectiveness, this highly interconnected environment encourages cooperation between humans and robots, opening the door for a more robust and adaptable industrial sector.

Another crucial area of focus is sustainability, as manufacturers use AI and IoT technology increasingly to lessen their environmental effect. Businesses are not only satisfying legal obligations but also satisfying consumer demand for greener practices by minimizing energy use, reducing waste, and putting circular economy principles into effect. Manufacturing is positioned as a leader in the shift to a more sustainable future due to the use of eco-friendly materials and renewable energy sources in industrial processes.

AI and IoT integration in manufacturing is expected to completely transform efficiency and productivity by 2030. Predictive maintenance, in which AI evaluates data from IoT devices to anticipate equipment breakdowns, and improved automation, which permits more intelligent, flexible production processes, are important themes. By moving toward intelligent systems, firms will improve resource allocation and reduce downtime, which eventually results in significant cost savings. Businesses may respond to shifts in demand and operating situations with previously unheard-of agility by using real-time data from connected devices to make well-informed decisions fast.

10.3 FUNDAMENTALS OF AIoT IN ROBOTICS

The IoT and AI have powerfully converged to form the AIoT. Fundamentally, AIoT gives connected devices intelligence, enabling them to evaluate, analyze, and make wise decisions in real time in addition to gathering enormous volumes of data. AIoT uses AI algorithms (such as ML and deep learning) to generate actionable insights, forecast trends, and automate responses, in contrast to traditional IoT, which concentrates on data collection and connectivity. This, in turn, leads to more effective operations, better human-machine interactions, and improved data management. The fundamental elements of an AIoT system include edge computing for localized processing that lowers latency and improves real-time decision-making, IoT devices (sensors, actuators, and cameras) that collect data from the physical environment, and AI algorithms for processing and analyzing this data.

Additionally, cloud computing is essential for improved analytics and larger data storage. Significant benefits of AIoT in the current market include improved operational efficiency through automation and optimization, predictive maintenance to avoid expensive downtime, improved decision-making skills because of real-time data analysis, and the development of customized experiences across multiple industries. For companies implementing AIoT solutions, these advantages translate into significant cost savings, enhanced sustainability, and a competitive advantage.

10.3.1 Architectural Overview of AIoT in Robotics

A meaningful change in the existing paradigm is brought about by the incorporation of AIoT into robotics, which turns robots from programmed machines into sentient, intelligent, and flexible beings. Each layer is essential to enable intelligent robotic activities, and the architectural overview of AIoT in robotics usually consists of multiple layers. A variety of IoT sensors, including cameras, LiDAR, ultrasonic sensors, temperature sensors, force sensors, encoders, and actuators, are integrated into the physical robotic system itself, which forms the device layer at the base. The internal condition of the robot (such as motor currents and joint locations) and its surroundings are continuously monitored by these sensors, which gather enormous volumes of raw data. Additionally, this layer contains the robot's embedded processing units, which may be used for local control and preliminary data filtering.

The internal condition of the robot (such as motor currents and joint locations) and its surroundings are continuously monitored by these sensors, which gather enormous volumes of raw data. Additionally, this layer contains the robot's embedded processing units, which may be used for local control and preliminary data filtering. Data can be delivered to an edge computing layer or straight to the cloud, depending on the bandwidth availability and latency needs of the application. The edge computing layer is a critical component for AIoT robotics, as it allows for real-time processing and analysis of data closer to the source (the robot). This localized intelligence eliminates the delay that comes with cloud connectivity, allowing for quick decision-making—essential for tasks like dynamic path planning, obstacle avoidance, and precise manipulation that call for quick reactions. Frequently, lightweight AI models trained for tasks are housed at this layer.

Finally, a centralized platform for high-performance computing, large data storage, and the training of more sophisticated and computationally demanding AI models (such as deep learning models for advanced perception or predictive maintenance) is offered by the cloud layer. This layer manages long-term trend monitoring, fleet management for several robots, global optimization, and ongoing learning for robotic systems. This infrastructure is supported by the application layer, which offers user interfaces, dashboards for tracking robot performance, and interaction with other business systems. The concept is based on a continuous feedback loop: the robot's IoT sensors gather data, which is then sent to the edge or cloud for analysis and decision-making driven by AI. The actuators then receive instructions to perform actions, allowing for intelligent and adaptive behavior. AIoT-enabled robots may function independently, learn from their experiences, and improve their performance over time because of this never-ending loop of detecting, analyzing, deciding, and acting.

10.3.2 Benefits and Limitations of AIoT Integration

"Artificial Intelligence of Things," or AIoT, is a term that embodies the potent combination of AI with the IoT. Through this connection, IoT devices—which often gather enormous volumes of raw data—can become "smarter" by utilizing AI's analytical and decision-making powers. The main advantage is that basic data collection may be transformed into insightful conclusions and automated processes. This results in increased operational efficiency in a number of industries, from smart cities that optimize waste management and traffic flow to smart factories that use predictive maintenance. Additionally, AIoT enables proactive risk management and real-time monitoring, enabling systems to identify irregularities and take prompt remedial action to reduce expensive downtime and increase overall reliability.

Even with its revolutionary potential, integrating AIoT has serious drawbacks. Data security and privacy are key concerns since AIoT systems gather and analyze large volumes of sensitive data, which leaves them open to cyberattacks. Processing, storing, and guaranteeing effective network connectivity are all made more difficult by the sheer amount of data produced by IoT devices, particularly in large-scale deployments. A smooth integration can be difficult and expensive because of the wide variety of devices, platforms, and communication protocols that make up an AIoT ecosystem.

Furthermore, the intricacy of overseeing and preserving AIoT systems sometimes necessitates specific knowledge, which raises the initial outlay for hardware, software, and qualified staff. There are also constant difficulties with ethical issues, such as biases in AI systems and the requirement for open decision-making procedures. The significance of strong fail-safe procedures and transparent accountability is highlighted by the possibility of unexpected repercussions or a loss of control resulting from an over-reliance on AIoT systems without adequate human monitoring. For AIoT technologies to be widely and responsibly adopted, several constraints must be addressed.

10.3.3 Comparison with Traditional Robotic Systems

The integration of AIoT differs from traditional robotic systems by introducing "intelligence" and connectivity into the real environment. Even though traditional robots are highly effective at repetitive, preprogrammed activities, they frequently work alone or in carefully regulated surroundings. In certain static tasks, including assembly line production or simple material handling, they are exceptionally accurate and quick. Due to their lack of adaptive learning and real-time decision-making capabilities, which are characteristics of AIoT, they are limited in situations involving variability or unanticipated events. Since their intelligence is usually hard-coded, any changes to their preset programming must be manually reprogrammed.

Conversely, AIoT enables robotic systems to be much more independent, flexible, and cooperative. AIoT-enabled robots can gather and analyze enormous volumes of data from their surroundings in real-time by combining AI, IoT devices, and frequently edge computing. This enables them to anticipate foreseeable problems (such as equipment failure for predictive maintenance), learn from past mistakes, and make wise judgments without continual human involvement. In contrast to typical, fixed-path robotic systems, an AIoT-powered robot in a warehouse, for instance, can utilize real-time inventory data to optimize its course or dynamically reroute itself to avoid obstructions.

AIoT has numerous advantages over conventional robotics. In addition to specific robotic activities, AIoT enables proactive problem-solving and optimization across large ecosystems. Because of the ongoing learning and adaptation, operational efficiency is increased, downtime is decreased, and quality control is improved.

While conventional robots are designed to perform single, repetitive tasks, AI enables more complex, dynamic interactions with the environment and other interconnected systems. This opens the door to more advanced applications, like self-driving cars that can navigate unpredictable traffic or smart factories that can dynamically modify production in response to machine performance and demand in real time. The primary distinction and the path forward for robotics is this transition from strict automation to cognitive autonomy.

10.3.4 AIoT-Based Robotic System Framework

AI and the IoT are combined in an AIoT-based robotic system framework to produce intelligent, networked robotic systems. Fundamentally, this framework uses IoT devices (sensors, actuators, and cameras) that are integrated into the robotic system and its surroundings to gather a significant quantity of data in real time. A central processing unit, which is frequently a cloud platform or edge computing device, receives this data, which includes anything from operating parameters and visual information to ambient variables and robot telemetry. The robot can correctly sense its environment and receive instructions or updates from a distance thanks to the IoT component, which guarantees smooth connectivity and data flow. Distributed intelligence is made possible by this interconnection, which optimizes latency, bandwidth, and computing capacity at multiple levels, from a single robot to a central server.

The incorporation of AI reveals the framework's true intelligence. AI algorithms are used to process, evaluate, and interpret the data once it has been gathered through the IoT. Essential elements include ML models, such as reinforcement learning for the best possible decision-making, natural language processing for human-robot interaction, and deep learning for image identification. The robot can anticipate potential problems, adjust to changing surroundings, learn from its experiences, and make decisions on its own thanks to these AI capabilities. Path planning, object detection and manipulation, predictive maintenance, and even the execution of complex tasks without continual human interaction are just a few examples of the applications of AI. Because AI and IoT work together, the robot is more than just a data collector; it is an intelligent agent that can reason, learn, and act depending on the information it has gathered from its interconnected data streams.

Thus, the all-encompassing AIoT-based robotic system framework creates a potent ecosystem in which robots are essential components of a broader intelligent network rather than standalone devices. Improved capabilities including fleet management, collaborative robotics, remote monitoring and control, and advanced automation across multiple industries are made possible by this connection. AIoT robots, for instance, may connect with other devices and systems, improve production lines, and conduct quality checks in smart factories. They can remotely monitor patients, help with procedures, and provide medications in the medical field. Robotic systems that are highly adaptive, efficient, and truly autonomous and capable of intelligently operating in complex and dynamic situations are made possible by the framework's ability to integrate real-time data collecting with sophisticated analytical and learning capabilities.

10.3.5 Sensor Integration and Real-Time Data Acquisition

The core component of an AIoT-based robotic system is made up of sensor integration and real-time data collection, which serve as the robot's "eyes and ears." This entails giving robots a wide range of sensors, including cameras for vision, LiDAR for mapping, ultrasonic sensors for proximity detection, accelerometers for tracking motion, temperature sensors, and even specialized sensors for jobs like pressure monitoring or gas detection. Large streams of unprocessed data are continuously gathered by these sensors from the robot's internal condition and immediate surroundings. The "real-time" component is essential since it means that data is gathered and sent with the least amount of delay possible, allowing the AI to act promptly, adjust to changing circumstances, and make well-informed decisions. The AI processing units, which may be located on the robot itself (edge computing) or on a cloud platform, receive this constant stream of sensory data. It serves as the vital "fuel" that the AI algorithms need to comprehend, learn, and direct the robot's intelligent behavior.

Case 1: Paint is sprayed on car bodies using a robotic arm in an automotive painting station. To guarantee consistent paint thickness across various parts, the robot must instantly modify its spray parameters in response to sensor inputs.

Integrated sensors:

LiDAR: Determines the robot's distance from the vehicle's surface.

Speed sensor: Determines the robotic arm's speed.

Real-time paint thickness measurement is possible with a paint thickness sensor.

The goal is to minimize underpainting and overpainting by using sensor data to dynamically manage the paint flow rate and maintain a target paint thickness of 50 μm.

The work analyzes the correlation of paint thickness with distance and speed of robotic arm as mentioned in table 10.1.

Adjust flow rate dynamically using an inverse relation:

$$\textit{Flow Rate}\ \alpha\ \frac{1}{\textit{Distance} \times \textit{Speed}}$$

The applied paint thickness varies over the 10-second interval, as seen in the first output image, "Paint Thickness vs Time," (Figure 10.1) with several departures from the desired value of 50 μm (shown by the red dashed line). In particular, the paint thickness increases to 59 μm at timestamps 2–4 and 9–10 seconds, which is far higher than the aim. This relates to times when the robotic arm was either moving

TABLE 10.1
Sample from 10 Seconds

Time (s)	Distance (cm)	Speed (cm/s)	Paint Thickness (μm)
0	25	10	45
1	24.5	12	47
2	23	15	52
3	21	18	58
4	22.5	16	55
5	24	13	49
6	25.5	11	46
7	26	10	44
8	24	12	48
9	22	17	56
10	21.5	19	59

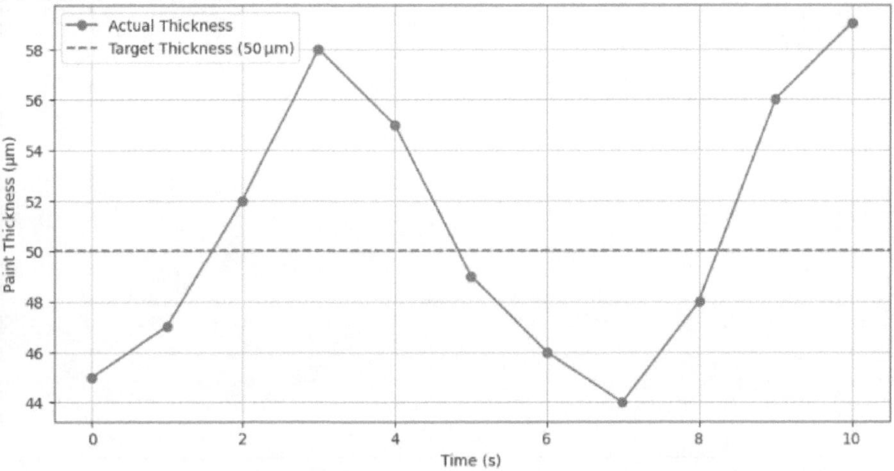

FIGURE 10.1 Paint thickness versus time.

more quickly or closer to the surface, suggesting that uncontrolled variations in these parameters result in uneven paint application. To reduce both underpainting and overpainting, this figure highlights the necessity of adjusting the paint spray in real-time based on live sensor feedback. This comprehension is further enhanced by the second and third output figures, 10.2 and 10.3. The correlation heatmap as shown in figure 10.2 indicates a strong negative relationship between paint thickness and both speed and distance; that is, as speed or distance increases, paint thickness drops.

Figure 10.3, which shows a distinct rising trend between the computed control factor (1 / (Distance × Speed)) and actual paint thickness, supports this conclusion.

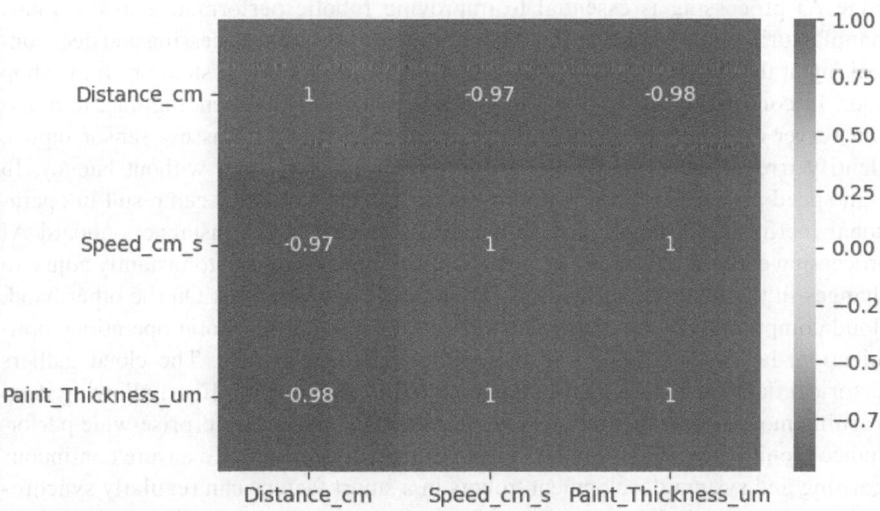

FIGURE 10.2 Correlation matrix.

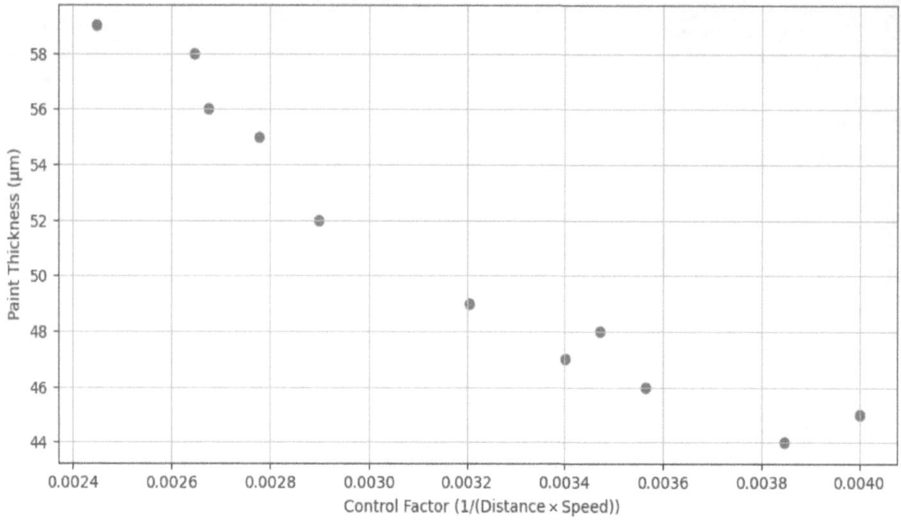

FIGURE 10.3 Effect of control factor on paint thickness.

The paint thickness increases in tandem with the control factor, demonstrating the effectiveness of this derived variable in predicting painting output. These observations support the use of control logic that dynamically modifies the paint flow in response to the combined effects of speed and distance, guaranteeing uniform surface coating and maximizing paint consumption.

10.3.6 Edge AI Processing and Cloud Computing Roles

Edge AI processing is essential to improving robotic performance in the smart manufacturing industry because it allows for real-time data processing and decision-making at the source, which can be either inside the robotic system or on the shop floor. In contrast to conventional techniques, which involve sending data to a distant server or cloud for analysis, Edge AI enables robots to assess sensor inputs, identify irregularities, and make modifications instantly and without latency. In high-speed production lines, where even milliseconds of delay can result in operational inefficiencies or product faults, this is essential. For instance, onboard AI processing enables a robot doing precision welding or painting to instantly adjust to changes in the material surface or surrounding circumstances. On the other hand, cloud computing offers a sturdy foundation for long-term robotic operations optimization, large-scale analytics, and centralized data storage. The cloud gathers historical data from various robotic units and manufacturing lines, allowing deep learning model training, predictive maintenance plans, and enterprise-wide performance monitoring, while Edge AI manages instant decisions. To ensure continuous learning and system development, robots in a smart factory can regularly synchronize with the cloud to update AI models, communicate operational metrics, and get software upgrades.

A hybrid architecture that strikes a balance between computational depth and responsiveness is produced by combining Edge AI with cloud computing. For example, Edge AI may spot faulty parts in a robotic quality inspection system instantaneously and notify operators, while the cloud can analyze batch trends to find systemic problems or recommend design changes. By increasing the autonomy and intelligence of robotic systems, this division of labor improves manufacturing's adaptability, efficiency, and resilience to disturbances.

Additionally, this synergy facilitates cooperation and scalability in international industrial networks. Robots may be remotely monitored, controlled, and coordinated across several locations thanks to cloud platforms, and Edge AI makes sure that every robotic unit runs precisely on its own. Because human-robot collaboration, customization, and sustainability are critical in Industry 5.0, this is especially advantageous. Automation, flexibility, and efficiency are increased in smart manufacturing by fusing real-time Edge AI decision-making with the analytical depth and scalability of the cloud.

10.4 METHODOLOGY AND IMPLEMENTATION

To facilitate intelligent decision-making, real-time monitoring, and adaptive automation, the approach and deployment of an AIoT-driven robotics system in smart manufacturing sectors entail the smooth integration of AI and the IoT. IoT-enabled sensors are installed on robotic systems to gather information on variables like motion, temperature, vibration, and pressure. To identify patterns, detect anomalies, and perform predictive maintenance, this data is sent to edge or cloud platforms where AI algorithms, such as ML and deep learning models, evaluate it. A centralized cloud for data storage and model training, edge computing for low-latency processing, and a feedback loop to actuators for self-reliant reactions are all parts of the system design. Strong connectivity, cybersecurity safeguards, scalable data infrastructure, and interoperability standards are necessary for implementation to guarantee dependable device-to-device communication. To meet the objectives of Industry 5.0, manufacturers can use AIoT to create intelligent, self-optimizing robotic systems that increase production efficiency, decrease downtime, and allow for mass customization.

10.4.1 Machine Learning Models for Anomaly Detection

ML models are essential to anomaly detection, which is the process of spotting odd patterns or behaviors that diverge from typical operations in AIoT-driven robotics systems utilized in smart manufacturing industries. These irregularities could be the result of unforeseen process deviations, cyberattacks, sensor malfunctions, or mechanical problems. Predictive maintenance, operational reliability, and intelligent automation are made possible by ML models that utilize massive real-time data streams from IoT-connected sensors integrated into robotic arms, conveyor systems, and control units. The purpose of data-driven intelligence-powered early warning systems is to increase productivity, save expenses, and limit downtime.

Let us look at an intelligent production line that employs collaborative robots, or cobots, for accurate assembly. IoT sensors on these robots allow them to track variables like vibration, temperature, torque, and motor current. Historical time-series sensor data that depicts typical operations is used to train a recurrent neural network model called a Long Short-Term Memory (LSTM) model. The LSTM uses historical sequences to anticipate the next set of values and learns temporal dependencies. The model indicates an anomaly when real-time values noticeably differ from expected values. For instance, the system might signal a probable motor wear or misalignment if the motor current suddenly surges over the typical range. Other methods, like autoencoders, try to recreate input data after compressing it into a lower-dimensional representation. An anomaly is indicated by a high reconstruction error. By randomly dividing the feature space and detecting the data points that need fewer partitions (likely outliers), isolation forests, on the other hand, isolate data points. Every model has advantages, such as isolation forests for sparse outliers, autoencoders for intricate non-linear patterns, and LSTMs for time-based anomalies. Self-adaptive anomaly detection and continuous learning are made possible by integrating these models into an AIoT ecosystem. Predictive intelligence and Industry 5.0 goals are supported by this combination of AI and real-time IoT data in robots, which turns manufacturing into a responsive, effective, and resilient environment.

10.4.2 Digital Twin Creation and Simulation Environment

In smart manufacturing industries, digital twin generation and simulation environments are essential for improving system performance, dependability, and decision-making in AIoT-driven robotics systems. A digital twin is a computerized model of a genuine robotic system that is constantly receiving data in real time from IoT sensors that are integrated into the actual asset. By simulating the robot's actions, surroundings, and behavior, this virtual model allows for defect diagnosis, performance optimization, and predictive analysis without interfering with real production. To assess results and direct in-the-moment decision-making, the digital twin uses AI algorithms to model different scenarios, such as workload shifts, component failures, or process modifications. Before deploying, engineers may test contemporary designs, streamline processes, and train robots in a risk-free virtual environment by integrating simulation environments with digital twins. A smart, flexible, and robust manufacturing ecosystem in line with Industry 5.0 is created by the convergence of AI, IoT, and digital twins. In this ecosystem, synchronized physical-virtual interactions enable real-time feedback and continuous development.

10.4.3 Predictive Maintenance Algorithm Design

Predictive maintenance algorithms are designed to anticipate equipment failures before they happen in AIoT-driven robotics systems within smart manufacturing sectors. This minimizes downtime and maximizes asset life. This is accomplished by gathering constant streams of sensor data from IoT-enabled robotic components, including temperature, vibration, current, and positional accuracy. To find patterns suggestive of wear or failure, ML techniques, such as Random Forests, Support Vector Machines (SVMs), and LSTM neural networks, are trained on both historical and real-time data. These models determine a component's remaining useful life (RUL) and send out maintenance scheduling reminders. In addition to deployment in edge or cloud contexts for real-time inference, the method design includes data preparation, feature extraction, model training, and validation. This initiative-taking strategy supports intelligent, self-monitoring robotics systems in accordance with Industry 4.0 and 5.0 efforts by lowering maintenance costs, ensuring operational continuity, and reducing unplanned outages.

10.4.4 Synchronized Multi-Robot Coordination Strategies

To achieve effective, adaptable, and high-throughput production processes in AIoT-driven robotics systems in smart manufacturing sectors, synchronized multi-robot coordination solutions are crucial. These tactics use AI algorithms and IoT connectivity to provide real-time data transmission, which allows many robots to coordinate cooperative efforts while working on shared or interdependent jobs. Workflow continuity, collision avoidance, and job allocation are all achieved through coordination strategies like behavior-based control, decentralized swarm intelligence, and centralized scheduling. Robots can adjust to unforeseen challenges or shifts in production demands thanks to AI models that evaluate sensor inputs and ambient data to enable dynamic decision-making. Robots can work together with the least amount of lag thanks to simulation and feedback loops that further improve synchronization through digital twin integration and edge computing. This clever coordination facilitates the flexible, adaptable manufacturing environments that Industry 5.0 envisions while increasing productivity and lowering energy use.

10.4.5 Use Cases and Industrial Applications

Modern manufacturing has been revolutionized by AIoT-driven robotics systems through a variety of significant use cases and industrial applications. Collaborative robots (cobots) with AI and IoT sensors conduct accurate welding, painting, and assembly jobs in the auto industry while continuously assessing quality and anticipating faults. In the semiconductor and electronics sectors, robotic arms controlled by AI vision systems manage fragile parts during fast printed circuit board (PCB) assembly, guaranteeing precision and minimizing flaws. Predictive analytics is used in pharmaceutical and food processing facilities to reduce contamination and equipment failures through the deployment of AIoT-enabled robots for sanitary packaging, inspection, and logistics.

AMRs in smart warehousing and logistics navigate warehouse layouts, monitor inventory, and optimize picking routes utilizing AI algorithms and IoT connectivity. In heavy industries, predictive maintenance systems are frequently used. Robots use sensor data to check their own health and notify maintenance staff before problems arise, minimizing downtime. Additionally, multi-robot coordination allows robots to interact and flexibly reorganize duties, allowing smart factories to manage specific production requirements with flexibility. These applications demonstrate the revolutionary potential of AIoT in building robust, effective, and adaptive manufacturing ecosystems that support Industry 4.0 and 5.0 objectives.

10.4.6 Case Studies from Manufacturing Sectors

BMW incorporated AIoT technologies for robot synchronization and automated visual inspection into its production lines. AI algorithms identify abnormalities in real time, while IoT sensors on robotic arms record information like torque, position, and vibration. AI-driven camera systems, for instance, confirm the accuracy of welding and fastening procedures. A centralized digital twin platform is used by robotic cells to coordinate, increasing line balance and cutting inspection time by 50%. This improves production precision and overall equipment effectiveness (OEE).

Programmable logic controller (PLC) production at Siemens' Amberg factory is highly automated through the use of AIoT-driven robotics. By simulating the behavior of genuine robotic systems, digital twins allow for real-time diagnosis and simulation-based optimization. To evaluate wear and initiate maintenance plans, predictive maintenance algorithms examine sensor data from robot joints and actuators. Consequently, unscheduled downtimes were almost eliminated and machine availability increased to nearly 99.8%.

Robotic assembly systems based on AIoT were integrated into Foxconn's smartphone manufacturing lines. Robots with IoT sensors and Edge AI units can adapt to different product designs on their own. Deep learning models use real-time data from force sensors and visual tracking systems to modify the robot's behavior for low-volume, high-mix output. Additionally, IoT analytics monitor energy consumption, allowing for power consumption optimization that lowers energy expenditures by 15% while improving assembly consistency and speed.

These case studies illustrate practical implementations of Industry 5.0 concepts by showing how AIoT in robotics improves quality, productivity, and sustainability in smart manufacturing.

10.5 RESULTS AND PERFORMANCE EVALUATION

AIoT-driven robotics is used in a smart manufacturing facility to perform automated assembly activities. The system incorporates AI algorithms for task scheduling, anomaly detection, and predictive maintenance with IoT sensors for real-time monitoring. Throughout 30 production cycles, performance is assessed using the following criteria:

- Time spent on the task (in seconds)
- The quantity of defects found
- Time spent idle by robot (in seconds)
- Energy use (measured in kWh)

Analyzing the system's dependability and efficiency in order to spot any patterns or problems is the aim.

The data set is mentioned in the Table 10.2. Based on this data of the table 10.2, the solution and performance metrics analysis for average task completion time, defect rate (total defects/total cycles), average robot idle time, average energy consumption,

TABLE 10.2
Sample for 10 Cycles

Cycle No	Task Completion Time (s)	Defects Detected	Robot Idle Time (s)	Energy Consumption (kWh)
1	120	2	15	1.8
2	115	1	10	1.7
3	130	3	20	2
4	118	0	12	1.6
5	125	2	18	1.9
6	110	1	8	1.5
7	122	0	10	1.7
8	117	1	11	1.6
9	128	3	22	2.1
10	115	1	9	1.5

and correlation between task completion time and energy consumption (efficiency indicator) are shown below.

- Average task completion time: 119.40 seconds
- Total defects detected: 32
- Defect rate per cycle: 1.07
- Average robot idle time: 12.70 seconds
- Average energy consumption: 1.72 kWh
- Correlation (task time vs energy consumption): 0.93

The graphical analysis as shown in Figure 10.4 offers valuable insights into how well the AIoT-powered robots system performs in the context of smart manufacturing.

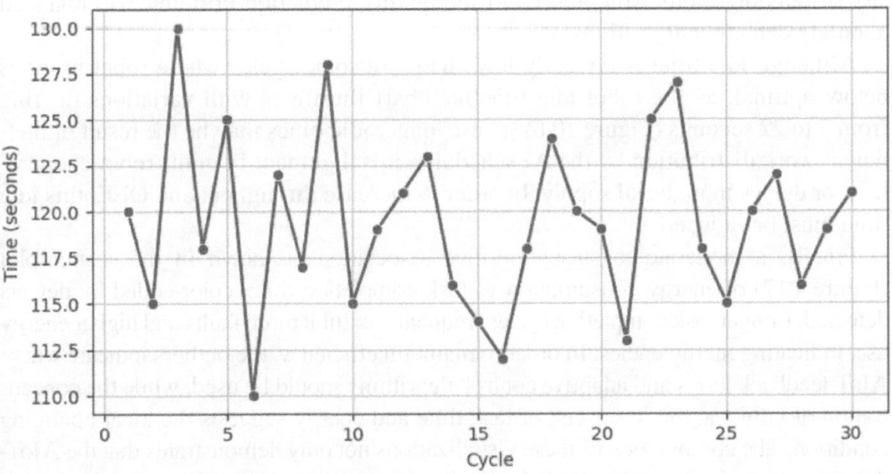

FIGURE 10.4 Task completion time per cycle.

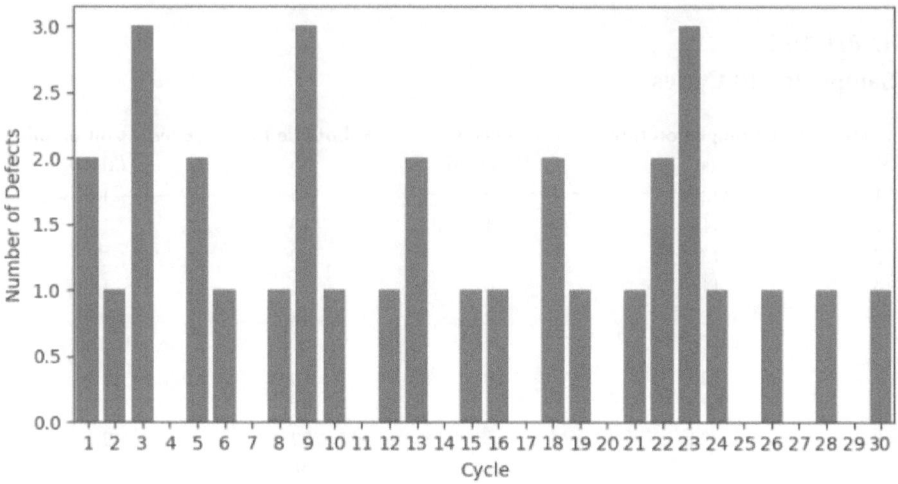

FIGURE 10.5 Defects detected per cycle.

With the majority of cycles taking between 115 and 125 seconds, the task completion time per cycle line plot shows a somewhat regular pattern. Occasionally, spikes that fall outside of this range indicate operational irregularities or an increase in workload, which suggests areas in which the system can profit from additional optimization or AI algorithmic prediction. However, the consistency shows that the robotic system can execute tasks in a generally stable and dependable manner.

The problems detected per cycle bar graph illustrated in Figure 10.5 shows that some cycles have up to three problems but the majority only record one or none at all. This uneven distribution raises the possibility that system irregularities or outside influences could be a factor in quality discrepancies. In order to enable manufacturers to proactively address the underlying reasons, these findings bolster the necessity of ongoing monitoring and real-time quality prediction utilizing AI-enhanced anomaly detection algorithms.

Although idle time is typically low, there are some cycles where robot usage is below optimal, as the robot idle timeline chart illustrates with variations ranging from 7 to 22 seconds (Figure 10.6). These longer idle times may be the result of inefficient work distribution by the AI scheduler, misalignment in multi-robot cooperation, or delays in material supply. In order to increase throughput and OEE, this idle time must be reduced.

Finally, a moderate to strong positive association is shown in the scatter plot (Figure 10.7) of energy consumption vs task completion time, color-coded by defects detected. Longer task completion cycles frequently exhibit more faults and higher energy use, indicating inefficiencies. In order to maintain efficiency, the outliers indicate where AIoT feedback loops and adaptive control algorithms should be used, while the concentration of points at the lower end of task time and energy suggests the ideal operating condition. The combination of these visualizations not only demonstrates that the AIoT-driven system offers a strong basis for intelligent automation but it also provides practical insights for future improvement through self-learning and real-time optimization.

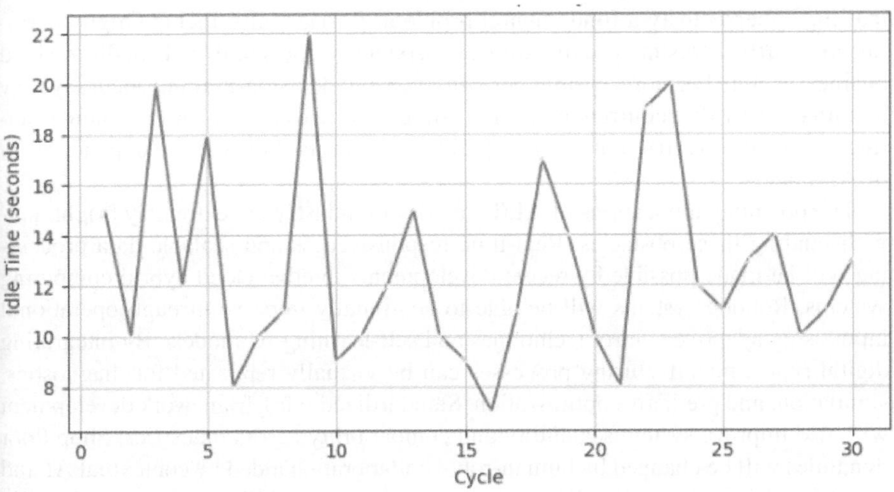

FIGURE 10.6 Robot idle time per cycle.

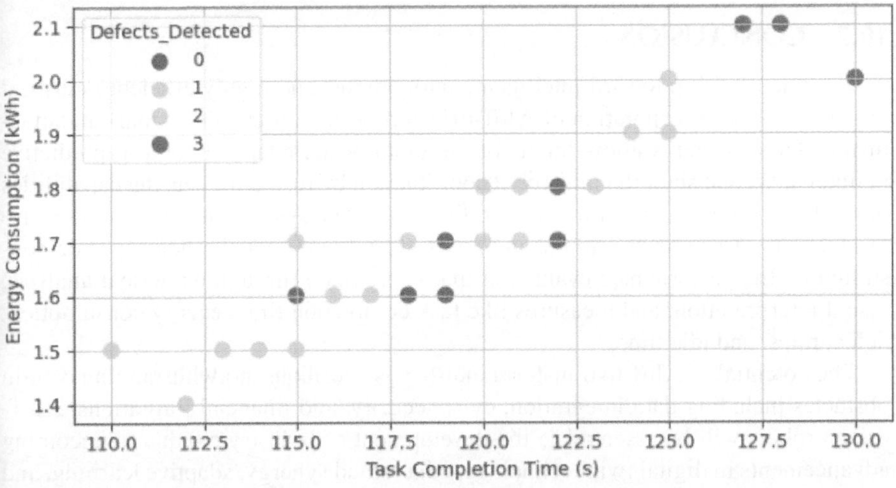

FIGURE 10.7 Energy consumption versus task completion time.

10.6 CHALLENGES AND FUTURE SCOPE

There are a number of significant obstacles to integrating AIoT-driven robotics in smart manufacturing. The intricacy of integrating real-time data from various IoT devices and industrial systems—which frequently lack standardization and interoperability—is one of the main problems. It is extremely challenging to guarantee data synchronization, accuracy, and consistency across sensors, devices, and AI models. Furthermore, heightened connections expose systems to risks including malevolent control, data breaches, and unauthorized access, making cybersecurity a fundamental problem. Barriers also include high operating and early investment expenditures, especially for small and medium-sized businesses. Furthermore, system performance and decision-making accuracy may be impacted by the requirement for real-time AI processing under low latency settings, as well as restrictions on the quality of training data and the dependability of AI models.

The potential applications of AIoT robotics in industry are extremely bright, notwithstanding these obstacles. Real-time responsiveness and scalable data processing will be made possible by recent developments in edge-cloud hybrid computing systems. Robotic systems will be able to continually improve through operational input using adaptive control techniques and self-learning AI models. By integrating digital twins, manufacturing processes can be virtually replicated for diagnostics, simulation, and predictive optimization. Standardized AIoT framework development will also improve system scalability and compatibility across industries. Shop floor dynamics will be changed by human-robot collaboration aided by contextual AI and user-friendly interfaces. Finally, advancements in sustainability, such as eco-friendly robotic designs and AI algorithms that consume less energy, will establish AIoT robotics as a major facilitator of intelligent and green production.

10.7 CONCLUSION

A revolutionary step toward intelligent, autonomous, and highly efficient industrial processes is the incorporation of AIoT-driven robotics systems into smart manufacturing. These systems allow for real-time monitoring, adaptive control, predictive maintenance, and smooth multi-robot coordination by fusing the sensing capabilities of the IoT with the decision-making ability of AI. The system's potential to increase productivity, decrease downtime, and boost overall operational efficiency is demonstrated by the performance evaluation that is provided through numerical analysis, visual interpretation, and measures like task completion time, energy consumption, defect rates, and idle time.

The potential of AIoT to transform industry is enormous, notwithstanding certain obstacles including data integration, cybersecurity, and financial limitations. AIoT-driven robots will be essential to the development of Industry 5.0 due to upcoming advancements in digital twin technology, edge-cloud synergy, adaptive learning, and sustainability-driven innovations. In addition to highlighting the present capabilities and outcomes of such systems, this work opens the door for additional study and industrial adoption with the goal of developing manufacturing ecosystems that are more intelligent, resilient, and sustainable.

REFERENCES

Abdulshaheed, H. R., Al Barazanchi, I., & Sidek, M. S. B. (2019). Survey: Benefits of integrating both wireless sensors networks and cloud computing infrastructure. Sustainable Engineering and Innovation, 1(2), 67–83.

Chander, B., Pal, S., De, D., & Buyya, R. (2022). Artificial Intelligence-Based Internet of Things for Industry 5.0. In Artificial Intelligence-Based Internet of Things Systems (pp. 3–45). Springer.

Feldt, J., Kontny, H., & Wagenitz, A. (2019). Breaking through the Bottlenecks Using Artificial Intelligence. In Artificial Intelligence and Digital Transformation in Supply Chain Management: Innovative Approaches for Supply Chains. Proceedings of the Hamburg International Conference of Logistics (HICL) (Vol. 27, pp. 30–56). epubli GmbH.

Haldorai, A. (2023). A review on artificial intelligence in Internet of Things and cyber physical systems. Journal of Computing and Natural Science, 3(1), 12–23.

Haleem, A., Javaid, M., & Khan, I. H. (2019). Current status and applications of Artificial Intelligence (AI) in medical field: An overview. Current Medicine Research and Practice, 9(6), 231–237.

Jayadatta, S. (2023). A study on latest developments in artificial intelligence (AI) and internet of things (IoT) in current context. Journal of Applied Information Science, 11(2), 21–28.

Kavitha, N., Nivethitha, T., Deebanchakkarawarthi, G., & Shakthipriya, P. (2025). Transforming Healthcare: Unleashing the Power of Artificial Intelligence of Things (AIoT). In AIoT (pp. 99–129). Auerbach Publications.

Khang, A., Rath, K. C., Satapathy, S. K., Kumar, A., & Agrawal, A. (2025). Robotic Process Applications and Tools for Manufacturing Sector. In Machine Vision and Industrial Robotics in Manufacturing (pp. 252–274). CRC Press.

Kim, Y. G., Donovan, R. P., Ren, Y., Bian, S., Wu, T., Purawat, S., & Li, G. P. (2022). Smart connected worker edge platform for smart manufacturing: Part 1—Architecture and platform design. Journal of Advanced Manufacturing and Processing, 4(4), e10129.

Lee, J., Ni, J., Singh, J., Jiang, B., Azamfar, M., & Feng, J. (2020). Intelligent maintenance systems and predictive manufacturing. Journal of Manufacturing Science and Engineering, 142(11), 110805.

Lele, A. (2018). Artificial Intelligence (AI). In Disruptive Technologies for the Militaries and Security (pp. 139–154). Springer Singapore.

Radanliev, P., De Roure, D., Van Kleek, M., Santos, O., & Ani, U. (2021). Artificial intelligence in cyber physical systems. AI & Society, 36(3), 783–796.

Rahmayanti, D. (2024). Artificial Intelligence of Things (AIoT) to improve efficiency and automation in industrial 4.0. Sci-tech Journal, 3(3), 255–270.

Rath, K. C., Khang, A., Satapathy, S. K., Kumar, A., Das, S. R., & Panda, M. R. (2025). Enabling the Future of Manufacturing: Integration of Robotics and IoT into Smart Factory Infrastructure in Industry 4.0. In Machine Vision and Industrial Robotics in Manufacturing (pp. 101–128). CRC Press.

Subramaniyan, M., Skoogh, A., Bokrantz, J., Sheikh, M. A., Thurer, M., & Chang, Q. (2021). Artificial intelligence for throughput bottleneck analysis – State-of-the-art and future directions. Journal of Manufacturing Systems, 60, 734–751.

Ullah, A., Younas, M., & Saharudin, M. S. (2025). Transforming manufacturing quality management with cognitive twins: A data-driven, predictive approach to real-time optimization of quality. Journal of Manufacturing and Materials Processing, 9(3), 79. https://doi.org/10.3390/jmmp9030079

Varriale, V., Cammarano, A., Michelino, F., & Caputo, M. (2025). Critical analysis of the impact of artificial intelligence integration with cutting-edge technologies for production systems. Journal of Intelligent Manufacturing, 36(1), 61–93.

Yao, X., Zhou, J., Lin, Y., Li, Y., Yu, H., & Liu, Y. (2019). Smart manufacturing based on cyber-physical systems and beyond. Journal of Intelligent Manufacturing, 30, 2805–2817.

11 Intelligent Process Automation through AI in Manufacturing Sector

Dina Darwish

11.1 INTRODUCTION

Intelligent process automation (IPA) is revolutionizing the manufacturing industries that combine with artificial intelligence (AI), Internet of Things (IoT), Industrial Internet of Things (IIoT), robotics, etc. IPA enables the automations in physical tasks followed by forecasting, inventory management, quality inspection, etc. Through IPA integration in manufacturing sectors, manufacturers are significantly enjoying the benefits in operational cost reduction and at the same time, they are availing accelerated production cycle by maintaining high level consistency and precision [Dalsaniya & Patel (2022)].

Smart manufacturing is a broad concept and is the advanced version in industrial revolution. It is not something that can be implemented in a production process by forcefully or directly. Rather, it is a sort of configuration and a system that combines several technologies and solutions together; if implemented in manufacturing ecosystem, it is termed as smart manufacturing [Lievano-Martínez et al. (2022)]. The advanced technology in upgradation with old manufacturing sectors will enable to optimize the entire manufacturing process and thus increases the production efficiency. Few prominent technologies that are adopted by various industries in their manufacturing systems are AI, IoT, IIoT, blockchain, robotics, etc.

All stake holders of smart manufacturing systems may be characterized typically in three types of companies, which can be broadly categorized as product and control solution provider, IT solution provider, and connectivity solution provider in the smart manufacturing system [Mathew et al. (2023)].

The IoT systems provide interrelated computing devices, mechanical machines, digital devices, and production environment interaction to communicate data over the setup network without human-to-human interaction. Whereas in the IIoT, various gadgets and processes are connected through data communication system. Here in each machine, electronic gadgets are embedded or connected with sensors which typically generate the relevant data. This is further transferred to software systems through data communication system that is setup in the smart industries [Waqar et al. (2024)]. Through the data analytics and analysis, a feedback mechanism will provide the feedback based on analysis to the production systems for the required corrective action.

AI is the ongoing and future technology in smart manufacturing sectors. Because the advanced manufacturing sectors are based on customer satisfaction with quality

 DOI: 10.1201/9781003503323-11

product and services, smart industries are interested to produce more volume of product with better quality and also want to maintain product quality with customer satisfaction. At the same time, the industries are on the way to enjoy the advanced technology for product and production efficiency. This suggests to industrial practitioner to develop and implement the technology in smart manufacturing environment that they manage to produce more and more products and identical parts. This process generates a bulk data and control the cost consumption. This can be achieved through AI algorithm. AI in manufacturing will build a modern world. But before AI, machines broke down without any warning to the system [Balasubramaniam et al. (2024)]. Now this AI technology and IIoT technology combined together and made a new form of technology called Artificial Intelligence of Things (AIoT). This technology provides maintenance for spot issues before fail at the earlier stage.

By integrating AI with IoT gadgets followed by advanced sensors, various manufacturers are capable to collect a large volume of data from manufacturing network system. These data further be analyzed by AI algorithm to provide a decision on the basis of data-driven mechanism. This data-driven mechanism will further enhance the production efficiency through optimization process. This alternatively increases the overall equipment effectiveness through predictive maintenance process. AI also supports to maintain quality control and supply chain management system [Chatterjee (2021)]. The machine learning (ML) and deep learning techniques are helping for self-optimization in the smart manufacturing system. AI also supports robotics adopted system and human-robot collaboration that enhance productivity and safety on the manufacturing shop floor. Hence, AI acts as the backbone of digital transformation technology and making the manufacturing system as smarter as possible.

11.2 TRANSFORMING MANUFACTURING WITH IPA TECHNOLOGIES

IPA technologies are the instrumental technologies that create smart factories. Through this, various machines and human collaborative mechanism play a vital role. These intelligent systems are monitoring the condition and status of connected gadgets and production line. Various inventories are well managed and adjusted production schedules of the system. IPA also helps to minimize wastes and optimize energy usages in compliance with environment regulation.

Process identification and assessment are well associated through IPA. Organizers can analyze their existing manufacturing workflows to identify rule-based, repetitive job and time-consuming tasks. This alternatively provides a proper understanding on interdependencies between different operations, resource utilization, and data flows. The main objective is to add value addition importance like quality inspection, supply chain management, prediction maintenance, etc.

AI, ML, and deep learning algorithms are capable to route and various structured tasks like data entry, equipment monitoring, inventory update, etc. These algorithms are well trained by using past real-time data as collected by various sensors connected with machines and enterprise network system. This algorithm has a valid potential to identify various anomalies and give insights for process optimization

[Rath et al. (2024)]. All the integrated technologies work in a balanced mechanism under the centralized IPA system.

11.2.1 Advantages of Intelligent Process Automation

Transferring manufacturing system through IPA is a great challenge in the manufacturing sectors to synchronize the demand of current digitalization system. It offers significant benefits to enhance operational efficiency, product quality, and business agility of smart industries. Few other benefits are noted down, as follows.

a. IPA technology in combination with AI, IoT, ML, and robotic process automation (RPA) are providing automation to handle repetitive and time-consuming tasks in manufacturing industries. As a result of this, the production cycles will be faster and minimize the lead time. The human intervention is also minimizing in the industries.
b. Through the AI-driven data analytics with real-time monitoring process, it is easy to identify the fault and defects of the system. This acts as early warning fault detection system in the manufacturing network. So, it is easy to take remedial step as decision process and ensure better quality control to reduce the number of defects.
c. The predictive maintenance is also clearly analyzed through IPA technology. This follows data-driven mechanism and decision process on real-time data as collected through sensor technology. It helps on schedule maintenance to reduce down time and will take care the health of equipment for better functioning.
d. Through process automation and optimization technique and algorithm, the IPA technology helps to reduce the labor cost and material wastage. Another advantage of IPA is to streamline the supply chain management system followed by reduction in inventory management to optimize the overhead cost.
e. High volumes of data are generated in manufacturing industries. IPA technology analyzes the real-time data and guide strategic decision-making. Data-driven techniques empower managers to make smarter and faster decision-maker to increase production and business performance.
f. IPA helps employees to focus on more strategic and value-added activities. This allows employees for their job satisfaction but encourages a culture of innovation in their work environment.
g. Improving quality with reduction on delay to cause customer satisfaction is caused by IPA by increasing the customized capabilities in the smart manufacturing sectors. This intelligent system also supports the customer need through tracking their order status in a short time and easy manner through proper tracking customized system to get their delivery.
h. IPA technologies also support sustainability by optimizing resource through proper use, minimizing wastages, and reducing energy consumptions through manufacturing environment monitoring process. This also monitors environmental impact in real time and adjusts the process to make an

eco-friendly system. This alternatively helps in meeting regulatory and corporate social responsibility to appeal environmentally conscious customers.

11.2.2 Challenges of Intelligent Process Automation

Various challenges associated with the implementation of IPA with manufacturing sectors are as follows.

a. Complexity of data integration and quality.
b. Managing organizational various challenges and also overcoming employee resistance.
c. Identification of right process automation and ensuring process mapping.
d. High initial investment costs and complexity technological selection.
e. Data security and skill gap within the manufacturing workforce.
f. Scaling IPA initiatives and ensuring optimization.
g. High strength network infrastructure.
h. Integrating AI with IPA through proper algorithm.
i. Implementation of worthy ML algorithm to improve manufacturing process.
j. Maintain proper standard IPA technology followed by strong protocol across the entire supply chain mechanism.
k. Flexibility of IPA with manufacturing environment as required.
l. Fixing of a culture for continuous improvement in the adopted IPA technology through optimization process.

11.2.3 Resolution of Challenges and Process Have to Follow

To overcome the IPA challenges, manufacturers have to upgrade their operation principle on the basis of digitalization technology through cutting-edge techniques.

a. Manufacturers have to start to conduct thorough assessment of their ongoing processes before and after adopting the IPA with their manufacturing system.
b. They have to maintain proper evaluation of adopted IPA technologies and solutions that meet the manufacturing needs.
c. Manufacturer should be alert for security though adopted technology and scaling the system through proper skills.
d. It is highly essential for manufacturer to go through thorough testing and validation of adopted IPA technology with their industries.
e. Phased implementation approach is preferable to check the stages of the process operated by IPA.
f. A skill-based technological familiarity training must have to conduct at regular interval for employees.
g. Manufacturers must have to identify track key performance indicators and conduct regular assessment that should meet business objectives.
h. Adoption of best practices for IPA with all stakeholders through clear roadmap and proper training with continuous monitoring and evaluating performance.

11.3 METHODOLOGY TO ADOPT INTELLIGENT PROCESS AUTOMATION IN MANUFACTURING SYSTEM

Adopting and implementing the IPA in various manufacturing sectors can be skilled up successfully through staged approach only. First, identify various bottlenecks, proceed with repetitive operations, and short out high-error areas. So that it would most value from automation by carrying out a systematic audit of current processes. To carry out this, contemporary actions must be mapped, and data must be surveyed for thinkable competence improvements. Second, choose the best IPA technologies which are correct for the weights that have been resolute. Examples of these technologies include process mining tools, AI and RPA, which includes ML techniques [Rath et al. (2025)]. Here manufacturers have to consider various factors like integration capabilities with legacy systems, vendor support, and scalability. Nest phase shows for pilot IPA solutions in a measured environment with low-risk and high-impact processes to establish value to collect feedback. This iterative approach consents for super fine-tuning and authentication of the automation strategy. Lastly, to assurance operative implementation and optimize return on deal, step up the effective pilot projects all the way through the production system by driving robust governance contexts in place, endlessly checking automated processes, and providing workforces with ongoing training. To overwhelm thinkable obstruction and encourage an incessant development culture, this entails generating a clear transformation administration plan [Rath et al. (2023)].

11.3.1 CASE STUDY 1: PHARMACEUTICAL INDUSTRY

There are many probabilities for IPA to encourage effectiveness and revolution in the pharmaceutical sector because of its sophisticated network of R&D, regulatory compliance, manufacturing, and distribution. Hetero Pharmaceutical, a renowned pharmaceutical firm in India, delivers a commanding design. The company's processes were overwhelmed by manual, error-prone procedures and outdated IT infrastructure. Quality control, passivity paperwork, and product assessments were overdue as a result of these problems, and the timely decision-making was troubled by the fragmented data. In specific, its drug safety and root cause analysis (RCA) followed by Annual Product Quality Review (APQR) techniques were precise resource-intensive and normally delayed.

So, Hetero Pharmaceutical is in progress through the digital transformation path to address these issues by putting in place a comprehensive IPA solution. Here the requirement is to implement RPA bots that are integrated with AI-driven automation and drive their infrastructure to smarter technology path. RPA bots are considered to automate processes as well as data entry, batch record review, and preliminary quality checks to make sure that all required paperwork will be avoided. This contracts the time requisite to meet and observe larger volumes of data, to turn three-month work procedure into a few days.

There were significant and wide-ranging outcomes from this IPA deployment. Significant financial savings were demonstrated by Hetero, who reported a surprising 30%–40% reduction in their total cost of ownership (TCO). More significantly, RPA automation in quality control alone resulted in over 13,500 hours saved per month, freeing up highly qualified staff to concentrate on more strategic and analytical work. While AI-powered pharmacovigilance decreased manual burden by 70%, AI-driven RCA cut investigation times by 80%, improving audit readiness. This scenario clearly demonstrates how IPA may maximize operational efficiency, lower costs, increase regulatory compliance, and speed up time-to-market for critical pharmaceutical goods by intelligently automating repetitive and data-intensive processes.

11.3.2 Case Study 2: Automotive Industry

Focusing on the order of progress on operational efficiency and maintain competitive edge, the automobile sectors are embraced with IPA now a days. A remarkable case is Hyundai's $7.6 billion EV Metaplant in Georgia, United States that incorporates cutting-edge AI and robotics technologies with its manufacturing system. By keeping a targeted goal in mind, the manufacturer is producing 500,000 electric vehicles (EVs) in a year. For this, they provide facility uses of 475 robotic arms and 300 autonomous guided vehicles (AGVs). Hyundai's dedication to manufacturing industry is always innovation and demonstrated by this automation technology. As a result of which the industry reduces human work and simultaneously increases efficiency.

Along with the technology facility, they adopted AI-powered inspection tools to maintain the record for quality control. The systems can assess up to 5,000 car bodies every day and use deep learning algorithms followed by high-resolution cameras to find every minute paint flaws on vehicle bodies. This automation has increased the accuracy in comparison to human inspectors and detect defect to improve the quality control procedure.

A significant advantage has been made with the incorporation of IPA in the automotive industry system. For example, Ford has concentrated on design time from hours to seconds by using AI agents and Nvidia Graphics processing units (GPUs) to gear up its engineering process and design procedure. This advances various facilitates in faster development and promotes flexibility by enabling the speedy manufacturing process of various models and reduces the manual predictions as shown in Table 11.1.

Figure 11.1 shows the development in fault detection accuracy over a 12-month period. Human inspectors collected data (as shown in figure) attain an accuracy of about 75%, whereas it progressively rises to 85% through AI adopted system. The

TABLE 11.1
Hyundai's EV Metaplant Automation

Metric	Value
Total investment	$7.6 billion
Autonomous guided vehicles	300
Robotic arms	475
Annual production target	500,000 vehicles
Jobs created on-site	8,500
Jobs created through suppliers	7,000

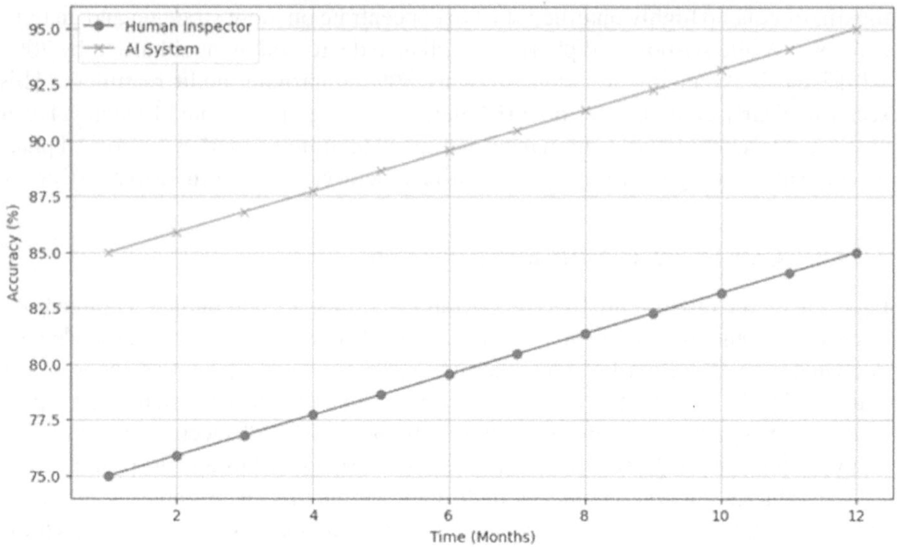

FIGURE 11.1 Defect detection accuracy over time.

AI system begins with an accuracy of 85% and increases to 95% accuracy during the same time frame work. This reflects the power of AI systems in quality control procedures and is more reliable in a consistent manner.

The Hyundai, Audi, and Ford industries have implemented the technology of IPA in their automotive industry sectors and became an example of AI technology revolutionizes smart manufacturing sectors to increase plant productivity. They maintain high-quality and creativity products in the world.

The above case studies are demonstrating the cutting-edge automation system in order to sustain a competitive edge in the dynamic automotive sectors.

11.3.3 AI in Production Process

AI provides the cutting-edge technology and makes industries in advanced manner from theoretical to significant applications. It has been seen that the AI rapidly changes industrial process in a multidirectional manufacturing. Generally, AI in production refers ML algorithms with AI technologies to improve through automation and optimize manufacturing operational processes in industry. The AI-integrated technology develops production systems with more intelligent, high effectiveness, and robust process in view to achieve various goals of Industry 4.0 in smart factories [Rath et al. (2025), Darwish, (2024)]. Over the production lifecycle, high volumes of data are generated and these data are analyzed by data analytics to find patterns, forecast outcomes, and carry out the requirements with minimal assistance from human effort.

The predictive maintenance in smart industries are using AI technology in manufacturing process to monitor and examine the collected sensor data from gadgets to anticipate various issues. These strategies are called as proactive process to lower the maintenance expenses and also support equipment lifespan. This also helps to minimize the downtime. Computer vision integrated with AI technology drives these smart industries to maintain standard quality of the process as intended in short duration of time. By anticipating various changes in industrial environment, the demand is fulfilled by logical algorithm adopted by AI technology to enhance optimizing logistics and various inventory control to minimize waste. It accelerates the delivery process and follows Just In Time (JIT) in some industries also. Several significant advantages are incorporated with AI technology when AI is adopted in production operations. Enterprise Resource Planning (ERP) with AI technology notably increases the operational potential of manufacturing environment in automation system [Biolcheva & Molhova (2022)]. This provides freeing up human workforce for creativity and speeding up the decision-making process.

There are several and significant advantages of incorporating AI into production operations. Businesses that use AI frequently see notable increase in operational efficiency as a result of the automation of repetitive and data-intensive operations, which frees up human workers for more intricate and creative duties. This automation provides a significant cost savings through optimum resource use, minimizing various wastes and controlling equipment failures. Additionally, intelligent systems maintain standards and will identify different anomalies AI helps to improve the quality of the process and products [Singh et al. (2024)]. Other significant benefit is to increase safety through AI-driven robots that can carry out various risky jobs.

AI has massive potential, but so many obstacles are also there to overcome before it is used in manufacturing sectors. These include suspicious data privacy and cybersecurity. The requirements are well-skilled staff to make and manage the AI systems in parallel with the significant honest deal in technology as well as the infrastructure [Ansari et al. (2022)].

However, this AI technology plays a vital role in shaping the direction of manufacturing sectors and becomes more adoptable. The rapid progress of AI creates a high potential that pushes various limits of industrial efficiency and productivity through advanced automation technology with optimum customized manufacturing technique and also provides fully autonomous factories to the manufacturer [Salem et al. (2024)].

11.3.4 Machine Learning Technique in Smart Manufacturing System

Through the data-driven technique, it is easy for decision-making process at faster rate. This also provides a better guidance for preventive maintenance and process optimization. All these can be possible through ML algorithm. As a result of this, the traditional manufacturing environment becomes more and more smart environment [Bajic et al. (2018)]. ML provides a good pattern and anticipates various equipment problems by modifying the operational settings in real-time monitoring and control system. It lowers the downtime and increases production efficiencies of the plant.

ML techniques as adopted by automobile sectors will improve quality control, optimize assembly lines, and also customize the output [Shahbazi & Byun (2021)]. To provide a qualitative product, computer vision-based ML algorithm technique is adopted and employed to find super fine critical faults also. ML in the advanced technology applications are well adopted through computerized module in industries like BMW, Tesla, etc. to automate various scheduling tasks and capable to detect machine breakdowns for preventive action. This alternatively reduces waste and operating costs to enhance the responsiveness of market demand.

BMW's smart factory is a leading industry and known for its well-set example of smart industry. This industry is integrated with ML technique in automotive industry arena. The entire automation system of this plant uses visual inspection systems to detect and classify various surface defects of paints on car body surface with high accuracy. In addition to this, BMW integrate the cutting-edge technology with ML algorithm in predictive maintenance to analyze the vibration, temperature, and performance of the system to forecast the breakdowns. This proactive strategy leads a significant reduction in unplanned downtime and accelerates overall equipment effectiveness of the facility of the plant. It also demonstrates the impactful role of ML in smart manufacturing in automotive industries.

11.3.4.1 Logic and Algorithm of ML Implementation

Step 1: Problem Definition

Clearly defined problem in automotive industry. In general, the problems encountered with predictive maintenance, autonomous driving, traffic signal recognitions, vehicle quality inspection, energy optimization, etc.

Step 2: Data Collection

Collect the manufacturing environment information as the data. The required data are collected from manufacturing area through sensors and cameras, ECU logs, GPS and telematics, manufacturing data, etc.

Step 3: Data Processing

The raw data will be cleaned and the transformation will be done before model training. Various steps included in this stage are noise removal, normalization, labeling, and handling missing values.

Step 4: Feature Engineering

Extraction of meaningful features.

Step 5: Model Selection

Have to choose suitable ML algorithm based on the problem type as classification, regression, clustering, and deep learning. This is used to identify objects from camera data in autonomous vehicles.

Step 6: Model Training

Train the model using historical data to split into training and validation sets, use appropriate loss functions and optimizers, and monitor metrics like accuracy, precision, and recall in this stage.

Step 7: Model Evaluation

Validate the model performance by using the test data through cross-validation, confusion matrix and ROC-AUC, etc.

Step 8: Model Deployment

Install the trained model into the automotive environment through edge, cloud-based data processing system, and Over-The-Air (OTA) updates the model time to time.

Step 9: Monitoring and Feedback Loop

Detect drift in data, collect feedback, and retrain or fine-tune models periodically.

TABLE 11.2
ML Algorithm and Its Benefits

Use Case	ML Algorithm	Benefit
Autonomous driving	CNN, RNN, deep reinforcement learning	Object detection, path planning
Predictive maintenance	Random Forest, gradient boosting	Reduce breakdowns, cost savings
Driver monitoring	Computer vision + CNN	Drowsiness or distraction alerts
Quality control	Image processing + ML	Detect defects in parts
Route optimization	Reinforcement learning	Save time and fuel

The purpose to use of ML algorithm in various cases of EV- industries are well noted in Table 11.2.

11.4 PREDICTIVE MAINTENANCE IN AUTOMOBILE INDUSTRY

Predictive maintenance uses AI to predict when a vehicle component fails in the automotive industry and allowing for suitable repairs. It also decreases the unplanned breaks. Various anomalies and trends indicate the wear and tear that is identified by adopted AI models through data gathered from various sources by sensors and previous maintenance records [Kundu et al. (2022)]. This data-driven strategy replaces reactive or scheduled maintenance with condition-based solutions, increasing vehicle reliability, reducing maintenance costs, and prolonging component life.

Data collection is the first step of the process by using IoT devices and the module will gather real-time data from engine sensors, brake systems, transmission units, and other crucial parts of the manufacturing unit. Then after, this data is pre-processed and eliminates various errors and noise encountered. To find correlations between sensor readings and component failures, ML algorithms—in particular, supervised learning models like Random Forests as well as deep learning techniques like Long Short-Term Memory (LSTM) networks—are well trained on labeled industrial datasets. These models are then strongly recommend about the suitable time schedule for maintenance and forecast the remaining useful life (RUL) of components or parts of machinery.

Case study: Proactively schedule maintenance by predicting an engine bearing's RUL from real-time sensor data.

Predictive maintenance is used to proactively identify cars that are at danger of engine failure. By examining past sensor data gathered from automobiles, the objective is to reduce maintenance expenses, increase operational uptime, and prevent malfunctions.

Part: Engine bearing.
Failure Mode: Wear and tear that causes vibration, increasing friction, and finally seizures.
Monitored Sensors:

- *Resource management* system *(RMS) Acceleration Vibration Sensor: A rise denotes deteriorating bearing health.*
- *Bearing Housing Temperature or Temperature Sensor: Higher temperatures signify more friction.*

TABLE 11.3
Various Parameters

Feature	Description
Engine temperature (°C)	Temperature of the engine
Oil pressure (bar)	Pressure of the engine oil
Vibration level (mm/s)	Level of vibration in the engine
Failure (1/0)	Whether the engine failed or not

- *A drop in the oil pressure sensor (at the bearing inlet) may be a sign of increased clearance or insufficient lubrication.*
- *Engine RPM: Wear is influenced by operating speed.*
- *Hours of Operation: Total utilization.*

Training an AI model to forecast the bearing's RUL in hours is the aim.

When the RUL was known at different times prior to a breakdown or planned replacement, it is required to use historical data.

Let us go to predict when an engine of car is likely to fail or require maintenance in automobile industry. Data are collected for various parameters (as shown in Table 11.3 and Table 11.4) such as engine temperature, oil pressure, vibration levels, etc.

The integrated AI and ML models provide the output as shown in Table 11.5. The Classification Report summarizes a model's performance using metrics like precision, recall, F1-score, and support for each class. It helps to evaluate classification accuracy and balance between correctly predicted and actual class instances.

Confusion matrix $\begin{bmatrix} 1 & 0 \\ 0 & 1 \end{bmatrix}$

Figure 11.2 shows about the confusion Matrix shows that the model correctly classified all samples, with one true positive and one true negative, resulting in perfect prediction accuracy (no misclassification).

TABLE 11.4
Parameter Data

Engine Temperature (°C)	Oil Pressure (bar)	Vibration Level (mm/s)	Failure (1/0)
90	10	0.5	0
95	12	0.7	0
100	15	1	1
92	11	0.6	0
98	14	0.9	1
91	10	0.4	0
96	13	0.8	1
94	12	0.7	0
99	16	1.1	1
93	11	0.5	0

TABLE 11.5
Classification Report

	Precision	Recall	f1-score	Support
0	1	1	1	1
1	1	1	1	1
Accuracy			1	2
macro avg	1	1	1	2
weighted avg	1	1	1	2

The Figure 11.3 represents about the feature importance and shows that Engine Temperature has the highest influence on the model's prediction, followed by Oil Pressure and Vibration Level, indicating their relative impact on system performance.

Figure 11.4 presents the predicted probabilities and illustrates the model's confidence levels for each class, showing how likely each observation belongs to a specific category based on the learned features.

The most important features are as follows:

- Engine temperature (°C): 0.37
- Oil pressure (bar): 0.35
- Vibration level (mm/s): 0.28

11.4.1 Interpretation

Accuracy (1.0): The adopted model is achieving an accuracy of 1.0, that means the model is correctly predicting the outcome (engine failure or not) for the given samples in the test set. In simple, it can be stated that the model provides no mistakes to predict about the engine performance (i.e., will fail or not).

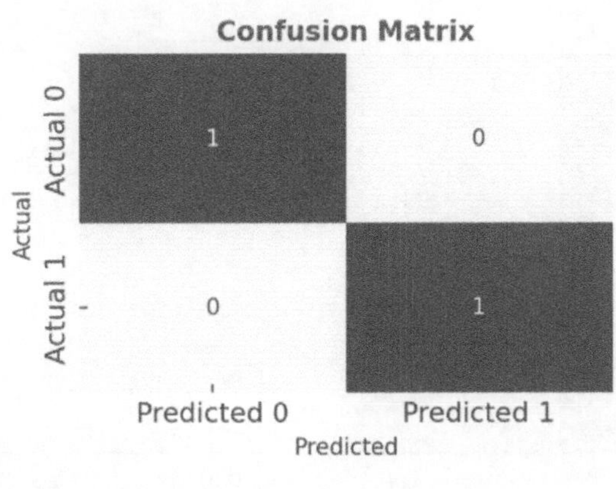

FIGURE 11.2 Confusion Matrix.

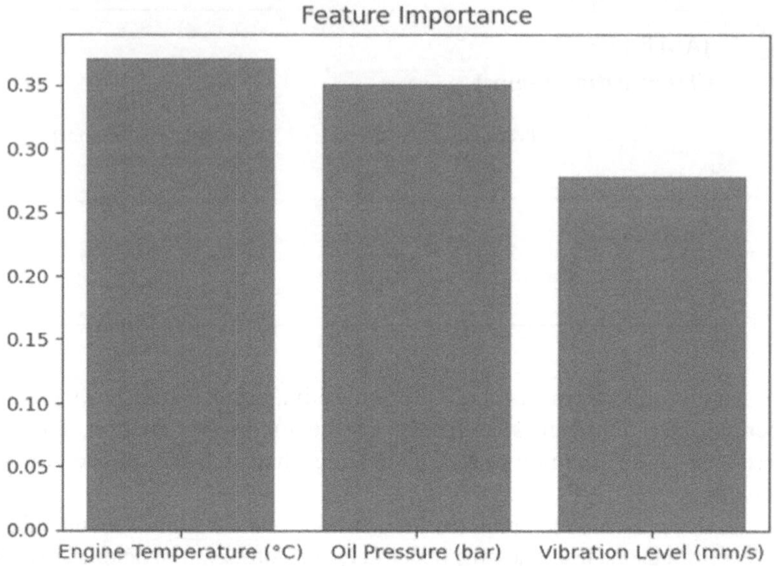

FIGURE 11.3 Feature Importance.

The classification report provides the following information.

Precision: *This deals the proportion of true positives (predicted engine failures) among all the predicted positives. A precision of 1.00 for both classes informed that the model is correctly identifying all engine failures and non-failures conditions.*

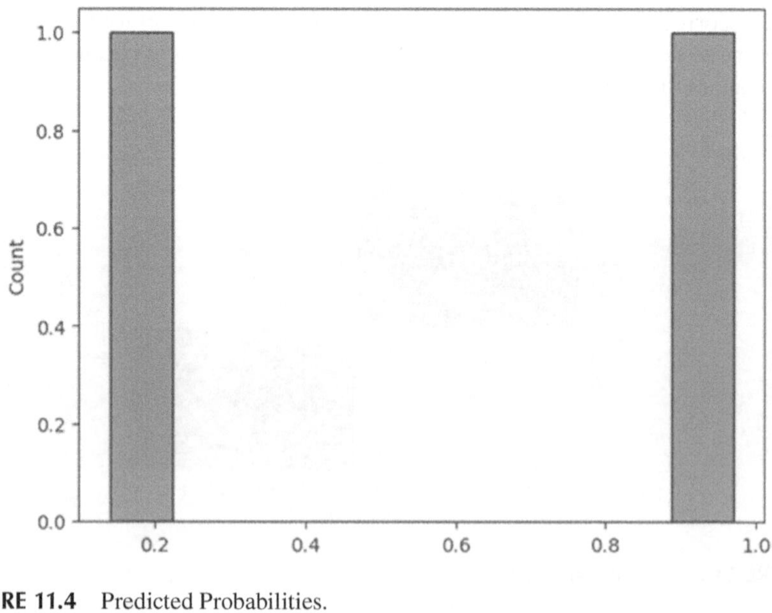

FIGURE 11.4 Predicted Probabilities.

Recall: *This method checks or measures the proportion of true positives among all actual positives through the data analysis system. The value for recall of 1.00 is meant for the both classes and suggesting that the model detects all actual engine failures and non-failures.*

F1-score: This is the harmonic mean of precision and recall. An F1-score of 1.00 for both classes indicates perfect performance.

11.4.2 Confusion Matrix

The matrix shows the number of true positives (TP), true negatives (TN), false positives (FP), and false negatives (FN). In this case, the matrix is: [[1 0]; [0 1]].

The overall interpretation shows that the model performs faultlessly on the test set, precisely predicting whether the engine would fail or not for every sample. Based on the input features (engine temperature, oil pressure, and vibration level), this shows that the model has mastered the art of spotting underlying patterns in the data and is capable of making precise predictions about engine failure.

11.4.3 Various Challenges and Their Overcomes

Predictive maintenance in the automotive industry refers to adopt various cutting-edge technologies to predict vehicle components status or breakdowns before any incidence happens. Manufacturers are capable to predict various possible breakdowns by analyzing vast sensor data, maintenance logs, and operating parameters as collected through IIoT technology with the combination of AI and ML techniques. This adopted technology will provide lower maintenance expenses, optimize vehicle performance, increase the related safety, and minimize downtime. AI/ML-integrated predictive models are supporting to increase the accuracy with time.

System adopted AI technology in predictive maintenance offers real-time fault detection as one of its prime benefits. IoT sensors are currently well installed in automotive industries manufacturing environment and with various gadgets to monitor different parameters and also predict the performances. AI algorithm analyzes the data and identifies various irregularity of the system to notify users before the system or gadget fails. ML models are able to predict engine problems, tire wear, battery deterioration, etc. easily through minute patterns and trends which are more difficult through human technicians. Car servicing and fleet management are being revolutionized by this move from reactive to proactive maintenance in well manner.

In spite of these advantages, there are a number of difficulties in applying AI and ML to predictive maintenance in the plant system. The accessibility and quality of data carry on being major challenges in the smart industrial technology system. Model accuracy may be affected by inconsistent or insufficient sensor data, and older designed cars have not advanced infrastructure with technology to support such systems facilities. Furthermore, standardization is challenging due to the diverse data formats produced by various car brands and models. Training and deploying ML models across many platforms is made more difficult by this data heterogeneity. The high expense and difficulty of implementing predictive maintenance systems present another difficulty. Investing heavily in sensors, data storage, and computing power is necessary to set an AI/ML infrastructure. Furthermore, these systems require trained workers to develop, manage, and upgrade. Smaller automakers or aftermarket service providers can find it difficult to meet these resource requirements. Predictive maintenance integration presents operational and technical challenges for current corporate systems like ERP or customer relationship management (CRM).

In order to overcome these obstacles, the industry is shifting toward open-source platforms that facilitate interoperability and standardized data formats. Cloud-based AI platforms are being used to boost scalability and save infrastructure expenses. Models learned on one set of cars can be converted to another using transfer learning and federated learning, which are also emerging as solutions to the problem of data scarcity. Tech companies and automakers are working together more and more to close the skills gap and jointly create solutions that meet their operational requirements.

By enabling safer, more intelligent, and more effective vehicle management, predictive maintenance using combined AI and ML is revolutionizing the automotive sector. Even if there are still issues with data, cost, and complexity, these are gradually being resolved by continued technical development and teamwork. The automobile industry will be able to more easily incorporate AI and ML technologies as they develop, opening the door for eventually completely autonomous maintenance systems.

11.4.4 Neural Network and Deep Learning Process for Engine Fault Detection in Automobile Industry

In the automotive sector, neural networks and deep learning are essential for identifying and diagnosing engine problems because of their capacity to evaluate large amounts of intricate data. Conventional diagnostic methods depend on preset guidelines and sensor thresholds, which could overlook nuances or developing problems. Neural networks, on the other hand, are able to identify patterns in enormous datasets gathered from car sensors, including temperature, vibration, RPM, and exhaust emissions. These models can detect early warning indications of possible problems that may be invisible to conventional diagnostic instruments by being trained on past engine fault data.

The ability to identify intricate patterns and relationships in data is improved by deep learning, a branch of ML that makes use of several layers in neural networks. To analyze time-series sensor data and even sound or vibration signals from engines, deep learning models—particularly convolutional neural networks (CNNs) and recurrent neural networks (RNNs)—are used in engine diagnostics. RNNs are useful for modeling the temporal behaviors and trends before an engine malfunction, whereas CNNs can be used to evaluate engine sound spectrograms and categorize particular fault kinds.

Additionally, by facilitating predictive maintenance, these solutions lower downtime and avoid expensive repairs. Through constant observation of engine performance data, a trained neural network can anticipate possible malfunctions before they happen. Both automakers and car customers save time and money by optimizing maintenance schedules and lowering the number of vehicle breakdowns. Deep learning makes it possible for adaptable models to get better with time as new data becomes available, which leads to more precise defect identification and prediction.

Furthermore, the development of linked automotive systems and autonomous vehicles depends heavily on neural networks and deep learning. They enable automated response plans and real-time engine health monitoring. For example, in an autonomous vehicle, a deep learning model may identify a serious engine issue in real time and safely stop the car, notifying the driver or remote assistance. From reactive diagnostics to intelligent, predictive, and automated fault management systems, these technologies are revolutionizing the car industry's approach to engine issue identification.

11.4.5 Benefits of Cutting-Edge Technology in Automation Systems

Advanced automation system technology has many advantages that greatly increase productivity and efficiency in a variety of sectors. Enhanced operational efficiency is one of the main benefits. Human workers can concentrate on more strategic, creative, or managerial duties when repetitive, time-consuming tasks are handled accurately and quickly by advanced automation technologies. By reducing manual intervention, industrial or service processes can operate more quickly and reliably by minimizing errors and increasing throughput.

Over time, cost reduction is another important advantage. Even while modern automation technologies can have a high initial cost, there are significant long-term savings. Large workforces are not as necessary with automated systems, which also reduce energy waste through streamlined operations and reduced maintenance costs thanks to predictive maintenance capabilities. The entire production process becomes more cost-effective with the aid of these solutions, which also reduce material waste and enhance inventory management.

The use of advanced automation technology also has the significant benefit of improved quality control. AI-driven monitoring tools, real-time data analytics, and high-precision sensors are all features of contemporary automation systems that guarantee constant product quality. These technologies enhance customer satisfaction and lessen the possibility of defective items by identifying irregularities and promptly resolving problems. Regulatory compliance is also guaranteed by the uniform use of standards and processes, particularly in highly regulated sectors like food processing and medicines.

Significant advantages also include increased scalability and flexibility. It is frequently difficult for traditional manufacturing systems to adjust to shifting consumer demands or evolving product designs. On the other hand, intelligent automation systems made possible by ML and the IoT can swiftly adapt to changing requirements or scale up or down production with no interruption. Because of this adaptability, companies can react to market demands more quickly and maintain their competitiveness in ever-changing contexts.

Advanced automation helps make better decisions based on data. Large volumes of real-time data are gathered and analyzed by these systems, which yield insightful information on client preferences, operational performance, and possible areas for development. This data can be used by decision-makers to better foresee trends, streamline procedures, and carry out strategic adjustments. All things considered, the incorporation of contemporary automation technologies is revolutionizing industries by stimulating creativity, enhancing output, and laying the groundwork for future expansion.

11.4.6 Various Modules and Gadgets of Cutting-Edge Technology in Automotive Industry

An overview of the technological innovations and devices used in the automobile sector are as follows:

a. **Advanced Driver Assistance Systems (ADAS)**: Lane-keeping assistance, automated emergency braking, adaptive cruise control, and traffic sign recognition are all examples of ADAS capabilities. These technologies help the driver maintain safety and control by monitoring the environment around the car using sensors, cameras, and radar. The primary purpose of automating some driving operations is to decrease accidents and enhance driver comfort.
b. **Infotainment and Connectivity Modules**: Modern cars include sophisticated infotainment systems that combine smartphone connectivity (through Apple CarPlay or Android Auto), music, hands-free communication, and navigation. Through the use of voice commands or touchscreen interfaces, these modules improve the driving experience by offering entertainment and necessary information.
c. **Telematics and Vehicle-to-Everything (V2X) Communication**: Vehicle monitoring systems, remote diagnostics, and GPS tracking are all part of telematics. Vehicle-to-pedestrian (V2X) communication is made possible by V2X. These technologies provide proactive communication between different traffic aspects, which enhances road safety, traffic efficiency, and real-time updates.
d. **Autonomous Driving Systems**: AI, LiDAR, radar, and video systems work together to help self-driving cars sense their surroundings, make choices, and steer without human assistance. These systems seek to improve traffic flow, lessen traffic accidents, and give those who are unable to drive mobility options.
e. **Electric Powertrains and Battery Management Systems (BMS)**: Electric motors and sophisticated batteries make up the powertrain of EVs. The BMS controls charging cycles, keeps an eye on battery health, regulates battery performance, and guarantees energy use is safe. The durability, efficiency, and long range of EVs depend on these technologies.

11.4.7 Real Implementation Model Difficulties and Process to Overcome

Real implementation model difficulties in the automotive industry and the processes to overcome them, each explained in a concise paragraph with examples:

Integration of Complex Systems: ADAS, infotainment, and telematics are just a few examples of the hardware and software modules that must seamlessly integrate in modern cars. Making sure these systems are compatible and communicate with one another is the difficult part. For instance, synchronized data flow and timing are required when integrating ADAS with navigation and braking systems. Modular architecture and standardized communication protocols, such as Ethernet or controller area network (CAN), are necessary to overcome this.

High Cost of Development and Deployment: The development of advanced technologies, such as electric powertrains or autonomous driving, requires significant R&D expenditures, testing facilities, and specialized personnel. Every year, for example, Tesla spends billions on battery and software technology. Automakers can cut expenses by utilizing shared platforms, collaborating with digital companies, or implementing open-source software solutions like Automotive Open System Architecture (AUTOSAR).

Cybersecurity and Data Privacy Concerns: Vehicles that are increasingly networked are more susceptible to cyberattacks. Breaking into braking or telematics systems could result in actual harm. The 2015 hack of the Jeep Cherokee, which revealed serious flaws, is one actual instance. Manufacturers are using cybersecurity standards like ISO/SAE 21434, encryption, and secure over-the-air (OTA) upgrades to lessen this.

Regulatory and Compliance Challenges: The laws governing safety, emissions, and autonomous systems in automobiles vary by nation. For instance, the General Safety Regulation (GSR) in Europe requires some ADAS characteristics but not all other countries do. Systems must be modular and flexible enough to accommodate local regulations, according to automakers. Early involvement with authorities and regulatory sandboxes facilitate more effective compliance navigation.

Customer Acceptance and Trust Issues: Consumers may be skeptical about adopting autonomous or fully EVs due to safety concerns or range anxiety. For example, while autonomous driving is technologically feasible, many drivers are reluctant to relinquish control. Building trust involves public education, transparent communication of benefits and risks, and real-world pilot programs (like Waymo's robotaxis in Phoenix).

Infrastructure and Ecosystem Readiness: Advanced vehicle technologies frequently rely on outside infrastructure, such as smart traffic systems for V2X communication or EV charging stations. EV adoption is slowed, for example, by the scarcity of fast chargers. Governments and businesses must work together to enable widespread implementation by constructing infrastructure, offering incentives, and guaranteeing backward compatibility.

11.5 CONCLUSION

In summary, by fusing the advantages of RPA, ML, AI, and data analytics, IPA driven by AI is transforming the industrial industry. Better decision-making, real-time monitoring, predictive maintenance, and more efficient production workflows are all made possible by this integration. Manufacturers can thereby attain increased productivity, lower operating expenses, better product quality, and a quicker time to market. Despite implementation obstacles including data integration, staff upskilling, and change management, IPA is a key factor behind Industry 4.0 because of its long-term advantages of agility, scalability, and innovation. Adopting AI-driven automation is now a strategic necessity for industrial companies that want to be prepared for the future.

11.5.1 Future Scope

With the ongoing development of AI, IoT, and edge computing, the potential applications of IPA in the industrial industry are numerous and bright. Future advancements will make it possible for factories to become completely autonomous, with self-optimizing machinery, real-time problem detection, and dynamic production line adaptation in response to resource availability or market demand. AI-driven simulations will enable zero-defect production and predictive planning with the incorporation of digital twins. Global smart manufacturing ecosystems will also become possible as 5G networks proliferate and improve remote automation and real-time data interchange. In the end, IPA will provide customization and sustainability at scale in addition to increasing productivity.

REFERENCES

Ansari, M. F., Dash, B., Sharma, P., & Yathiraju, N. (2022). The impact and limitations of artificial intelligence in cybersecurity: A literature review. International Journal of Advanced Research in Computer and Communication Engineering, 11(9).

Bajic, B., Cosic, I., Lazarevic, M., Sremcev, N., & Rikalovic, A. (2018). Machine learning techniques for smart manufacturing: Applications and challenges in industry 4.0. Department of Industrial Engineering and Management Novi Sad, Serbia, 29, 18.

Balasubramaniam, S., Prasanth, A., Kumar, K. S., & Kadry, S. (2024). Artificial Intelligence-Based Hyperautomation for Smart Factory Process Automation. In Hyperautomation for Next-Generation Industries (pp. 55–89). Wiley Publisher.

Biolcheva, P., & Molhova, M. (2022). Integration of AI supported risk management in ERP implementation. Computer and Information Science, 15(3), 37.

Chatterjee, A. K. (2021). Process Automation to Autonomous Process in Cement Manufacturing: Basics of Transformational Approach. In Intelligent and Sustainable Cement Production (pp. 79–98). CRC Press.

Dalsaniya, A., & Patel, K. (2022). Enhancing process automation with AI: The role of intelligent automation in business efficiency. International Journal of Science and Research Archive, 5(2), 322–337.

Darwish, D. (2024). *Hyperautomation in Business and Society.* IGI Global Scientific Publishing.

Kundu, P., Luo, X., Qin, Y., Cai, Y., & Liu, Z. (2022). A machine learning-based framework for automatic identification of process and product fingerprints for smart manufacturing systems. Journal of Manufacturing Processes, 73, 128–138.

Lievano-Martínez, F. A., Fernández-Ledesma, J. D., Burgos, D., Branch-Bedoya, J. W., & Jimenez-Builes, J. A. (2022). Intelligent process automation: An application in manufacturing industry. Sustainability, 14(14), 8804.

Mathew, D., Brintha, N. C., & Jappes, J. W. (2023). Artificial Intelligence Powered Automation for Industry 4.0. In New Horizons for Industry 4.0 in Modern Business (pp. 1–28). Springer International Publishing.

Rath, K. C., Khang, A., Mishra, S. K., Patnaik, P. K., Mohanty, G. K., & Dash, T. (2024). Integration of Artificial Intelligence and Internet of Things Technology Solutions in Smart Manufacturing. In Machine Vision and Industrial Robotics in Manufacturing (pp. 155–177). CRC Press.

Rath, K. C., Khang, A., Satapathy, S. K., Kumar, A., Das, S. R., & Panda, M. R. (2025). Enabling the Future of Manufacturing: Integration of Robotics and IoT into Smart Factory Infrastructure in Industry 4.0. In Machine Vision and Industrial Robotics in Manufacturing (pp. 101–128). CRC Press.

Rath, K. C., Kumar, R., & Mishra, B. K. (2023). Implementation of the Industrial Internet of Things (IIoT) Is the Tool of Digital Transformation Technology. In Internet of Things (pp. 61–84). Apple Academic Press.

Rath, K. C., Panda, L. P., Rath, S. K., Khadanga, S. P., & Olasoji, O. T. (2025). Enhancing Production Efficiency and Sustainability through Smart Manufacturing. In Sustainable Advanced Manufacturing and Logistics in ASEAN (pp. 11–32). IGI Global Scientific Publishing.

Salem, A. H., Azzam, S. M., Emam, O. E., & Abohany, A. A. (2024). Advancing cybersecurity: A comprehensive review of AI-driven detection techniques. Journal of Big Data, 11(1), 105.

Shahbazi, Z., & Byun, Y. C. (2021). Smart manufacturing real-time analysis based on blockchain and machine learning approaches. Applied Sciences, 11(8), 3535.

Singh, H., Veeraiah, V., Khan, H., Singh, D. K., Talukdar, V., Anand, R., & Sindhwani, N. (2024). Investigating Scope and Applications for the Internet of Robotics in Industrial Automation. In Robotics and Automation in Industry 4.0 (pp. 132–151). CRC Press.

Waqar, M., Bhatti, I., & Khan, A. H. (2024). AI-powered automation: Revolutionizing industrial processes and enhancing operational efficiency. Revista de Inteligencia Artificial en Medicina, 15(1), 1151–1175.

12 Lean Principles in Manufacturing for Resource Efficiency

Mallikarjun Koripadu, K. Venkatasubbaiah, and Teresa Ptaszynska

12.1 INTRODUCTION

In today's competitive landscape, businesses face constant pressure to optimize resources and minimize waste wherever possible. Lean manufacturing principles offer a proven standard approach and framework for achieving these goals, which enable every organization to create more value with fewer resources. These can be applied across different manufacturing and services sectors with a minimal difference of application of lean principles and toolset [Abualfaraa and Salonitis (2020)]. By systematically identifying and eliminating waste, lean practices provide a way for increased efficiency, improved quality, and a more sustainable approach to product, process deliverables that meet the customer requirements and expectations, in turn help retain customers and gain more customers [Abolhasani and Sabbagh (2017)].

Lean manufacturing is a structured, systematic approach to streamline production workflow, process workflow by minimizing waste identification and minimizing or improving the efficiency within a manufacturing system without sacrificing productivity and quality of the end deliverables to the user. Lean management primarily concentrates on increasing efficiency as well as effectiveness by conducting detailed analysis to classify and remove non-value-added (NVA) activities, thus optimizing resource use across the departments and cross-utilize the resources effectively [Anvari et al. (2011)]. The founder of the Toyota group during 1902, Sakichi Toyoda, wanted to do something differently to improve the efficiency of the deliverables and resources, who invented an automated loom that used to stop every time a thread broke as a notification to the operator. This helped in reduction of quality defects, and a single operator was able to monitor more than one machine at one time as this has improved the resource time and efficiency [Bhuiyan and Baghel (2005)]. After many years, one more attempt by Taichi Ohno, who was working as a production engineer at the Toyota Company applied these concepts and he thought to remove the wastes in the workplace and NVA activities. Furthermore, at every fault to stop production line, where he introduced another concept, Just-in-time (JIT). These two Jidoka and JIT together made a decision-making pillar of the Toyota Production System (TPS) that supported by a foundation of level loading, which is the basis of lean. Originating in Japan in the TPS during 1930s, lean process and manufacturing has become a widely adopted methodology for achieving process and

DOI: 10.1201/9781003503323-12

operational excellence focuses to bring improved value to the customers with smaller number of resources [Bhamu and Sangwan (2014)]. There are several principles of lean manufacturing that include continuous improvement, respect for people, cultural and behavioral, and the eradication of activities that do not add any value to the final product or service, which will be discussed in detail in the upcoming sections. Lean manufacturing really helps to produce a more value for customers' deliverables while using fewer or limited resources along with the continuous process improvement approach and methodology to remove waste and enhance productivity [Bon and Mustaffa (2013)].

12.2 UNDERSTANDING LEAN PRINCIPLES

Lean principles are all about making work easier and more efficient by cutting out unnecessary steps. The idea is to focus on what's important to the customer and remove anything that does not help the outcome. This could be extra work, delays, or fixing mistakes. Lean started in manufacturing, but it can be used in any type of work to save time, reduce costs, and improve quality [Boyer (1996)]. The goal is to keep improving how things are done so that the process runs smoothly and delivers better results.

12.2.1 Lean Manufacturing

Lean manufacturing, in simple terms "lean," is a simple method to using simple and proven systematic approach that was developed to minimize or remove or eliminate waste in the process of a manufacturing system while productivity remains constant or improved [Chiarini (2014)].

Lean manufacturing is a philosophy that highlights creating more value to the end users and customers with fewer resources by improving the flow of information, product or services, and information through the entire process or product value chain [Ciarapica and Matt (2016)]. This approach not only aims to enhance production efficiency but also improves product quality and reduces lead times. Lean manufacturing is all about use of tools and align with the mindset, beliefs, and change; when properly implemented, can lead to transformative developments in productivity (saving time), quality (reduced defects), and sustainability (improved continuous improvement). The core philosophy of lean is to create more value for customers with fewer resources by optimizing workflows and reducing inefficiencies.

12.2.2 Key Lean Principles

For any strategy, the principles are the key and also for lean, the implementing and practicing lean effectively centers on knowing how to apply lean principles successfully in the business environment is very important. Here the point to emphasize is that, lean management principles focus on teams; their behaviors to drive continuous improvement culture from bottom-up and top-down [Cudney and Elrod (2011); Duarte and Machado (2011)]. Once the lean approach is implemented effectively, these principles create a self-supporting feedback loop between the team members, processes, and customers that each one of the team members [Dues et al. (2013)], leadership involved in the complete end-to-end process delivers a product or service to the customer. James P. Womack and Daniel T. Jones, in the book named "Lean Thinking," mentioned its essence of the TPS, and in another book of Womack and Jones, namely "The Machine That Changed the World," they have outlined the five key principles of lean manufacturing that can help achieve the organizational objectives and goals. The listed principles function as a guide for improving overall efficiency and quality of the final product or service. To take you all back to manufacturing days, as mentioned earlier, lean concepts were initially applied in manufacturing; however, in the recent past (may be early 1990–2000), lean approach has changed the environment of expertise work and management strategies and perceptions and embedded in the culture of many manufacturing (as shown in figure 12.1) and service along with medical and pharma companies. At the same time, lean manufacturing nurtures to prepare for the continuous improvement approach based on the structural solution and ideation of respect for human and the continuous change management, which is the key for bringing the change [Esmemr et al. (2010)].

From Figure 12.1, the attention here goes to:

- "JIT": What customer wants, when customer wants it. This is a powerful strategy that aligns inventory management closely with production needs and customer demand, leading to cost savings, reduced waste, and enhanced operational efficiency [Fahmi et al. (2019)]. This is a manufacturing and inventory management strategy pointed to improve efficiency and lowering waste by accepting goods only as they are needed in the production process. The JIT concept has been widely used in multiple different organizations

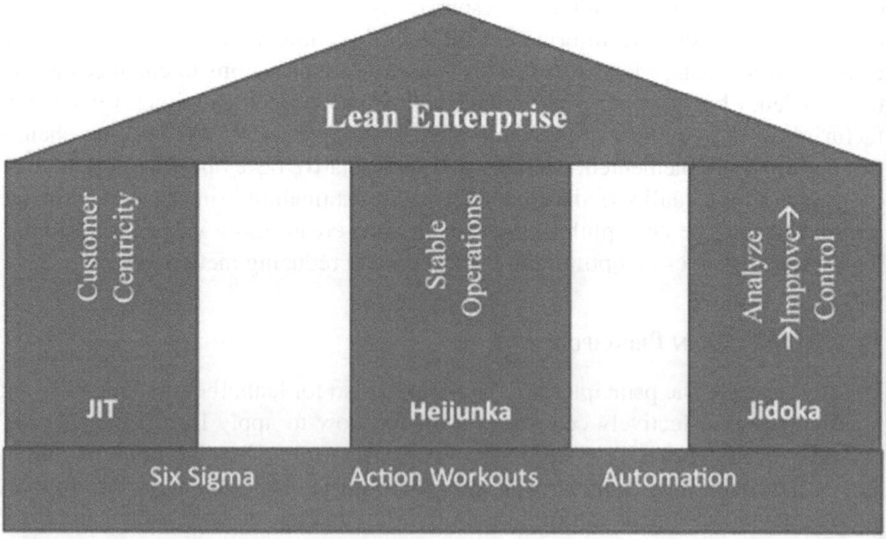

FIGURE 12.1 House of lean enterprise.

apart from manufacturing, including retail, healthcare, pharma, and food service providing the delivery of products and services in the agreed timelines, is most critical.

- "Heijunka": Sequencing and level loading. This is a Japanese term that means "production leveling" or "production smoothing" [Furlan et al. (2011)]. The primary purpose of "Heijunka" is to reduce the unevenness in the processes by level loading the type and quantity of production over a given period. There are two major types:
  - Production leveling: Production schedule to meet customer demand consistently rather than producing in large batches and create the inventory that is again an NVA.
  - Volume leveling: Balancing the production volume based on time and ensuring products are produced at a steady pace, which avoid overproduction and long lead times.

    One of the applications of Heijunka is that balancing the production of various models and components to avoid excess inventory and reduce lead times in the mobile and electronic components.
- "Jidoka": Pause at every issue; machines are designed with human intelligence. Basically, this is an automation with the human touch. This is a combination of automated detection with human involvement to ensure defects are identified and corrected immediately to improve efficiency and quality of deliverables. One of the examples in healthcare is settings to halt processes when errors are identified, ensuring patient safety and quality care.

The core principles of lean manufacturing include the following:

The Figure 12.2 illustrates the five core principles of Lean, which aim to enhance efficiency, eliminate waste, and deliver maximum value to customers.

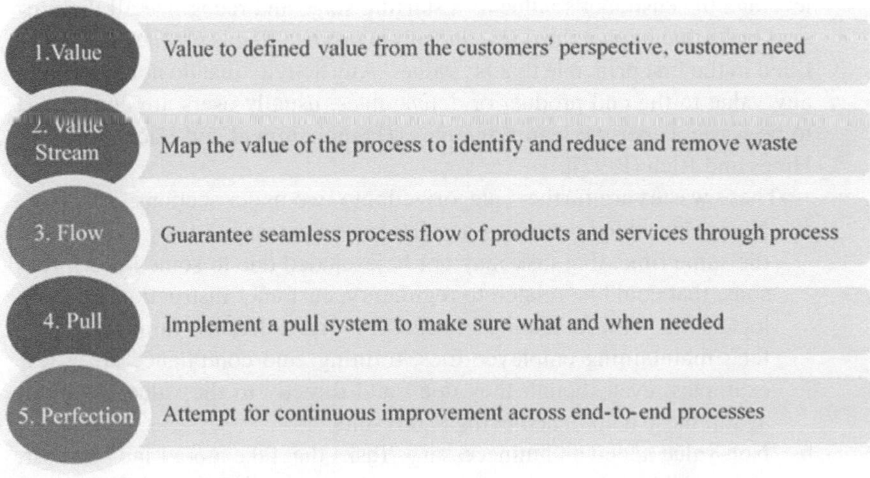

FIGURE 12.2 Core principles of lean.

From Figure 12.2, let us discuss these principles in detail:

1. **VALUE**: Let us start with the "value," where the customer is willing to pay or use the product or service. This is one of the most important to determine the actual requirements asking the clarifying questions to the customer. Sometimes, they may not be able to clearly articulate what is the ask or want, or they may unaware of it at the beginning when the conversations started. To gain more information on what the customer needs, there are many techniques such as:
    a. Interviews: Ask a series of questions to gather detailed information from customers about their requirement. These could be structured, semi-structured, and unstructured based on the need.
    b. Surveys: To gather quantitative data and can be employed to understand what, when, and where the deliverables to be completed with having some questions. These could be online survey, telephonic/web survey, and face-to-face survey.
    c. Demographic information: This helps in audiences understanding target markets and developing targeted strategies suitably.
    d. Web analytics: This helps in doing some web-based analysis on the customer requirement before the meeting with the customer to make some analysis about the services and deliverables.

    The above should help identify or define what customers find value of the deliverables. By using all or a few of the techniques listed above, which could be of qualitative and quantitative techniques, one should be able find out "what is the ask from customer?," "how the customer want the product or service to be delivered?," and "at what price the customer is willing to buy or afford?."
2. **VALUE STREAM**: The most important one to help detecting and plotting the complete value chain of the given process. Here, the main objective is to leverage the customer's value as a starting stage and recognize all the process tasks that impact the end value of the product that is planned and identified in the first principle that is "value." Any activity that do not contribute any value to the end product or deliverables, usually users are considered to be waste as per the lean principles [Hajmohammad and Vachon (2014); Hines and Rich (1997)].

    These wastes are further categorized into two major sections:
    a. Non-value added but necessary: Tasks that doesn't add any value but at the same time, that task may not be excluded due to some known reasons, that could be related to regulatory, customer instruction, and any local laws. This is required because of risk tolerance limit and "buffer"; maintaining employee files, training, and compliance are a few examples, even though they don't add directly to the value of output again; these help in achieving end results.
    b. Non-value added and unnecessary: Tasks that take more time and cost, along with employees but create no value to the customer. A few examples include manual entry of data, movement of the employees for the

printing and fax, unnecessary approvals, holding frequent meetings, and so on; like this there are many examples.

The non-value-added and redundant tasks should be removed from the process while the NVA but necessary should be reduced to maximum possible in the whole process so that the overall time of the deliverables is reduced and improved quality. Eliminating redundant processes or steps ensures customers receive exactly what they need while simultaneously reducing the cost of delivering the product or service.

To perform the above value-added and non-value added tasks recognition, the "value stream mapping (VSM)" for an organization's war on waste. VSM helps the organization recognize how value-added and NVA activities influence their products and processes as it progresses from raw or semi-finished material to final material or service to the customer [Hofer et al. (2012)]. VSM results in identifying wastes in the operational processes triggering elimination or reduction of the wastes.

Waste Exists in every process, Eliminates IT, or Reduces IT. To address this waste in the process, it is important that one should know how to identify these wastes in the process. To make that possible, there has been a method called seven wastes from the modern age and in the recent past, added one more waste to seven wastes. That has become eight wastes in the process that make the process inefficient and more of quality issues across the end-to-end process.

The typical seven lean wastes are explained using the TIMWOOD approach in many of the books, they all are explained as follows:

a. Transportation (T): The transfer of goods from one location to another is referred to as transportation. In manufacturing, this might mean performing different tasks in different locations. For example, producing product parts in one location and shipping them to another location to assemble as depicted in Figure 12.3.
b. Inventory (I): If we maintain the goods at the manufacturing site or warehouse after producing the goods to get them sold, that mean to "Inventory." Which is a waste happening when your product is sitting there waiting to be sold. Takes valuable space, requires moving, losses due to damage, ties up capital, multiplies quality problems and may depreciate.
c. Motion (M): To perform the activities in any of the manufacturing site, if people, machines, or items are moved unnecessarily, that is termed as "Motion," that do not add value is the product hence that is termed as the waste and leads to ergonomic concerns, safety of the employees as shown in Figure 12.4.
d. Waiting (W): When an employee requires raw materials, semi-finished, or finished goods to complete their work but remains idle without taking any action, this is considered waste, known as "Waiting." This waste leads to increased costs due to delays in final product delivery timelines. Consumes valuable resources, slows response to customers, increases work in process.

FIGURE 12.3 Unnecessary movement of material.

FIGURE 12.4 Unnecessary motion of men.

e. Over Processing (O): In some cases of the manufacturing, the product is delivered to the client with additional steps that are not listed as part of the standard operating procedure; however, to please, the customer generally provided with additional features that is termed as "Over-processing"; they may include adding more features to a product than the customer will use or requires. Consumes valuable resources, creates delay, and opportunity for more defects.
f. Over Production (O): Anything that produced more than required quantities and reports is termed as "Over Production" and this is not inefficient way of working. That consumes valuable resources time and effort that is not needed immediately. Sometimes this can be considered as an inventory also, and that is one more waste as per the lean principles.
g. Defects (D): This is the most common word that we use in many deliverables in manufacturing, that is, a "Defect." This is the one that is probably the most obvious waste type as defects are scrap products and they don't meet the customer specifications and requirements that are caught as shown in Figure 12.5.
h. Finally, the additional and last and eighth waste, in lean is Talent or Skill, that are not utilized to fullest and not utilized the right resources for the right activities or the work. It is also known as the waste of

FIGURE 12.5 Performing quality errors.

"talent" or "underutilized" talent that leads to the failure to fully utilize the skills, knowledge, and abilities of employees. The other seven wastes listed above focus physical and process inefficiencies, whereas eighth waste concentrates on human aspect of the waste. Hence, leveraging the employees with the right skills for the right job is very important to have the effective and efficient process that can deliver the required results.

The above-listed different wastes are explained in Table 12.1 that helps understand more about the seven wastes.

3. **FLOW**: The third principle is to create a smooth flow that eliminates process bottlenecks, optimizes efficiency, and reduces the overall lead time for the entire process. As mentioned in Table 12.1, once the process is assessed and identifying different types of process wastes and then removing the wastes from the value stream of the process, several strategies can be employed to ensure a seamless flow, such as establishing cross-functional

TABLE 12.1
Types of Waste in Lean

Waste Type	Definition	Manufacturing	Service	Software
Transportation	Something farther than the necessary	Shipping product from plant to warehouse	An invoice with errors are further processed with billing	A software with bug being mailed for further processing
Inventory	Any supply more than customer requirement	Finished goods lying in store waiting for customer	Processes invoice undelivered to customer	Software developed waiting for customer
Motion	People with equipment walking are moving more than required	Moving a welder from one job to another	Carrying a greater number of hits in the process than required	Carrying a greater number of hits in the process than required
Waiting	Waiting for the next production step to start	Work in progress waiting for the process to start	Waiting for manager to start the meeting	A software with bug waiting for fixing
Over Processing	Processing to the extent customer does not require	Producing a shaft of +/−2 mm tolerance though customer asked for +/−5 mm	Preparing the presentation that was not requested by customer	Adding additional features that were not in the ask
Over Production	Producing more quantities than required	Producing 100 tons of stainless steel per month against the order of 50 tons per month	Producing 60 invoices per day against the target of 40 invoices per day	A software with bug being mailed for further processing
Defect	Producing more than required with bad quality	Delivering 1000 nuts and quality defects were for 100 nuts	Delivering a report with mistakes in the report	Delivering a dashboard with data misses and errors

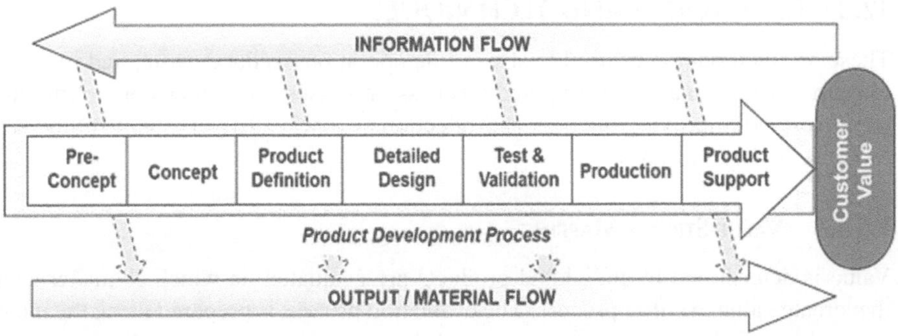

FIGURE 12.6 Lean flow method.

departments, balancing the workload, work breakdown structure, training employees to be multi-skilled, and more [Hu et al. (2015)]. This enables improved production time lines, reduced inventories at each stage of the process, and increased responsiveness to customer demand.

4. **PULL**: In production environment, pull approach (Figure 12.6) is applied on true demand and not the forecasted demand keeping the future demand in mind. This means that start producing products only when they are needed, against the customer orders, and not pushing products into the market based on forecasted demand. Hence, JIT, Kanban helps achieve this to control the inventory which is considered one of the biggest wastes in any production system. The process listed in the value stream using the second lean principle and working backward through the production system, one can make sure that the products produced satisfy the customers' requirements. Pull helps in reduced inventory costs, reduced waste, and a very well-responsive and flexible production process that can be adoptable very quickly to the changes in customer demand [Karim and Arif-Uz-Zaman (2013)].
5. **PERFECTION**: Perfection in lean manufacturing is the activity of continuous improvement to achieve the ideal state where there is no waste, processes are optimized, and customer satisfaction is maximized. However, that require too much coordination, cooperation, and collaboration among various individuals, departments, and regulatory to achieve. It involves consistently seeking alternative ways to refine processes, eliminate defects, and improve overall quality of the products. This is possible to achieve by bringing some methods like introducing mistake-proof processes, predictability by processes in control, and continuously improve by fine-tuning processes.

In a whole, by integrating the lean values basically, flow, pull, and perfection, organizations can create a more efficient, customer-focused production system that reduces waste, responds dynamically to customer needs, and continuously improves over time. These principles work together to form the foundation of lean thinking at the grassroots level that helps drive sustainable and operational excellence [King and Lenox (2001)].

12.3 LEAN TOOLS AND TECHNIQUES

These are methodologies used to streamline operations, reduce waste, and increase efficiency within manufacturing and business processes. They focus on optimizing resources and improving the flow of production or service delivery. Some of the key tools are listed as follows.

12.3.1 Value Stream Mapping

Value is determined by individual products are evaluated on which characteristics that create value to entire process. The evaluation process is prepared using the internal and external customer standpoints while mapping the process. Precise understanding of customer requirements and delivering the products to the customer on time, at the right process, as prescribed by the user or customer which is the key to implement any lean principles at the organization [Li et al. (2016)].

Value stream is the combination of tasks that include both value-added and NVA process steps along with value enables, mandatory to bring a product from order to delivery. VSM is a string of all activities required to process data, material, and information from the customer that can be transformed the product that is on its way to the customer as per the requirements.

This is the very important step to remove all the types of redundancy in the processes across organization. VSM helps the organization to spend enough time and effort to recognize how the value-added and NVA tasks affect the products and processes as it transforms raw or semi-finished material to finished material that is ready to handover. This results in identifying wastes in the operational processes triggering elimination or reduction of the wastes.

VSM is a tool that is leveraged to visualize the data, material, and information flow as an end product and makes its path through the complete chain of the process.

As depicted in Figure 12.7, VSM is created and are similar to "Flow Charts" with few additional elements and icons. This is a very simple process diagram of each step

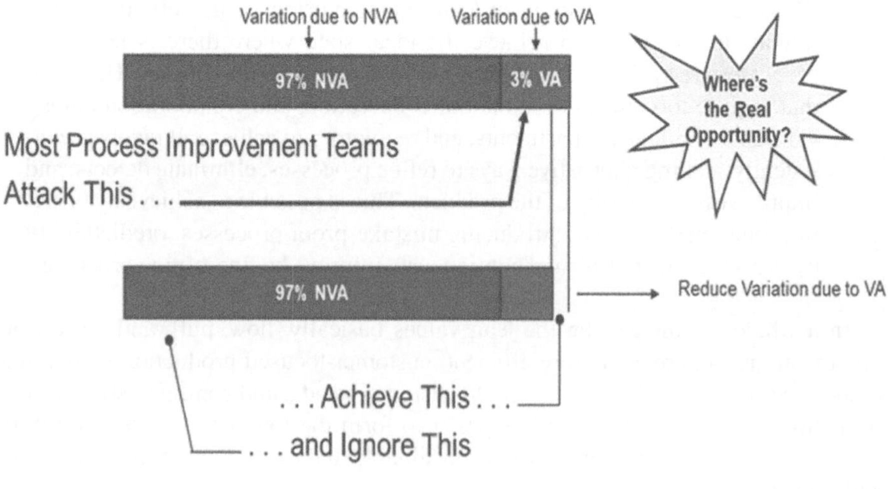

FIGURE 12.7 Value stream mapping benefits.

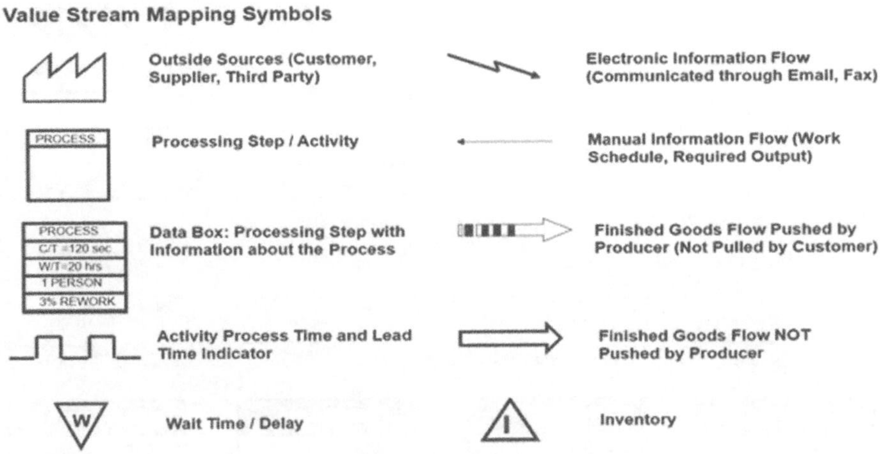

FIGURE 12.8 Symbols used for VSM.

listed from value stream is depicted with the time it takes to complete that step and the time gap between any two consecutive steps.

Figure 12.8 gives a list of few common symbols used during mapping the value stream map but not limited to the listed.

One sample is depicted in Figure 12.9, with all the mentioned time, resources, and other details in value stream map.

12.3.2 5S Methodology

5S is a key concept in lean methodology, offering a structured approach to workplace organization. Its goal is to enhance efficiency, safety, and productivity by eliminating waste and maintaining a clean, well-organized environment. This was originated from foundational component of lean manufacturing and is used to create a workplace where everything has its place, processes are streamlined, and performance is maximized.

This can be applied to physical workspaces, databases, table spaces, shop floors, or any other areas prone to accumulating waste, which leads to decreased efficiency.

The 5S stands for five Japanese terms that describe the steps of this process: seiri (sort), seiton (straighten), seiso (shine), seiketsu (standardize), and shitsuke (sustain) as explained with some examples in Figure 12.10.

12.3.3 Kaizen (Continuous Improvement)

This is a Japanese terminology as depicted in Figure 12.11; Kaizen emphasizes small, incremental changes repeatedly made to improve efficiency and effectiveness and "change for the better" or "continuous improvement." It involves every employee in the organization at every level from top management to the shop floor staff. Kaizen is an intensive, focused approach to process improvement through small incremental changes. Kaizen means, "To take it apart and put back together in a better way." The goal of Kaizen is to remove waste, standardize work, and deliver right product at right

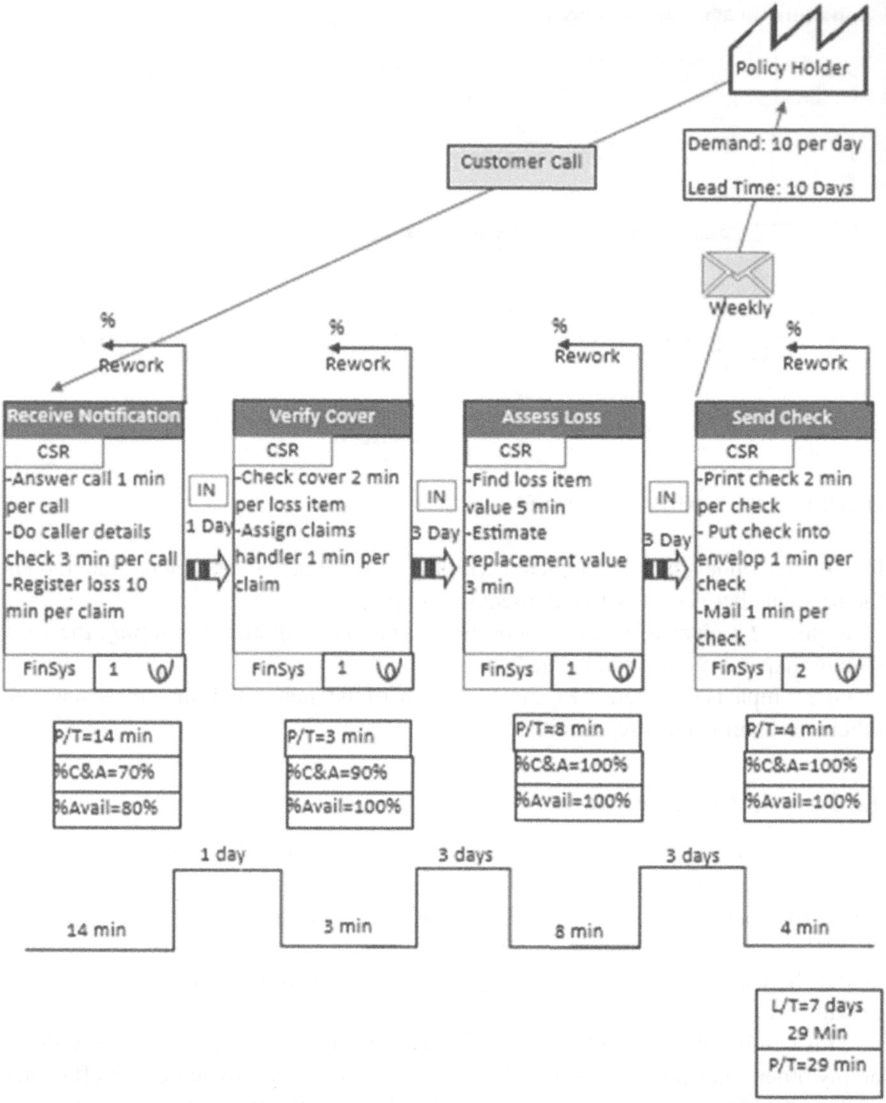

FIGURE 12.9 Example of VSM in making the policy.

time with optimum resources to customers. Figure 12.11 illustrates the concept of Kaizen, a Japanese philosophy that emphasizes continuous improvement in all aspects

Kaizen follows the following rules:

1. Work on process and results (not just results).
2. This approach involves systemic thinking through the Plan-Do-Check-Act (PDCA) cycle, offering a structured method for continuous improvement

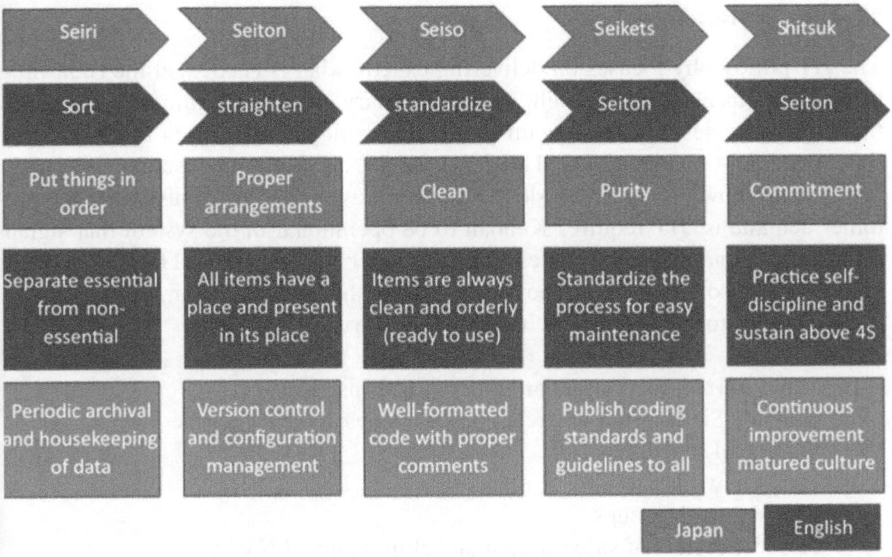

FIGURE 12.10 5S examples.

while considering the broader context rather than focusing on a limited perspective.

3. No-judgmental and no-blame game as blaming is wasteful.
4. All the stakeholders irrespective of their level should participate and contribute.
5. Team to continuously analyze and evaluate the activities of their operational process.
6. Any activity with scope of improvement to reduce cycle time will trigger Kaizen.

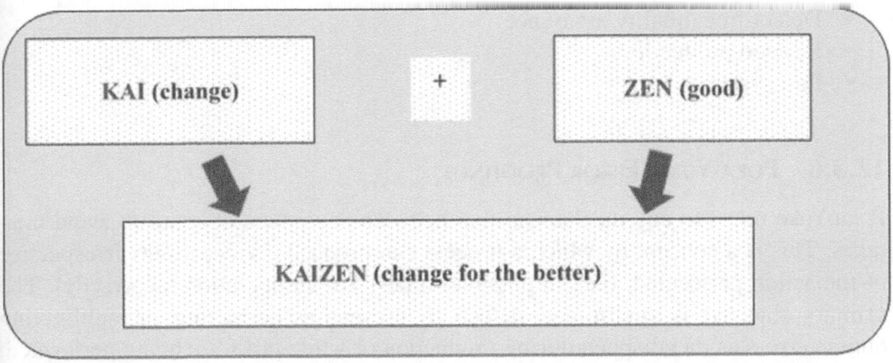

FIGURE 12.11 Kaizen (change for the better).

12.3.4 Just-in-Time

The JIT philosophy focuses on delivering exactly what is needed, at the right time, in the right place, and at the right cost. Its primary goal is to eliminate NVA activities from processes. Eliminating unwanted steps will naturally result in shorter cycle time, better on-time delivery (OTD) performance, and optimum resource utilization. JIT is also known as inventory-less production as the output is just what the customer demand is. JIT requires Kanban to be operational in the system that signals information from upstream processes on customer requirements. JIT also requires a balanced load so that there are no bottlenecks in the process stream.

Some ways to implement JIT in service industry are as follows

- Standardized Work-Standard Operational Instructions
- Kanban: Signal from upstream process steps at the right time
- Stabilized schedules
- Uniform workload
- Focused work groups
- Periodic review of value stream and elimination of NVAs
- Kaizen: Continuous measurement and improvement in processes

12.3.5 Kanban

Kanban applies a form of visual control to the movement of work units, which is understood by all. Nothing is manufactured unless there is a "signal" to manufacture. This contrasts with a push-manufacturing environment where production is continuous.

In service and IT environment, electronic communication and agreement are essential before initiating any work (e.g., purchase order and requirement sign-off).

There are two types of Kanban – Production and withdrawal for initiating or stopping the work.

There are three prerequisite conditions that need to be met before you can introduce the Kanban effectively. The Kanban can be a deadly weapon that hinders your efforts to achieve goals if it is implemented improperly.

- Defect-free (quality assurance)
- Leveled production
- Pull system

12.3.6 Poka-Yoke (Error Proofing)

Poka-Yoke refers to any mechanism that helps an equipment operator to avoid mistakes. This is a concept by which a process can produce correct results irrespective of the action performed. (i.e., impossible to perform an operation incorrectly). The primary objective is to eliminate defects by preventing, correcting, or highlighting human errors as they happen during production or while tasks are being performed. Figure 12.9 represents the concept of Poka-Yoke, a Japanese term that refers to error-proofing or mistake-proofing techniques used in manufacturing and service processes.

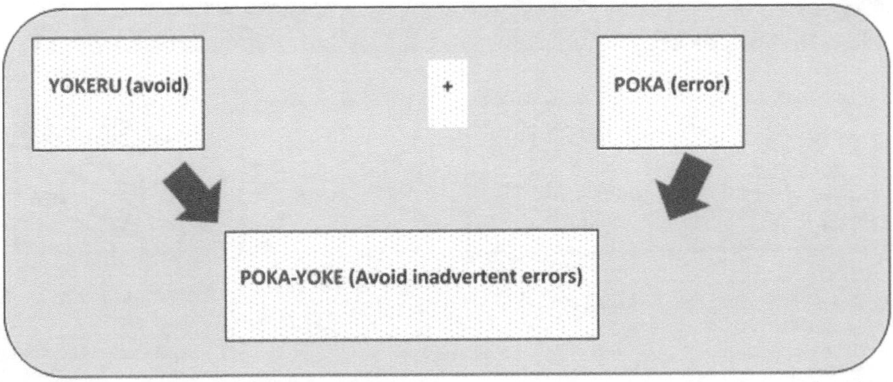

FIGURE 12.12 Poka-Yoke classification.

The available classifications of Poka-Yoke are depicted in Figure 12.12.

1. Prevention-based Poka-Yoke
   a. Control method
   b. Warning method
2. Detection-based Poka-Yoke

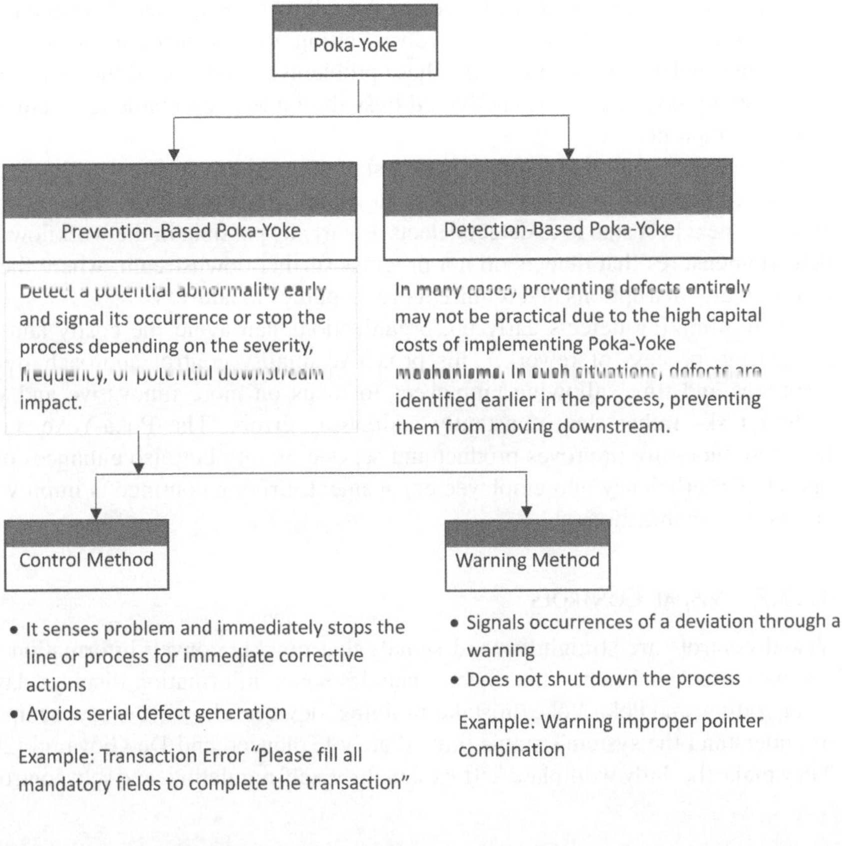

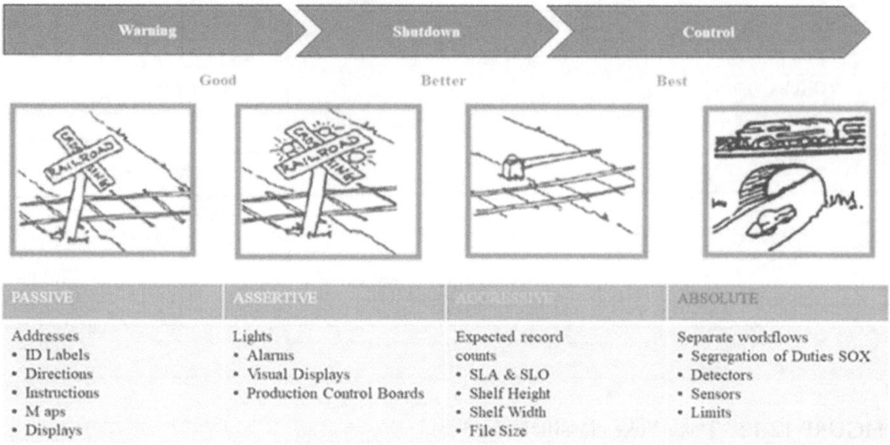

FIGURE 12.13 Poka-Yoke examples.

If mistakes never happen in the first place, there needs to be no defect fixing with examples in Figure 12.13.

Most errors in organizations can be significantly reduced or eliminated through the implementation of standard processes that allow for early detection and prevention of defects. A robust approach involves setting up clear procedures for identifying when problems occur, analyzing these problems to understand their root causes, and applying corrective actions that address the underlying challenges rather than just the symptoms.

The primary goal of Poka-Yoke is to design processes in such a way that prevents defects from occurring in the first place. However, if complete prevention is not feasible, the next best step is to detect defects as early as possible in the workflow. Early detection ensures that defects do not progress further downstream, where they can cause greater disruptions and require more resources to address.

By minimizing defects early on, organizations can avoid the costly and time-consuming process of rework. This proactive quality control approach frees up resources and time, allowing employees to focus on more innovative and value-adding tasks rather than constantly addressing errors. The Poka-Yoke method, therefore, not only improves product and service quality but also enhances overall operational efficiency and employee engagement, driving continuous improvement across the organization.

12.3.7 Visual Controls

Visual controls are straightforward signals that quickly convey information about a situation or condition. This system includes signs, information displays, layouts, color coding, and Poka-Yoke (mistake-proofing) devices, allowing everyone involved to understand the system's status immediately [Trabucco and De Giovanni (2021)]. They make the daily workplace self-explanatory, self-regulating, and self-controlling,

FIGURE 12.14 Production status display.

FIGURE 12.15 Fault message display.

enabling the management of variances in processes and products at the source during production:

- Process conditions (includes machines, materials, manpower, and methods)
- Product outcomes (that are quality, delivery, and cost)

Figures 12.14 and 12.15 illustrate visual management tools in Lean manufacturing, where the Production Status Display shows real-time performance metrics and the Fault Message Display alerts operators to equipment issues for timely corrective action.

12.4 BENEFITS OF LEAN MANUFACTURING

Lean manufacturing offers a wide range of benefits to businesses by focusing on reducing waste, improving efficiency, and enhancing value. Below highlighted some of the key benefits.

12.4.1 Improved Efficiency

Lean manufacturing significantly improves the efficiency of the process by removing waste and optimizing all aspects of production. Streamlined processes lead to faster production times and reduced cycle times. Methods such as VSM are used to analyze and optimize each step of the production process. For example, in cellular manufacturing, grouping similar processes near reduce movement and improve workflow, significantly boosting overall efficiency.

12.4.2 Cost Reduction

By reducing waste, lean manufacturing helps lower production costs. This includes saving on materials, labor, and overhead costs. Companies can achieve significant cost savings by minimizing excess inventory, reducing scrap, and improving resource utilization. Some of the examples referring to the previous sections, excess inventory, rework, scrap, and operational inefficiencies are reduced that help cost saving allows companies to invest more on business. Another example, a furniture manufacturer reduces warehousing costs by receiving materials only as they are needed for production, minimizing storage costs and the risk of obsolete inventory

12.4.3 Enhanced Quality

Lean principles lead to higher product quality by focusing on defect prevention and continuous improvement. Techniques like Poka-Yoke and quality at the source ensure that defects are caught and corrected early in the production process. By applying Poka-Yoke in packaging, line uses sensors to ensure all items are correctly packed; if an item is missing, the line stops automatically, reducing the risk of errors reaching customers. One more example of using Jidoka, in electronics manufacturing, machines are equipped with sensors to detect abnormalities, like missing components, halting the process to fix the issue immediately [Larson and Greenwood (2004)]. This is because, when a problem is detected in the work process and the workflow that need to prevent errors from moving further down the line.

12.4.4 Faster Delivery Times

Lean manufacturing enables faster response to customer demands through efficient workflows and reduced cycle times. JIT and Kanban systems help ensure that products are produced just in time to meet customer needs, reducing lead times and improving customer satisfaction. A few examples of application of Kanban for car parts supplier uses Kanban cards to signal when new parts are needed on the production line, ensuring a smooth and continuous flow without excess inventory buildup.

12.4.5 Greater Employee Satisfaction

Lean promotes employee participation in problem-solving and process enhancements, resulting in increased job satisfaction and motivation. A culture of continuous improvement fosters teamwork and empowers employees to contribute to the organization's success. This helps using the Kaizen events by having the regular small group meetings where employees at a factory identify inefficiencies and propose solutions, leading to incremental but impactful changes.

12.5 LEAN MANUFACTURING AND RESOURCE EFFICIENCY

Lean manufacturing focuses on eliminating waste and optimizing processes, leading to more efficient use of resources such as time, materials, and labor, while delivering maximum value to customers.

12.5.1 Reducing Waste

Lean manufacturing identifies eight types of waste (muda) discussed in the previous sections, including overproduction, waiting, transportation, over-processing, inventory, motion, defects, and the underutilization of talent, which is recognized as the eighth type of waste.

12.5.2 Energy Efficiency

Implementing lean principles can lead to more energy-efficient production processes by optimizing machinery use and reducing idle times that reduce waste, lower costs, and minimize environmental impact. Lean practices such as preventive maintenance and equipment efficiency improvements contribute to lower energy consumption.

- Energy audits: Performing routine energy audits helps pinpoint opportunities for reducing energy consumption.
- Equipment efficiency: Upgrading to energy-efficient machinery and ensuring equipment is well maintained.
- Process optimization: Streamlining processes to decrease energy utilization, such as by adjusting the heating, cooling, and lighting systems.

A few techniques on how the energy efficiency is aligned with the listed lean principles and lean manufacturing:

- Switching to energy-efficient LED lighting and using motion sensors or timers to turn lights on only when needed can significantly reduce energy consumption in any of the manufacturing or warehousing. This is currently used across the streetlight where the light gets switched on at particular time and switched off at particulate time.
- Designing layout of the plant in such a way to minimize travel distance and improve workflow that just not reduces motion waste but also reduces energy use, as conveyors and transport vehicles operate less frequently.
- Currently in many metropolitan cities, the traffic light management is completely based on the peak hours and non-peak hours and also based on the number of vehicles the time will be decided.

Hence, by integrating energy efficiency into lean principles, many organizations achieve their goals of waste reduction sand operational resilience.

12.5.3 Material Efficiency

Lean encourages the efficient use of materials by minimizing excess inventory and reducing scrap and rework. Techniques like JIT and continuous flow ensure that materials are used effectively, reducing waste and lowering costs. Material waste refers to any unnecessary use of raw materials, excess inventory, scrap, defects, over-processing, and any other tasks that utilizes resources.

- Material audits: Conducting audits to track material usage and identify opportunities for reduction.
- Recycling and reuse: Implementing recycling programs and reusing materials where possible.
- Design for manufacturability: Designing products in a way that minimizes material use and facilitates efficient production processes.

A few techniques on how the material efficiency is aligned with the listed lean principles and lean manufacturing:

- In industries like textiles or metal fabrication, optimizing cutting patterns can significantly reduce scrap material by automating the entire process.
- If you relate the recycle, implementing recycling programs for scrap materials (such as mobiles, plastic, etc.) can turn waste into a valuable resource.
- Implementing key performance indicators (KPIs) related to material usage, scrap rates, and inventory levels helps track status and identify areas opportunities.

Hence, material efficiency in lean manufacturing is a crucial aspect that not only reduces costs but also supports sustainable operation.

12.5.4 Water and Resource Conservation

By streamlining processes, lean manufacturing can also contribute to conserving water and other resources, ensuring that manufacturing processes are more sustainable. Efficient production processes reduce the need for water and other inputs, contributing to environmental sustainability.

- Water use audits: Regular audits to identify areas where water use can be reduced.
- Efficient water use: Implementing water-efficient technologies and practices, such as closed-loop water systems.
- Waste reduction: Minimizing waste generation through process optimization and resource recovery initiatives.

One of the examples includes implementing water recycling systems in manufacturing processes to reuse water, reducing overall consumption. For instance, a factory producing automotive parts might use a closed-loop cooling system to recycle water used in cooling machinery. Another example, in a chemical manufacturing plant, precise measurement of chemicals can prevent excess and waste.

12.5.5 Environmental Sustainability

Lean manufacturing principles contribute to overall environmental sustainability by reducing the environmental footprint of production processes. This includes lower emissions, reduced waste, and more efficient use of resources.

- Sustainable practices: Incorporating sustainable practices into production, such as using renewable energy sources and sustainable materials.
- Environmental management systems: Implementing systems to monitor and manage environmental performance.
- Compliance and certification: Achieving certifications such as ISO 14001 for environmental management.

One of the examples, designing processes to prevent environmental contamination that was used in a semiconductor manufacturer might use error-proofing techniques to ensure that hazardous materials are handled and disposed of properly, minimizing the risk of pollution.

12.6 IMPLEMENTING LEAN PRINCIPLES

Lean principles involve systematically eliminating waste, improving processes, and fostering a culture of continuous improvement to enhance productivity and deliver greater value to customers.

12.6.1 Steps to Implement Lean

1. Assess current state: Evaluate the existing processes by conducting the detailed analysis that can help identify areas for improvement in the entire process or a sub-process that can improve the part of the process. This is possible by using the tools like VSM or process analysis to create a baseline of current operations that was carried out by the operational teams.
2. Set goals: Define clear objectives and goals (e.g., reducing lead time, cutting costs, and improving quality) for implementing lean principles. Ensure that the goals are specific, measurable, achievable, relevant, and time-bound (SMART) to measure and achieve the results as planned that are aligned in the beginning of the project.
3. Engage employees: Assemble and engage the team responsible for leading and executing lean initiatives. Involve employees at all levels of the organization in the lean transformation journey and not just an operational team and not just the leadership team. Encourage a culture of continuous improvement and open communication across team members.

4. Train and educate: Provide training on lean tools and techniques. Ensure that all employees understand the principles of lean and how they can contribute to the implementation process. Make sure that all staff members are involved in the lean transformation journey.
5. Pilot projects: Once the solutions and improvements are identified, build the solutions and implement as a prototype so that one can identify any further enhancements in the improvement are needed or ready to roll out. Start with pilot projects to test and refine lean implementations. Choose small, manageable projects that can demonstrate the benefits of lean principles.
6. Measure and adjust: Continuously measure performance and adjust as necessary. Make sure to apply the KPIs to measure improvements in areas of improvements with respect to lead time, quality, defect rate, and efficiency. Regularly review progress and make adjustments as needed and improve further.

12.6.2 Overcoming Challenges

Most common challenges include resistance to change, lack of training, lack of management support, inadequate training and knowledge sharing, insufficient resources, inadequate objectives, and no constant lean deployment approach. Overcoming these challenges requires a proactive and strategic approach, including effective communication, training, resource allocation, and continuous monitoring that can successfully implement lean manufacturing and achieve long-term benefits.

12.7 CONCLUSION

In the search of operational excellence and resource efficiency, lean manufacturing develops as a transformative approach that not only streamlines processes but also significantly enhances the utilization of resources. The principles of lean, deeply rooted in the quest for waste elimination and continuous improvement, offer a robust framework for achieving greater efficiency and effectiveness in manufacturing.

In a nutshell, the application of these lean manufacturing practices and principles help achieve:

- Holistic resource optimization
- Enhanced process efficiency
- Long-term benefits and competitive advantage
- Continuous improvement and innovation
- Sustainable practices

In summary, lean manufacturing principles provide an effective framework for maximizing resource efficiency within the manufacturing industry. By prioritizing waste reduction, process enhancement, and ongoing improvement, organizations can significantly boost their operational performance, advance their sustainability objectives, and ensure sustained success. Adopting lean principles not only revolutionizes resource management but also fosters a culture of excellence, guiding organizations toward greater efficiency and resilience in the future.

Before concluding the chapter, would like to relate the lean philosophy with a story of "The Blind Men and the Elephant." This story is often used to illustrate the concept of perceptions of an individual and the limitations.

Once there were six blind men who stood by the roadside every day and often heard of elephants, but they had never seen one as they could not see, and wanted to know how the elephant could be. One day, somehow, they were able to manage to feel the elephant requesting the driver. The challenge is they can't see the elephant, but they can feel the elephant with their hands. Each individual began to feel the elephant. The first man, who touched its leg, exclaimed, "The elephant is like a tree!" The second man, feeling the tail, responded, "Oh, no! It resembles a rope." The third man, who touched the trunk, remarked, "It feels like a large snake." The fourth man, upon touching the ears, said, "It's like a big hand fan." The fifth man, who touched the body, stated, "It feels like a huge wall." Lastly, the sixth man, who touched the tusk, declared, "It is like a spear," as shown in Figure 12.16.

Figure 12.16 illustrates how the six blind men's differing perceptions of an elephant symbolize the importance of combining diverse viewpoints to achieve a complete understanding of reality. Once the elephant moved from there, the six blind men sat together, eager to share their experiences. Each man was convinced that he understood what the elephant truly looked like based on his own touch. Though their interpretations differed, leading to a lively discussion, this encounter highlighted the importance of considering multiple perspectives to gain a fuller understanding. As discussed in the above story, this leads to confusion and conflict, as each man believes his partial view represents the whole truth. This fable serves

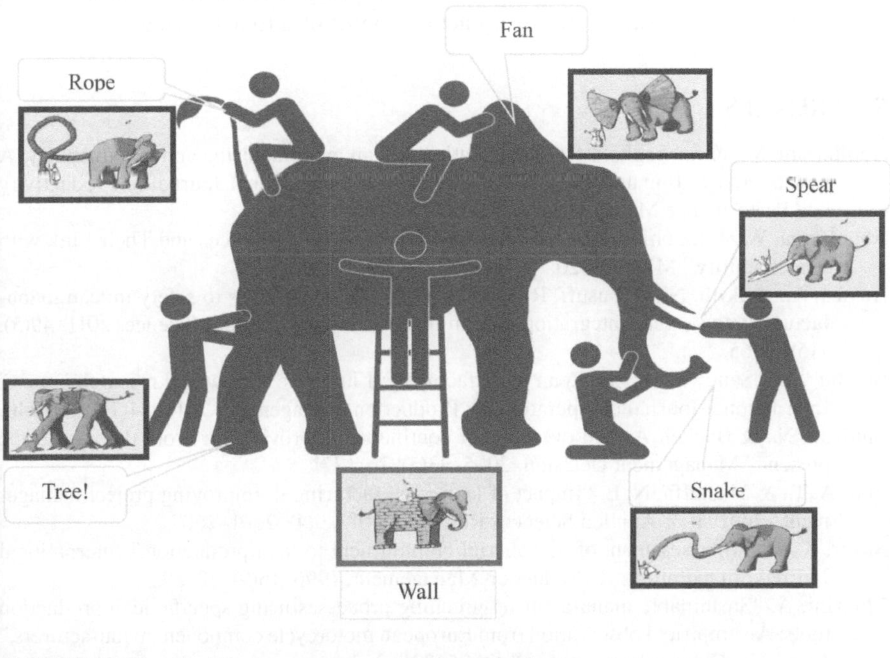

FIGURE 12.16 Six blind men perspective about elephant.

as a powerful metaphor for lean manufacturing, highlighting the importance of holistic understanding instead of the silos and collaboration across different areas of an organization.

As explained in the chapter above on the lean principles, a quick relation to the story with the lean principles is listed as follows:

1. Look at the holistic process view: In lean manufacturing, it is very critical to look at the entire value stream rather than focusing on individual or isolated processes. Just like the blind men only understood a small part may miss the inefficiencies and opportunities across the whole system.
2. Collaboration between and across functions: At every stage, lean emphasizes the need for collaborating and working across teams by sharing insights and data from various parts of the organization. This collaboration allows everyone having a full understanding of processes, leading to better decision-making, much like the blind men would have needed to share their observations to truly understand what an elephant is.
3. Continuous learning and improvement: Lean manufacturing depends on continuous improvement (Kaizen) and the ongoing identification opportunities for perfection. This requires an openness to learn from all parts of the process; acknowledging perspectives by integrating different viewpoints, organizations can continuously improve and refine their processes.
4. Understanding the customer viewpoint: In lean, the goal is to deliver value to the customer. Each department or team may have their own view of what adds value, but like the blind men, they might be missing the broader customer experience. Lean encourages teams to align their activities with customer needs, ensuring that every action contributes to delivering.

REFERENCES

Abolhasani, M., & Sabbagh, R. "The integration of lean manufacturing and sustainability: A new approach to manufacturing improvements." International Journal of Productivity and Performance Management, 2017, 66(7), 1006–1022.

Abualfaraa, W., & Salonitis, K., "Lean-Green Manufacturing Practices and Their Link with Sustainability." MDPI, 2020.

Anvari, A., Zulkifli, N., & Yusuff, R. M. "Evaluation of approaches to safety in lean manufacturing and safety integration in lean implementation." Safety Science, 2011, 49(5), 1158–1165.

Bhamu, J., & Sangwan, K. S. "Lean manufacturing: Literature review and research issues." International Journal of Operations & Production Management, 2014, 34(7), 876–940.

Bhuiyan, N., & Baghel, A. "An overview of continuous improvement: From the past to the present." Management Decision, 2005, 43(5), 761–771.

Bon, A. T., & Mustaffa, N. E. "Impact of lean manufacturing to improving project management." Journal of Applied Sciences Research, 2013, 9(4), 2601–2607.

Boyer, K. K. "An assessment of managerial commitment to lean production." International Journal of Operations & Production Management, 1996, 16(9), 48–59.

Chiarini, A. "Sustainable manufacturing-greening processes using specific lean production tools: An empirical observation from European motorcycle component manufacturers." Journal of Cleaner Production, 2014, 85, 226–233.

Ciarapica, F. E., & Matt, D. T. "Lean thinking and resource efficiency: Simulation models for the optimal design of lean green plants." International Journal of Sustainable Engineering, 2016, 9(2), 102–114.

Cudney, E., & Elrod, C. "Incorporating lean concepts into supply chain management." International Journal of Six Sigma and Competitive Advantage, 2011, 6(3), 156–171.

Duarte, B. M., & Machado, V. C. "Exploring lean and green supply chain performance using balanced scorecard perspective." International Journal of Business Performance and Supply Chain Modelling, 2011, 3(3), 240–263.

Dues, C. M., Tan, K. H., & Lim, M. "Green as the new lean: How to use lean practices as a catalyst to greening your supply chain." Journal of Cleaner Production, 2013, 40, 93–100.

Esmemr, S., Cinar, M., & Demirdogen, R. E. "Application of lean manufacturing techniques in the Turkish furniture industry." African Journal of Business Management, 2010, 4(10), 1873–1884.

Fahmi, M., Arifin, H., & Fahmi, R. "Implementation of lean manufacturing to reduce waste at Batik industries: A case study in Indonesia." Journal of Engineering and Applied Sciences, 2019, 14(7), 2341–2345.

Furlan, A., Vinelli, A., & Dal Pont, G. "Complementarity and lean manufacturing bundles: An empirical analysis." International Journal of Operations & Production Management, 2011, 31(8), 835–850.

Hajmohammad, S., & Vachon, S. "Lean manufacturing and environmental performance: Exploring the impact and relationship." Production and Operations Management, 2014, 23(12), 1862–1880.

Hines, P., & Rich, N. "The seven value stream mapping tools." International Journal of Operations & Production Management, 1997, 17(1), 46–64.

Hofer, C., Eroglu, C., & Rossiter Hofer, A. "The effect of lean production on financial performance: The mediating role of inventory leanness." International Journal of Production Economics, 2012, 138(2), 242–253.

Hu, Q., Mason, R., Williams, S. J., & Found, P. "Lean implementation within SMEs: A literature review." Journal of Manufacturing Technology Management, 2015, 26(7), 980–1012.

Karim, A., & Arif-Uz-Zaman, K. "A methodology for effective implementation of lean strategies and its performance evaluation in manufacturing organizations." Business Process Management Journal, 2013, 19(1), 169–196.

King, A. A., & Lenox, M. J. "Lean and green? An empirical examination of the relationship between lean production and environmental performance." Production and Operations Management, 2001, 10(3), 244–256.

Larson, T., & Greenwood, R. "Perfect complements: Synergies between lean production and eco-sustainability initiatives." Environmental Quality Management, 2004, 13(4), 27–36.

Li, W., Alvandi, S., Kara, S., Thiede, S., & Herrmann, C. "Sustainability cockpit: An integrated tool for continuous assessment and improvement of sustainability in manufacturing." CIRP Annals, 2016, 65(1), 5–8.

Trabucco, A., & De Giovanni, P., "Artificial Intelligence in Lean Manufacturing." Emerald Insights, 2021.

13 Improvement of Titanium Alloy (Ti-15-3) Wear Resistance and Hardness by Solution Treatment and Aging to Pioneer Sustainable and Environmentally Friendly Manufacturing

Poppy Puspitasari, Hadaq Maulana Muji Burrochman, Diki Dwi Pramono, Muhammad Ilman Hakimi Chua Abdullah, and Abdul Munir Hidayat Syah Lubis

13.1 INTRODUCTION

Titanium alloys have become the focus of interest in recent years because of their excellent resistance to corrosion, biocompatibility, microhardness, and strength. Titanium alloys are extensively used in automotive industry for making crucial components such as exhaust system components, heat exchangers, chassis elements, and engine parts. In the marine industry, medium-strength titanium alloys are used to manufacture products such as propeller shafts, main housings, and hydraulic pump parts. Titanium alloys with high strength are used in manufacturing thin-walled tubes specifically designed for condensers, heat exchanger tubes, and diesel engine exhaust systems in submarines (Pervukhin et al. 2017). Additionally, high-quality titanium alloys, capable of withstanding temperatures above 600°C, are predominantly used in the aerospace industry for building supersonic fighter aircraft, including compressor blades, fan housings, casing parts, and turbine disks for jet engines (Puspitasari et al. 2023a).

Despite its widespread use and numerous advantages, titanium alloy has some drawbacks, particularly its lower hardness, with Vickers hardness values typically

DOI: 10.1201/9781003503323-13

ranging from 250 to 350, and poor wear resistance (Chen et al. 2024). These deficiencies may be alleviated by using suitable heat treatments. Presently, a variety of treatment techniques and technologies have been developed to enhance the properties and long-lasting nature of titanium alloys, one of which is solution treatment. The wear resistance of titanium alloys is influenced by factors such as microstructure, phase composition, and tribological response. Solution treatment and aging (STA) are two commonly used techniques for enhancing the mechanical characteristics of titanium alloys. The objective of solution treatment is to achieve a homogeneous microstructure by dissolving the alloy in a solvent, whereas aging enhances the strength and toughness of the alloy by causing the precipitation of phases (Chen et al. 2024).

Heat treatment methods are generally used for all kinds of titanium alloys to tailor and modify its mechanical properties, including STA or solution treatment and aging (Totten 2016). The tensile strength is 20% higher after STA compared with mill annealing. This procedure is also used to improve the mechanical characteristics of aeronautical components in order to gain adequate ductility. Heat treatment of the microstructure in a dual α-β phase region may improve tensile and fatigue properties of titanium alloys. STA is an effective and useful method to improve the wear resistance of titanium alloys (Su et al. 2022).

Su et al. (2022) investigated the Ti-6Al-4V titanium alloy wear-resistant behavior for several surface treatment methods including STA, direct solution, and direct aging. The results of their studies showed that STA treatment has better wear resistance than direct solution treatment and direct aging. Ganesh et al. (2012) studied on the influence of solution aging on wear property of wrought Ti-6Al-4V alloy. The samples were treated with a solution and aged, and the researchers found that these showed better wear resistance. The improvement was attributed to the development of a protective oxide film over the entire testing process. Sharma et al. (2021) conducted a study and investigated the wear properties of Ti-6Al-4V alloy generated via electron beam melting and their response to solution treatment. The study discovered that, when water quenched, solution treated the alloy on the β phase, it had superior wear resistance than before. Elshaer et al. (2023) studied the effectiveness of annealing, solution treatment, and STA treatment at 925°C for 30 min with different cooling medium, i.e., furnace cooling, water cooling, and air cooling. Their research demonstrated that the most effective treatment for improving wear resistance and mechanical properties such as tensile strength, toughness, and hardness was shown to be the STA treatment.

The temperature in the STA process is crucial since it profoundly affects the microstructure of the titanium alloy, hence impacting its properties (Vahidshad and Khodabakhshi 2022). Titanium and their alloys experience exhibiting allotropic transition around specific temperatures. During this period, the α-phase, characterized by a face-centered cubic (FCC) structure, transforms to the β phase, which has a body-centered cubic (BCC) crystal structure. According to Joshi (2006), the β-transformation temperature may be influenced by the presence and amount of alloying elements. If α-stabilizing elements are present, the temperature can rise, but if β-stabilizing components are present, the temperature can drop. In their study, Ju et al. (2021) examined how solution temperatures within the range of 760°C–1060°C affect the wear characteristics of laser powder bed fusion (LPBF)-processed titanium alloy Ti-6Al-4V in the context of additive manufacturing. The experiment demonstrated that submitting the material to solution treatment at 1060°C produced in the greatest degree of wear resistance, irrespective of the applied load. Additionally, this treatment led to the lowest coefficient of friction (CoF) and wear rate.

This study aims to investigate the effect of STA temperatures at 950°C, 1000°C, and 1050°C on the wear resistance of titanium alloy (Ti-15-3). The findings will provide vital understanding into the capacity of this method to enhance sustainability and ecological compatibility in the production of titanium alloys. By optimizing the wear resistance of Ti-15-3, this research may contribute to the development of more environmentally friendly and sustainable manufacturing practices, ultimately reducing the environmental impact of titanium alloys in various industrial applications.

13.2 LITERATURE REVIEW

13.2.1 Titanium

The discovery of titanium in England was made by Reverend William Gregor in 1790. Gregor first presented it in 1791, providing a detailed account of the chemical composition and qualities of a kind of magnetic black sand that he discovered on the South Cornish coast. Titanium has been used in industrial applications for over six decades. In the first stages of attempting to get pure titanium from its compounds, the outcome typically included the creation of nitrides (TiN), carbides (TiC), or carbonitrides (TiCN) due to the common misinterpretation of its metallic sheen as a different metal (Froes 2015).

In 1887, L.F. Nilson and O. Pettersson successfully obtained a titanium product with a purity of 97.4% by using sodium to reduce titanium tetrachloride within a sealed steel cylinder. In 1906, M.A. Hunter, an employee at General Electric Company in the United States, successfully created the first almost entirely pure titanium metal. Hunter improved upon the technique used by Nilson and Pettersson by successfully eliminating air from the process, which led to the production of titanium metal with very few impurities (Froes 2015).

In 1925, A.C. Van Arkel and J.H. de Boer from the Netherlands devised a technique to create titanium by subjecting titanium tetrachloride to heat breakdown. While this procedure resulted in titanium of exceptional purity, it was also excessively expensive. Subsequently, W.J. Kroll's research resulted in the synthesis of titanium metal by using magnesium in an atmosphere saturated with an inert gas (argon) inside a sealed system, obviating the need for elevated pressure. Xu et al. (2021) state that titanium has gained significant popularity in the aviation, aerospace, and shipbuilding sectors owing to its remarkable characteristics, such as robust corrosion resistance, high specific strength, low thermal conductivity, and broad temperature adaptability. Due to scientific and technological progress, high-performance titanium has been successfully used in several fields such as autos, sports equipment, medical devices, and light industrial. Table 13.1 highlights the benefits of titanium in comparison to other metals, showcasing its better qualities.

Titanium may be classified into two categories, with the first form being pure titanium (cpTi). Titanium in its pure form is a shiny white metal that has a low density, great strength, and outstanding resistance to corrosion. The second kind denotes a metallic mixture consisting of titanium with a composition of 6% aluminum and 4%

TABLE 13.1
Comparison of Titanium Properties with Other Metals

	Metals				
Properties	**Ti**	**Al**	**Cu**	**Fe**	**Mg**
Melting point (°C)	1953	933	1084	1811	923
Density (kg/dm²)	4.51	2.7	9.84	7.784	1.738
Thermal conductivity (W/m °K)	21.9	237	3.98	80.4	156
Specific heat (J/kg K)	522	903	386	447	1046
Modulus elasticity (MPa)	10.7	6	11	211	45
Yield strength (MPa)	240	28	69	130	–
Tensile strength (MPa)	330	69	200	260	–

Source: Campbell (2006).

vanadium. This alloy has superior strength compared to pure titanium. Titanium metal has a density that falls between that of aluminum and steel. Titanium is the ninth element with a high abundance in the earth's crust and is known for its plentiful metal formations. Titanium exhibits allotropy, meaning it can exist in two distinct crystal forms at varying temperatures. Titanium has a hexagonal close-packed (HCP) crystal structure in its pure state when it is at ambient temperature. The structure is referred to as the alpha phase and remains stable until it reaches a temperature of 1620°F (882°C), at which point the crystal structure undergoes a shift. At elevated temperatures, the arrangement of atoms in the crystal changes to a BCC crystal structure, which is often known as the β phase.

The $\alpha + \beta$ titanium alloy and β metastable titanium alloy have attracted many scientists to explore and develop titanium. Alpha-beta titanium alloy consists of aluminum material content as α stabilizer and a combination of β stabilizers such as chromium, iron, molybdenum, and vanadium, and there are tin and zirconium as solid solution strengthening elements for replacement. Such alloys have good strength and creep resistance at high temperatures. Pure titanium can undergo allotropic transformation. Allotropic transformation is a change in the crystal structure of a material caused by a change in temperature; discussion in terms of pure titanium, at 885°C, titanium undergoes a transformation from a tightly packed hexagonal crystal structure (α) to a BCC structure (β). The BCC structure is more easily deformed than the α structure because there are more slip systems available for deformation. The strength and formability of many of these alloys can help optimize when given heat treatments (Froes 2015).

13.2.2 Classification of Titanium Alloys

Titanium alloys are composite materials composed of titanium and other metals, including aluminum, vanadium, molybdenum, and niobium. The mechanical characteristics of titanium alloys are affected by several variables, including chemical composition, manufacturing method, and ambient conditions. Titanium alloys have exceptional tensile strength, with values ranging from 441 to 960 MPa. Furthermore, titanium's physical characteristics, including its low elastic modulus, low paramagnetic susceptibility, low thermal expansion, short radioactive half-life, and excellent ballistic qualities, are also used in the aerospace industry. Titanium alloys provide the benefit of possessing both high strength and low weight, making them highly ideal for use in industries that need robust and lightweight materials, such as aerospace (Froes 2015).

The alloys of titanium are often used in the aerospace sector for the production of various aircraft parts, such as engines, wings, and aircraft bodies. Titanium alloy is advantageous as a material for making aviation components due to its exceptional strength and lightweight properties, which effectively alleviate the aircraft's overall weight burden. Furthermore, titanium alloys possess a consistent degree of strength and exhibit resistance to corrosion, so enabling them to last for extended periods and guarantee flight safety. Titanium is classified into two distinct kinds, namely:

a. Pure titanium

   Pure titanium conducts heat and electricity well and has a low density. Pure titanium crystallizes into a hexagonal shape at low temperatures. Up to a beta transus point of 885°C (1625°F), this structure can exist. The lowest equilibrium temperature at which 100% of the alloy is in the beta phase is known as the beta transus. A tiny portion of the oxygen present in pure titanium helps to produce the exterior TiO_2 layer, while more than 50% of the metal is added to create oxygen diffusion regions (Warlimont 2018).

b. Titanium alloy

   Titanium is used to make a variety of alloys, including Tiα alloy, Tiβ, and Ti α + β; further information about titanium alloys is provided as follows:

   1. Titanium alloy Tiα

      Tiα alloy is a single-phase alloy that has good properties and is stable at high temperatures. Tiα alloy contains 5% Al and 2.5% Sn. When annealing process at high temperatures right in the region β using rapid cooling, it will produce grain structure α-Widmanstätten and produce good resistance to fatigue. Room cooling will produce a plate-like α structure that provides good resistance to creep.

   2. Titanium alloy Tiβ

      Although the addition of large amounts of vanadium or molybdenum will create a β structure at room temperature, there are number of β alloys in this condition, but actually in Tiβ, 100% alloy has a lot of β stabilizer content in it. This causes rapid cooling which can result in a meta-stabilized structure consisting entirely of β. Strengthening can be achieved by using large amounts of solvent saturated strengthening agents and aging the metastable β structure, which forms alpha precipitates. Applications of this metal include strong fasteners, beams, and connectors for use in spacecraft.

   3. Titanium alloy Ti α + β

      Alpha-beta titanium alloys possess a crystal structure that comprises both alpha and beta crystal types at normal room temperature. The beta phase of the alloy improves its capacity to be formed, whereas the alpha phase acts as the binding matrix. Conversely, alpha-beta alloys possess well-balanced mechanical characteristics. This alloy is typically weldable and has excellent corrosion resistance at room temperature. It also possesses great malleability and stays stable up to a temperature of 4250°C. When beta stabilizer elements, such as molybdenum, vanadium, tantalum, and niobium, are introduced to pure titanium, the beta phase is inclined to augment at ambient temperature, while the alpha stabilizer elements will enhance the alpha phase.

13.2.3 Titanium Alloy (Ti-15-3)

Ti-15-3 is a titanium alloy that is beta metastable and offers benefits in terms of weight reduction when compared to other technical materials. Ti-15-3 has exceptional cold formability while in the solution treated state and possesses significant strength when aged. The alloy is often used in crucial applications within the aviation and space sectors. Ti-15-3 has an elevated operational temperature of 550°F (288°C) and is offered in strip widths ranging from 0.016 inches (0.4 mm) to 0.094 inches (2.4 mm). The metastable β-titanium alloys are extensively used in the aircraft industry is attracted to the extraordinary characteristics of these features, such as high specific strength, low modulus of elasticity, and corrosion resistance. Some examples of these alloys are Ti-15-3 and other recently developed alloys that have great promise for many applications. These alloys possess desirable qualities, such as a maximum tensile strength of around 1200–1500 MPa and an elongation range of 8%–18%. The primary source of strengthening in these two kinds of alloys comes from the formation of the secondary α (alpha) phase, The precipitation occurs from the metastable β (beta) phase as a result of the ageing process in the two-phase (α + β) region. Studies on the titanium alloy Ti-15V-3Cr-3Al-3Sn indicate that subjecting it to pre-aging at temperatures of 250°C and 350°C prior to high-temperature aging may result in the formation of uniform and small precipitates, hence enhancing the dispersion of these tiny precipitates throughout the alloy (Ji et al. 2021).

The Ti-15-3 alloy's polished smooth surface reduces the susceptibility to the formation of metastable and stable spots, hence enhancing the corrosion resistance. Polishing the surface of titanium may enhance its resistance to corrosion in tropical marine aircraft environments (Xie et al. 2021). Further investigations have shown that when subjected to slow strain testing, the Ti-15-3 alloy that has undergone solution annealing has remarkable resistance to corrosion when exposed to 3.5% NaCl and simulated saltwater. Titanium grade 15-3, in its solution annealed form, exhibited superior resistance to deterioration compared to titanium grade Ti-6Al-4V in the annealed state (Tiyyagura et al. 2018).

To enhance the properties of the Ti-15-3 alloy, it is essential to carry out solution processing and heat treatment to induce phase precipitation from the matrix. The physical features of Ti-15-3 are formed by several aspects, such as cooling rate during treatment, the temperature of the solution treatment, the temperature and length of the aging process, and the heating rate of the material. The experiment used the Ti-15-3 alloy, which underwent a solution treatment at a temperature higher

TABLE 13.2
Elements of Ti-15-3

Element of Ti-15-3	Percentage (%)
Vanadium, V	15
Titanium, Ti	76
Chromium, Cr	3
Tin, Sn	3
Aluminum, Al	3

Source: Maurotto et al. (2012).

than the β transus temperature. This treatment resulted in the formation of a microstructure consisting only of the β phase. Slow strain rate testing was conducted on Ti-15-3 material in its solution annealed state. At a temperature of 850°C, researchers observed the creation of a 250 nm thick metallurgical bond due to TiC diffusion which resulted in excellent elongation. In contrast, samples subjected to a temperature of 950°C formed a diffusion layer that was 400 nm thick, composed of TiC and FeTi compounds. This resulted in the highest tensile strength. During the recent test, when the application of a 1050°C treatment was used, the diffusion thickness increased, resulting in lower tensile strength measurements.

Table 13.2 shows the composition contained in Ti-15-3 which is made of titanium as the main material with a composition of 76%. There are also variant materials given, namely vanadium with a composition of 15%; Ti-15-3 also has additional ingredients such as chromium, tin, and aluminum with each having a composition of 3%. Looking at the content, it should be noted that the composition will affect the mechanical characteristics of titanium. Changes in mechanical properties are shown by Li et al. (2012) in Table 13.3.

TABLE 13.3
Mechanical Properties of Ti-15-3

	Unit	
Properties	**Metric**	**English**
Hardness (Rockwell)	95	95
Ultimate tensile strength	790 MPa	115000 psi
Yield strength	770 MPa	112000 psi
Elongation at break	22%	22%
Modulus of elasticity	82 GPa	119000 ksi
Poisson's ratio	0.33	0.33
Shear modulus	30.8 GPa	4470 ksi

Source: Li et al. (2012).

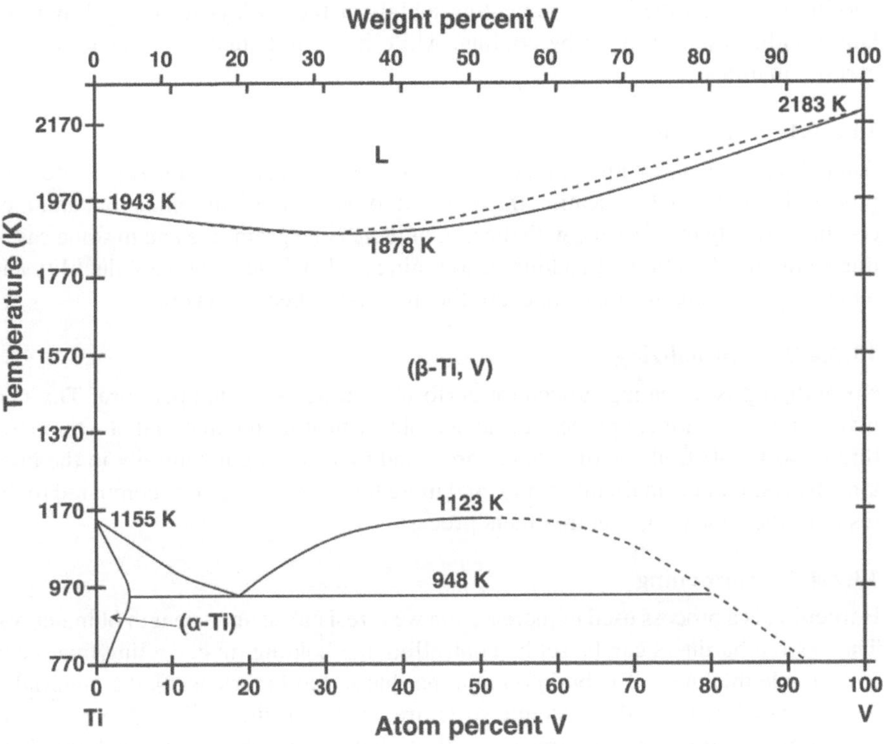

FIGURE 13.1 Phase diagram of Ti–V.

The Ti–V diagram is a representation of the equilibrium phase between titanium and vanadium. This diagram consists of a mixture of titanium and vanadium mixed in certain proportions. In Figure 13.1, we can see the mixture gap in the BCC phase (β-Ti, V) and the monotectoid reaction that occurs at 948 K, namely (β-Ti) which can change to (α-Ti) + (V). At 1155 K, titanium and vanadium can dissolve completely in solid solution (β-Ti, V), but the solubility of vanadium in the HCP phase (α-Ti) is limited to a maximum of 2.7% at a composition V.

The microstructure of β-titanium alloys is very vulnerable to the heat treatment procedure. The precipitation of various phases in Ti-15-3 titanium alloy is influenced by the temperature and time of the heat treatment, leading to changes in its mechanical characteristics. Through the examination of heat treatment settings and their impact on the resulting phases and properties, we are able to manipulate the microstructure and characteristics of the alloy, and assess its conditions for safe use.

13.2.4 Solution Heat Treatment

Solution heat treatment is a process that involves heating a metal sample in a furnace to its recrystallization temperature for a certain duration, followed by chilling it using various cooling media such as air, water, brine, and oil, each with a distinct cooling rate. This method is subjecting metals or their alloys to thermal manipulation, either by heating or by cooling, while in a solid state, in order to achieve certain qualities.

13.2.4.1 Annealing

Annealing is a heat treatment method used to produce coarse but soft pearlite. The process is carried out by heating the material to the point of austenization, and then cooling it slowly in a furnace with the aim of increasing grain size and in some cases, improving machinability. In addition, softening is also done to increase ductility and reduce internal stresses that can cause the material to become brittle.

13.2.4.2 Normalizing

Normalizing is a heating procedure performed at austenite temperature. The normalizing heat treatment process is carried out by heating the material at a temperature of 10°C–40°C above the critical area, and then cooling it naturally in the open air. The result of normalizing is finer and more homogeneous grains compared to the result of the annealing heat treatment process.

13.2.4.3 Hardening

Hardening is a process used to increase the wear resistance and strength of materials. The desired hardness can be set by controlling the holding time, cooling treatment, and sample thickness. To obtain optimal hardness (hard martensite), the material is heated until it reaches the austenite structure. However, the hardening process can have problems if other structures are still present during heating. The solution is to perform quenching to obtain a structure that consists mainly of martensite.

13.2.4.4 Tempering

Tempering is a heating process carried out after the quenching process at a tempering temperature, which aims to achieve the desired ductility in the material. Tempering at low temperatures, between 150°C and 230°C, will only result in an insignificant reduction. However, if the tempering temperature is increased, martensite can break down faster. At temperatures around 315°C, martensite phase change occurs rapidly.

13.2.4.5 Quenching

Quenching is a rapid cooling process to prevent grain growth in the material, which can occur when using slow cooling. The quenching rate is influenced by several factors, such as specific heat, heat of vaporization, viscosity, thermal conductivity of the medium, and flow of the cooling medium. There are several ways to perform the quenching process, such as direct quenching, single quenching, and double quenching.

13.2.4.6 Aging

The aging process is generally carried out at low temperatures and for a long period of time, ranging from 4 to 48 hours. Determining the duration of the aging process must be carefully considered because the time used can affect the type, volume fraction, size, and distribution of precipitate particles, which ultimately affects the final structure and mechanical properties. According to Zhang et al. (2002) Al-Si-Mg can reach a peak tensile strength of 90% after undergoing solution treatment at 540°C for 10 minutes and several hours. Increasing the solution treatment holding time to 30 minutes is enough to change the microstructure to 90% elongation.

The aging process aims to remove the alloying elements from the solid solution and form precipitates that can gradually increase the strength of the alloy. The strengthening mechanism occurs through precipitate shearing and dislocation bypass (or Orowan looping). In precipitate shearing, dislocations will move through the precipitate by cutting the precipitate. This mechanism generally occurs in particles that are easily deformed, such as in small-sized precipitates.

13.3 METHODOLOGY

Titanium alloy (Ti-15-3) was treated with solution treatment with temperatures of 950°C, 1000°C, and 1050°C for 1 hour with an aging process using a temperature of 450°C for 4 hours, as shown in Figure 13.2.

Titanium alloy (Ti-15-3) specimens that have been treated with STA were subjected to Vickers hardness using ESEWAY Vickers Hardness Tester type EW-412AAT with force of 1000 g and dwell of 10 s, Ogoshi wear resistance test using Ogoshi universal wear testing machine type OAT-U with distance of 15 m and a load in the test of 6.36 kg, microstructure test using Nikon optical microscope with 650X magnification, and conducting a surface roughness test with the Mitutoyo Surftest type SJ-301. The research scheme is shown in Figure 13.3.

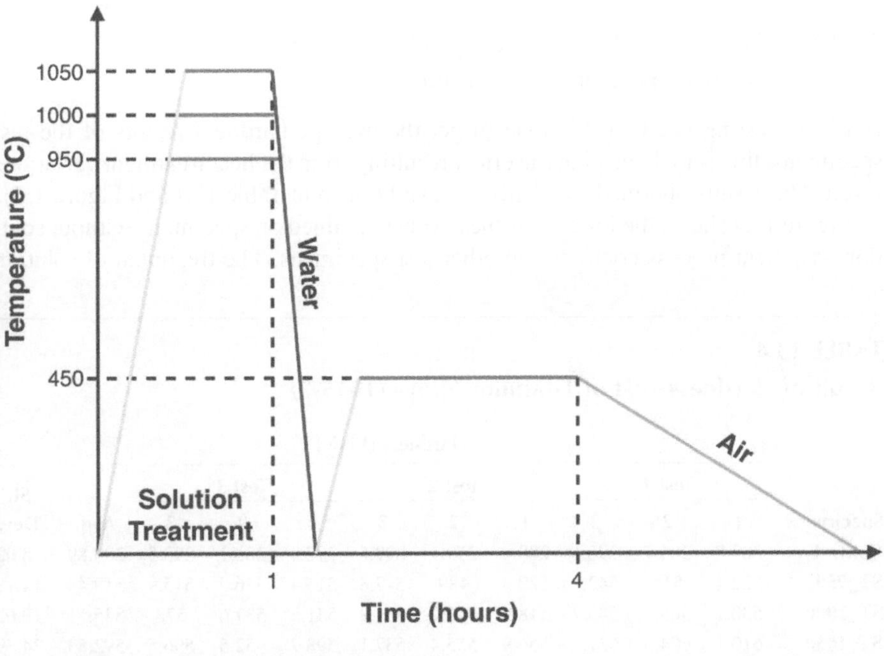

FIGURE 13.2 Heat treatment process diagram.

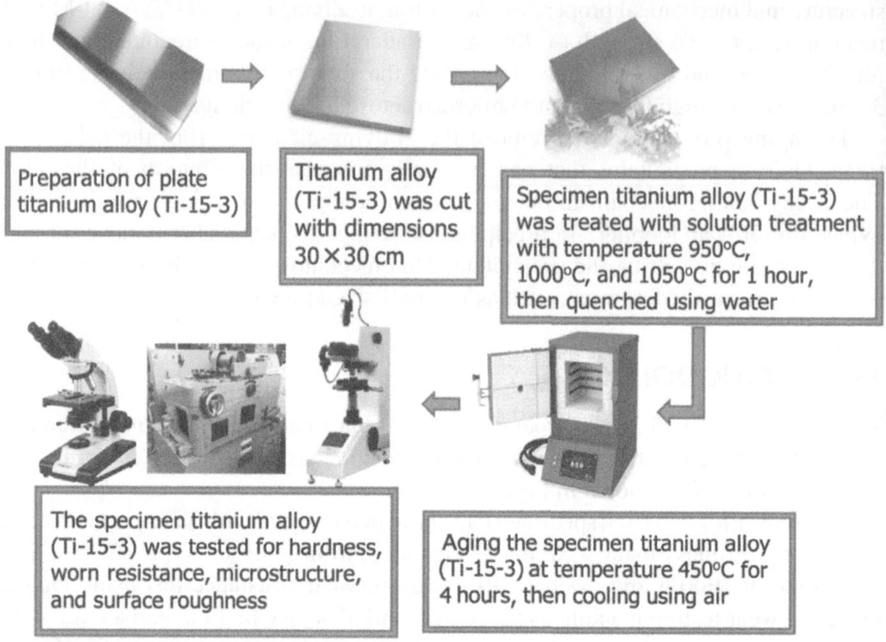

FIGURE 13.3 Schematic of the research.

13.4 RESULTS AND DISCUSSION

13.4.1 Hardness of Titanium Alloys (Ti-15-3) with Solution Treatment and Aging

Hardness testing needs to be done to get the average hardness results of the test specimens; this is to determine the effect resulting from the heat treatment variations given. The results obtained from this test can be seen in Table 13.4 and Figure 13.4.

Figure 13.4 shows the lowest hardness value obtained on specimens without solution treatment process compared to other test specimens. The treatment of solution

TABLE 13.4
Result of Hardness Test of Titanium Alloys (Ti-15-3)

	Hardness [VHN]										
	Test 1			Test 2			Test 3				
Specimens	1	2	3	1	2	3	1	2	3	Avg.	St. Dev.
Control	303.8	284.7	296.1	288.4	279.1	299.5	292.8	280.1	293.5	290.89	8.02
ST_950	522.4	515	542.2	520.4	489	517.8	515.4	536.9	515.9	519.44	14.18
ST_1000	530.1	505.1	545.7	518.9	527.2	531.9	541.5	537.6	578	535.11	19.03
ST_1050	610.1	604.3	621.3	556.9	555.4	537.1	598.7	652.5	596.5	592.53	34.39

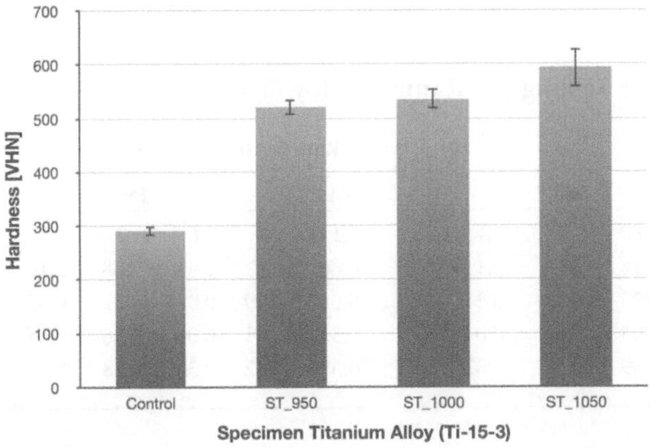

FIGURE 13.4 Hardness of titanium alloy sample (Ti-15-3).

treatment variations with a temperature of 1050°C gets the highest average hardness value when compared to other tests, namely with a value of 592.52 VHN. The process of giving higher temperatures makes an increase in the average graph of surface hardness (Da Rocha et al. 2006).

The hardness of the specimen progressively increases as the temperature applied to it rises. If the specimen is heated to the austenite temperature and then rapidly cooled with a cooling rate that above its critical cooling speed, it will lead to the creation of martensite and result in a substantial enhancement in hardness. By subjecting the test specimen to a higher temperature during the solution treatment procedure, the hardness of the specimen may be further increased by quenching cooling treatment. The use of cooling media can also affect the speed in cooling the test material which contributes to the formation of martensite structure from austenite transformation (Huang et al. 2013). A significant increase in hardness value occurs because the martensite structure that was initially formed finely through rapid cooling in water, then changes into a more rounded shape and the area of the α phase fraction formed is also getting bigger as a result of the process. The data is also reinforced by other studies; the martensite growth process is specifically affected by the quick cooling process (Wang et al. 2019).

13.4.2 Wear Resistance of Titanium Alloys (Ti-15-3) with Solution Treatment and Aging

Wear testing is intended to identify the wear rate of titanium alloy (Ti-15-3) specimens after solution treatment with different temperatures. After testing, the wear results obtained the average value which is then displayed in the form of a graph in Table 13.5 and Figure 13.5.

Figure 13.5 shows that the control titanium Ti-15-3 specimen has the highest average wear value of 0.001658 mm^3/kg.m. The average wear value obtained from specimens with solution treatment at 950°C for 1 hour is 0.000132 mm^3/kg.m. The average

TABLE 13.5
Result of Wear Testing of Titanium Alloy (Ti-15-3)

	Wear Rate (×10⁻³) [mm³/kg.m]										
	Test 1			Test 2			Test 3				
Specimens	**1**	**2**	**3**	**1**	**2**	**3**	**1**	**2**	**3**	**Avg.**	**St. Dev.**
Control	1.14	2.01	2.19	0.50	0.66	1.27	2.57	2.38	2.19	1.6567	0.730
ST_950	0.29	0.21	0.24	0.08	0.08	0.09	0.08	0.07	0.05	0.1322	0.083
ST_1000	0.05	0.08	0.07	0.37	0.32	0.31	0.06	0.06	0.05	0.1532	0.129
ST_1050	0.59	0.51	0.32	0.37	0.39	0.32	0.36	0.38	0.32	0.3756	0.104

wear value obtained in the solution treatment specimen with a temperature of 1000°C is 0.000153 mm³/kg.m; the data is not higher than the specimen with treatment without aging. The solution treatment at 1050°C has the highest average wear value in the wear assessment with a value of 0.000377 mm³/kg.m, but this value does not exceed the value of the specimen without solution treatment. Wear on Ti-15-3 with solution treatment increased in wear value as the temperature increased. Based on this, it shows that the STA temperature affects the wear value of titanium (Puspitasari et al 2023b). The wear of Ti-15-3 without solution treatment produces the highest value this is because in the previous hardness test results, the hardness value without solution treatment obtained the lowest value, so the wear obtained the highest.

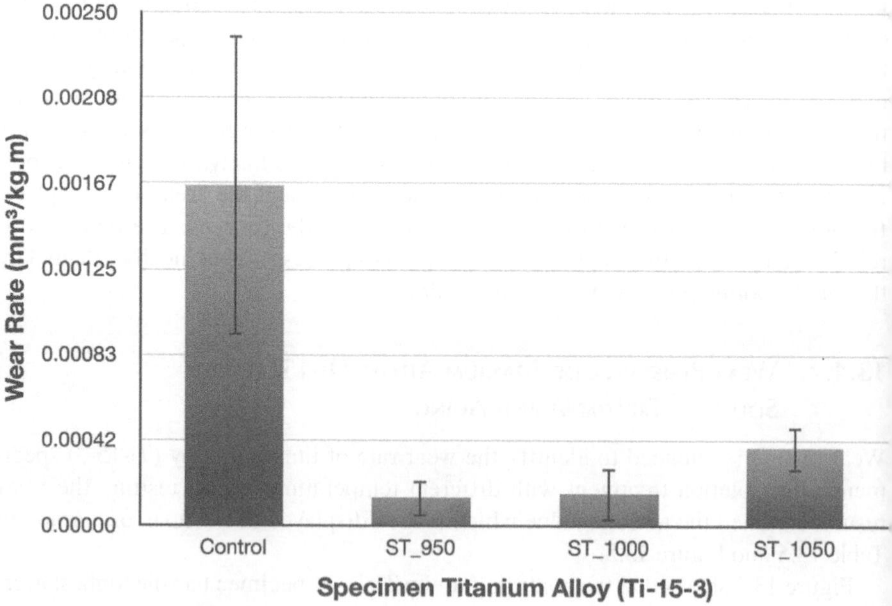

FIGURE 13.5 Average wear value titanium Ti-15-3.

13.4.3 Microstructure of Titanium Alloys (Ti-15-3) with Solution Treatment and Aging

Microstructure testing was carried out using a Nikon 59520 optical microscope; the following are the results of the microstructure test of the Ti-15-3 titanium alloy specimen.

Figure 13.6(a)–(d) is the result of microstructure tests on titanium alloy (Ti-15-3) specimens. Figure 13.6(a) controls specimen titanium alloy (Ti-15-3); there is martensite in the microstructure results although there is only a little arising. This is due to an increase in the hardness value which makes the martensite slowly change into a more rounded shape. Figure 13.6(b) results from testing using 950°C solution treatment shows the metallographic structure after STA but tends to the equiaxed structure due to the results obtained from previous studies that show the same results. This phenomenon is caused by the martensite decomposition process in which recrystallization of α phase occurs evenly at the upper and lower boundaries of α phase, as well as at the internal twin interface of α′. This results in the loss of the needle-like structure of α′ and produces a uniformly dispersed α + β phase structure, which significantly contributes to the strengthening effect (Wang et al. 2019). Figure 13.6(c) explains the results of the solution treatment test with a temperature

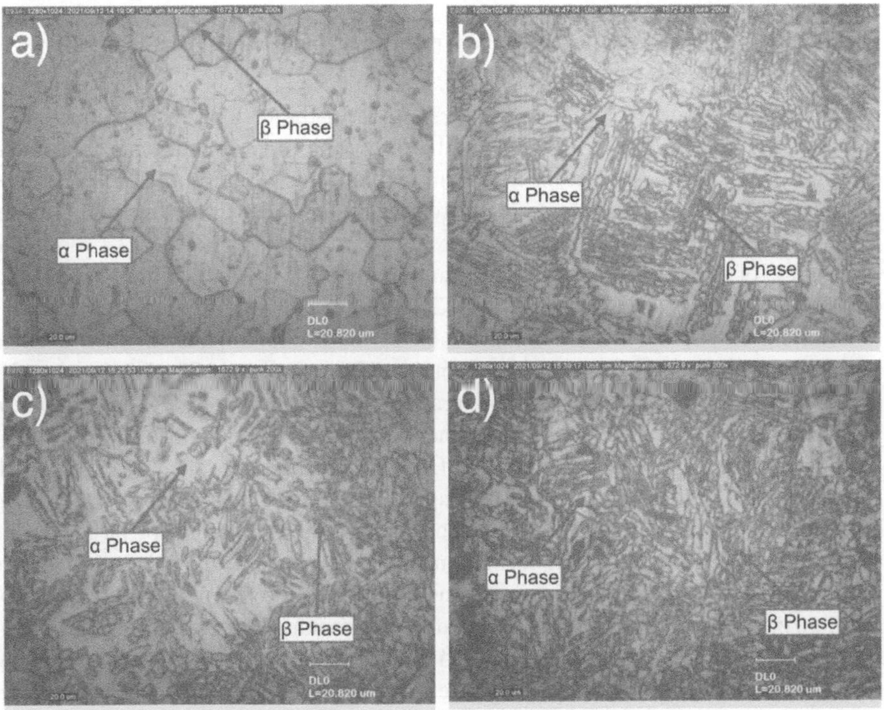

FIGURE 13.6 Microstructure of titanium alloy (Ti-15-3): (a) control, (b) solution treatment 950°C, (c) solution treatment 1000°C, and (d) solution treatment 1050°C.

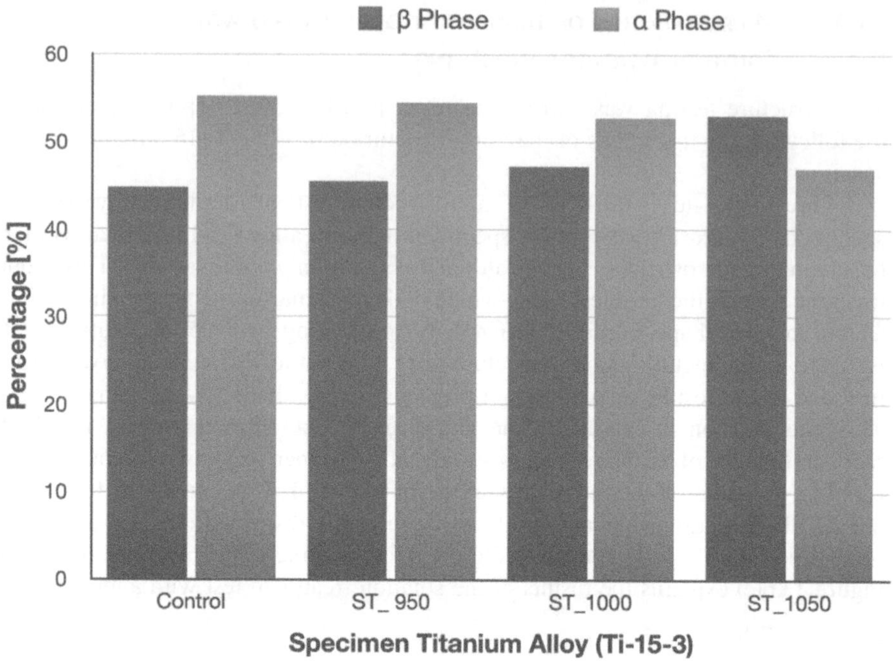

FIGURE 13.7 Percentage of α phase and β phase in titanium alloy (Ti-15-3).

of 1000°C showing the complete transformation of the α′ martensite phase into a stable α + β lamella phase. The results that show the highest beta phase are found in Figure 13.6(d) with a solution treatment temperature variation of 1050°C. Based on these results, it proves that solution treatment can affect grain size and have an impact on the increase of β phase or the reduction of α phase. Figure 13.7 shows the percentage of α phase and β phase of titanium alloy (Ti-15-3) specimens with solution treatment analyzed using OriginPro 2021 software.

Figure 13.7 displays the percentage value of the α (alpha) phase and the average percentage value of the β (beta) phase obtained from the tested Ti-15-3. Figure 13.7 demonstrates that altering the solution treatment temperature of the Ti-15-3 titanium alloy leads to a decrease in the α phase and an increase in the β phase. This is attributed to the varying grain sizes resulting from the solution treatment process, which are influenced by the temperature levels applied to the Ti-15-3 specimen, improving the temperature during solution treatment leads to the production of larger grains (Sridevi et al. 2024). The transformation of tiny grains into larger grains occurs due to the capacity of the solution treatment procedure to regulate the stability of the matrix and the dimensions of the grains. Specimens subjected to solution treatment exhibit enhanced visibility of the β phase, suggesting that solution treatment has an impact on the size of the β phase. Several variables that impact heat treatment include the temperature provided during heating, the duration of keeping the material at that temperature, the cooling medium used, and the ambient conditions present. The characteristics of titanium alloys are significantly affected by their

TABLE 13.6
Result of Surface Roughness Test of Titanium Alloys (Ti-15-3)

	Surface Roughness [μm]				
Specimens	**Test 1**	**Test 2**	**Test 3**	**Avg.**	**St. Dev**
Control	0.17	0.35	0.18	0.2333333	0.0825967
ST_950	0.21	0.74	0.65	0.5333333	0.2315647
ST_1000	0.5	0.73	0.47	0.5666667	0.1161417
ST_1050	0.86	0.51	0.53	0.6333333	0.1604854

microstructure, which is determined by the size, organization, and volume percentage of the α and β phases.

13.4.4 Surface Roughness of Titanium Alloys (Ti-15-3) with Solution Treatment and Aging

The surface roughness of the titanium alloy (Ti-15-3) specimen was tested using the Mitutoyo Surftest type SJ-301 instrument. The measured surface roughness value can be found in Table 13.6 and Figure 13.8.

Figure 13.8 shows the surface roughness value of titanium alloy (Ti-15-3) specimens without solution treatment (control) is lower than the treatment variation

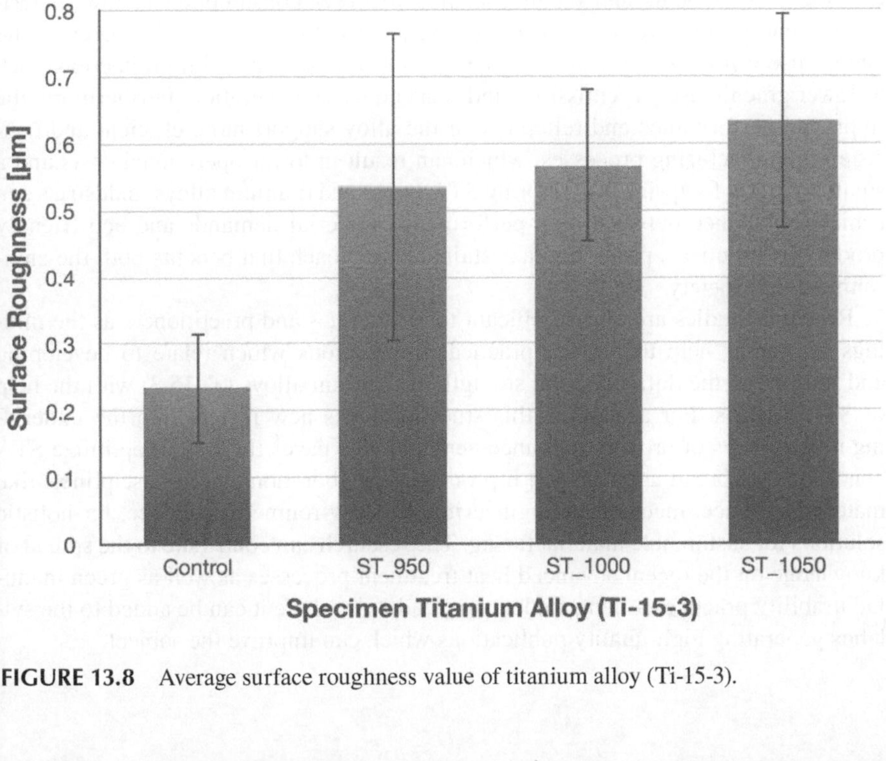

FIGURE 13.8 Average surface roughness value of titanium alloy (Ti-15-3).

using solution treatment with temperature variations of 950°C, 1000°C, and 1050°C. An increase in the roughness value of titanium specimens treated with solution treatment, this is due to specimens that have undergone a heating process with higher temperatures, will experience an increase in surface roughness value. Providing temperature variations during heating can also increase the surface roughness value. During the solution treatment process, the specimen is given with a predetermined temperature variation. The higher the temperature given, the higher the roughness value. This roughness value will have an impact on specimens that are resistant to corrosion. The surface roughness value will directly correlate with the temperature rise; these findings are consistent with the data collected from prior study, which indicates that subjecting the solution treatment procedure to larger temperature swings might enhance surface roughness and improve the test material's resistance to corrosion (Puspitasari et al. 2024). The research's findings indicate that the treatment of the test material and the duration of the testing procedure have a significant impact on the results obtained from the test material (Das et al. 2014).

13.4.5 The Study's Contribution

The study on the improvement of titanium alloy (Ti-15-3) wear resistance and hardness through STA significantly contributes to sustainable manufacturing and socio-environmentally friendly production processes. By enhancing the durability and mechanical properties of the titanium alloy, components made from this material exhibit a longer lifespan and reduced wear and tear. This longevity translates to fewer replacements and repairs, leading to lower consumption of raw materials and energy resources over time. Consequently, this reduction in material and energy use minimizes the environmental impact associated with production, such as lower greenhouse gas emissions and reduced waste generation. Furthermore, the improved performance and reliability of the alloy support more efficient and consistent manufacturing processes, which can result in lower operational costs and a smaller carbon footprint. By adopting STA-enhanced titanium alloys, industries can achieve a balance between high-performance material demands and eco-friendly production practices, promoting a sustainable approach that benefits both the environment and society.

Research studies are also significant for academics and practitioners as the findings of studies help to provide practical implications which relate to developing and improving the durability and strength of titanium alloys (Ti-15-3) with the help of STA process. For academia, this study provides new insight into the underlying mechanisms of property enhancement and also paves the way to optimize STA conditions for broad applications. It promotes collaboration among disciplines, like materials science, mechanical engineering, and environmental science, for holistic solutions for sustainable manufacturing. The research can contribute to the spread of knowledge on the recent advanced heat treatment processes as well as green manufacturability practices among students in academia where it can be added to the syllabus generating high-quality publications which can improve the subject.

Practitioners gain from robust STA at the forefront of technology acceptance hrough titanium alloys (Ti-15-3) improved performance and component life, thereby educing operational and maintenance costs. STA processes can be implemented along the lines of the practitioner input, which allows for a higher efficiency and ower environmental footprint due to lower material and energy expenses. Therefore, STA implementation aids companies to sustain growing environmental regulatory compliance standards and increases market reputation as it shows the commitment toward sustainable manufacturing practices. In addition, this gives us a strategic edge over competitors and serves to improve product quality and reliability, which leads to customer satisfaction.

13.5 CONCLUSION

The resulting hardness value of Ti-15-3 specimens with STA increases in hardness value when given a higher temperature. This can show that the STA process influences the hardness of titanium materials. The hardness value continues to increase with increasing temperature given. These results can conclude that solution treatment can increase the hardness value of titanium materials, and the hardness value can exceed that of test materials without solution treatment.

The wear rate of titanium alloy (Ti-15-3) specimens with solution treatment decreased, which can show that the solution treatment process has an effect on the wear value of titanium materials. The increase in wear value continues to increase with increasing solution treatment temperature; this can conclude that the provision of the solution treatment process can reduce the wear value of titanium materials and the provision of variations in solution treatment temperature can increase the wear value but not to exceed the wear value obtained in the test material without solution treatment.

The percentage of β phase continues to increase after solution treatment. Various microstructures can be obtained through the heat treatment process, so the results of the heat treatment process on Ti-15-3 depend on the parameters and controls used.

The surface roughness increased as the solution treatment temperature increased. This can prove that the increase in roughness value can be influenced by solution treatment. Providing variations in solution treatment temperature is a factor due to the increase in the roughness value of the Ti-15-3 surface. Evidenced by the increasing temperature variations given, the roughness value also increases.

The solution treatment process can significantly enhance the wear resistance and hardness of titanium alloy (Ti-15-3). Therefore, the improved wear resistance of titanium alloy (Ti-15-3) can add up to an extended lifespan. The extended duration of its service life results in a reduced need for replacements and maintenance, leading to lower consumption of raw materials and energy resources over time. As a result, this reduction in material and energy use minimizes the environmental impacts associated with production, such as lower greenhouse gas emissions and reduced waste generation. This directly supports sustainable manufacturing principles by minimizing resource use and environmental impact.

REFERENCES

Campbell, Frederick Charles. Manufacturing Technology for Aerospace Structural Materials Elsevier, 2006.

Chen, Yazhou, Honggang Zhang, Bitao Wang, Jianyong Huang, Meihong Zhou, Lei Wang Yuntao Xi, et al. "A Review of Research on Improving Wear Resistance of Titanium Alloys." Coatings 14, no. 7 (2024): 786.

Das, Sudhansu Ranjan, Amaresh Kumar, Debabrata Dhupal, and Kali Charan Rath. "Estimating the Effect of Machining Parameters on Surface Roughness During Machining of Hardened EN24 Steel Using Coated Carbide Inserts." Industrial Engineering 4, no. 2 (2014): 10.

Elshaer, Ramadan N., Shimaa El-Hadad, and Adel Nofal. "Influence of Heat Treatment Processes on Microstructure Evolution, Tensile and Tribological Properties of Ti6Al4V Alloy." Scientific Reports 13, no. 1 (2023): 1–20.

Froes, F.H. "Principles of Alloying Titanium." In Titanium: Physical Metallurgy, Processing and Applications, pp. 51–74. ASM International, 2015.

Ganesh, Barathi K.C., N. Ramaniah, and P.V.C. Rao. "Effect of Heat Treatment on Dry Sliding Wear of Titanium-Aluminum-Vanadium (Ti-6Al-4V) Implant Alloy." Journal of Mechanical Engineering Research 4, no. 2 (2012).

Huang, Chien-lung, Yu-jia Gao, Tao-liang Chuang, Vincent K S Hsiao, and Tair-i Wu "Study of the Behavior of Titanium Alloys as the Cathode for Photovoltaic Hydrogen Production." International Journal of Photoenergy 2013, no. 857491 (2013): 1–6.

Ji, Xuanming, Panpan Ge, Song Xiang, and Yuanbiao Tan. "Effects of Double-Ageing Heat Treatments on the Microstructure and Mechanical Behaviour of a Ti-3.5Al-5Mo-4V Alloy." Materials 14, no. 1 (2021): 1–12.

Joshi, Vydehi Arun. Titanium Alloys. CRC Press, 2006.

Ju, Jiang, Chunlei Zhao, Maodong Kang, Jingjing Li, Lin He, Chenchen Wang, Jinlong Li, Hanguang Fu, and Jun Wang. "Effect of Heat Treatment on Microstructure and Tribological Behavior of Ti–6Al–4V Alloys Fabricated by Selective Laser Melting." Tribology International 159 (2021): 106996.

Li, Jianfeng, P.A. Agyakwa, and C.M. Johnson. "Interfacial Reaction in Cu/Sn/Cu System during the Transient Liquid Phase Soldering Process." Acta Materialia 59, no. 1 (2012): 1–10.

Maurotto, Agostino, R. Muhammad, A. Roy, V.I. Babitsky, and V.V. Silberschmidt. "Comparing Machinability of Ti-15-3-3-3 and Ni-625 Alloys in UAT." Procedia CIRP 1, no. 1 (2012): 330–335.

L.B. Pervukhin, D.B. Kryukov, A.O. Krivenkov, and S.N. Chugunov. "Structural Transformations and Properties of Titanium–Aluminum Composite during Heat Treatment." Physics of Metals and Metallography 118, no. 8 (2017): 759–763.

Puspitasari, Poppy, D.D. Pramono, M.N.A. Habiby, P. Prabowo, A. Jaelani, M.I.H.C Abdullah, and A. Suyetno. "Experimental Evaluation of Biolubricant with Additive Nanoparticle Calcium Carbonate ($CaCO_3$) from Scallop Shell Waste as Cutting Fluids Using Minimum Quantity Lubrication (MQL) in CNC Milling Process." FME Transactions 52, no. 2 (2024): 319–334.

Puspitasari, Poppy, Diki Dwi Pramono, Chandra Setiawan Putra, Avita Ayu Permanasari Abdul Munir Hidayat Syah Lubis, and Muhammad Ilman Hakimi C. Abdullah "Optimization of Performance Carbon-Based Nanolubricant on Ti6Al4V Pin on Disc Testing Using Taguchi Method." In 8th International Conference on Electrical Electronics and Information Engineering (ICEEIE), pp. 1–5, IEEE, 2023a.

Puspitasari, Poppy, Muhammad Raffli Putra Wardana, Diki Dwi Pramono, Abdul Munir Lubis, Avita Ayu Permanasari, Muhammad Ilman Hakimi Chua Abdullah, and Puput Risdanareni. "Tribology Properties of Titanium Alloy (Ti-6Al-4V) at Various Temperature on α/β Solution Treatment and Aging Condition." In Lecture Notes in Mechanical Engineering, pp. 367–372. Springer, 2023b.

Rocha, Sicknan Soares Da, Gelson Luis Adabo, Guilherme Elias Pessanha Henriques, and Mauro Antônio De Arruda Nóbilo. "Vickers Hardness of Cast Commercially Pure Titanium and Ti-6Al-4V Alloy Submitted to Heat Treatments." Brazilian Dental Journal 17, no. 2 (2006): 126–129.

Sharma, Deepak, Mohammed Kamran, Nilesh Kumar Paraye, and Ramkishor Anant. "Insights into the Wear Behaviour of Electron Beam Melted Ti–6Al–4V Alloy in the As-Built and the Heat-Treated Conditions." Journal of Manufacturing Processes 71 (2021): 669–678.

Sridevi, G., Kali Charan Rath, Sujit Mishra, Santosh Patro, and P. Srinivasa Rao. "Optimization of EDM Process Parameters with Fuzzy Logic Technique for SS-316 Steel and Investigation of Microstructural Characteristics of EDM Machining Surface Evolutionary Manufacturing." In Evolutionary Manufacturing, Design and Operational Practices for Resource and Environmental Sustainability, pp. 245–256. Scrivener Publishing, 2024.

Su, Jinlong, Haiming Xie, Chaolin Tan, Zili Xu, Jin Liu, Fulin Jiang, Jie Tang, Dingfa Fu, Hui Zhang, and Jie Teng. "Microstructural Characteristics and Tribological Behavior of an Additively Manufactured Ti-6Al-4V Alloy under Direct Aging and Solution-Aging Treatments." Tribology International 175 (2022).

Tiyyagura, Hanuma Reddy, Suman Kumari, M.K. Mohan, Bhanu Pant, and M. Nageswara Rao. "Degradation Behavior of Metastable β Ti-15-3 Alloy for Fastener Applications." Journal of Alloys and Compounds 775 (2018): 518–523.

Totten, George E. "Principles of Heat Treating of Nonferrous Alloys." In Heat Treating of Nonferrous Alloys, pp. 3–31. ASM International, 2016.

Vahidshad, Yaser, and Amir Hossein Khodabakhshi. "Effect of Solution Treatment and Aging Temperature on A′ and Ti3Al(A2) Phase Formation and Mechanical Properties of Water-Quenched Ti–6Al–4V." Metallography, Microstructure, and Analysis 11, no. 1 (2022): 59–71.

Wang, Wenyan, Miaofei Zhang, Yazhou Liu, Jingpei Xie, Yingying Wu, and Saisai Yang. "Effect of Heat Treatment on Microstructure and Properties of Ti-6Al-4V-0.5Si Alloy." Procedia Manufacturing 37 (2019): 592–598.

Warlimont, Hans. Titanium and Titanium Alloys. Springer Handbooks, 2018.

Xie, Longfei, Wenlong Zhou, and Shiwen Zou. "Pitting Behavior of Ti-15-3 Titanium Alloy with Different Surface in Salt Spray Studied Using Electrochemical Noise." Journal of Materials Research and Technology 14 (2021): 2865–2883.

Xu, Ying, Guoli Zhao, and Yanqing Cai. "Preparation of Titanium by Electro-deoxidation of CaTiO3 in a Molten CaCl2-NaCl Salt." International Journal of Electrochemical Science 16 (2021): 211022.

Zhang, Deliang, Li Zheng, and D.H. Stjohn. "Effect of a short solution treatment time on microstructure and mechanical properties of modified Al–7wt.%Si–0.3wt.%Mg alloy." Journal of Light Metals 2 (2002): 27–36.

Index

Q

R

S

T

V

W

For Product Safety Concerns and Information please contact our EU representative GPSR@taylorandfrancis.com Taylor & Francis Verlag GmbH, Kaufingerstraße 24, 80331 München, Germany